DEEP LEARNING THROUGH SPARSE AND LOW-RANK MODELING

Computer Vision and Pattern Recognition Series

Series Editors

Horst Bischof   Institute for Computer Graphics and Vision, Graz University of Technology, Austria

Kyoung Mu   Department of Electrical and Computer Engineering, Seoul National University, Republic of Korea

Sudeep Sarkar   Department of Computer Science and Engineering, University of South Florida, Tampa, United States

DEEP LEARNING THROUGH SPARSE AND LOW-RANK MODELING

Edited by

ZHANGYANG WANG
Department of Computer Science and Engineering,
Texas A&M University,
College Station, TX, United States

YUN FU
Department of Electrical and Computer Engineering and
College of Computer and Information Science (Affiliated),
Northeastern University,
Boston, MA, United States

THOMAS S. HUANG
Department of Electrical and Computer Engineering,
University of Illinois at Urbana-Champaign,
Champaign, IL, United States

ACADEMIC PRESS
An imprint of Elsevier

Academic Press is an imprint of Elsevier
125 London Wall, London EC2Y 5AS, United Kingdom
525 B Street, Suite 1800, San Diego, CA 92101-4495, United States
50 Hampshire Street, 5th Floor, Cambridge, MA 02139, United States
The Boulevard, Langford Lane, Kidlington, Oxford OX5 1GB, United Kingdom

Notices

Knowledge and best practice in this field are constantly changing. As new research and experience broaden our
understanding, changes in research methods, professional practices, or medical treatment may become necessary.

Practitioners and researchers must always rely on their own experience and knowledge in evaluating and using any
information, methods, compounds, or experiments described herein. In using such information or methods they
should be mindful of their own safety and the safety of others, including parties for whom they have a professional
responsibility.

To the fullest extent of the law, neither the Publisher nor the authors, contributors, or editors, assume any liability for
any injury and/or damage to persons or property as a matter of products liability, negligence or otherwise, or from
any use or operation of any methods, products, instructions, or ideas contained in the material herein.

Library of Congress Cataloging-in-Publication Data
A catalog record for this book is available from the Library of Congress

British Library Cataloguing-in-Publication Data
A catalogue record for this book is available from the British Library

ISBN: 978-0-12-813659-1

For information on all Academic Press publications
visit our website at https://www.elsevier.com/books-and-journals

Working together
to grow libraries in
developing countries

www.elsevier.com • www.bookaid.org

Publisher: Mara Conner
Acquisition Editor: Tim Pitts
Editorial Project Manager: Ana Claudia A. Garcia
Production Project Manager: Kamesh Ramajogi
Designer: Matthew Limbert

Typeset by VTeX

CONTENTS

Contributors   *xi*
About the Editors   *xiii*
Preface   *xv*
Acknowledgments   *xvii*

1. Introduction   **1**
Zhangyang Wang, Ding Liu

    1.1. Basics of Deep Learning   1
    1.2. Basics of Sparsity and Low-Rankness   2
    1.3. Connecting Deep Learning to Sparsity and Low-Rankness   3
    1.4. Organization   4
    References   4

2. Bi-Level Sparse Coding: A Hyperspectral Image Classification Example   **9**
Zhangyang Wang

    2.1. Introduction   9
    2.2. Formulation and Algorithm   12
       2.2.1. Notations   12
       2.2.2. Joint Feature Extraction and Classification   12
       2.2.3. Bi-level Optimization Formulation   15
       2.2.4. Algorithm   15
    2.3. Experiments   17
       2.3.1. Classification Performance on AVIRIS Indiana Pines Data   20
       2.3.2. Classification Performance on AVIRIS Salinas Data   23
       2.3.3. Classification Performance on University of Pavia Data   23
    2.4. Conclusion   26
    2.5. Appendix   27
    References   28

3. Deep ℓ_0 Encoders: A Model Unfolding Example   **31**
Zhangyang Wang

    3.1. Introduction   31
    3.2. Related Work   32
       3.2.1. ℓ_0- and ℓ_1-Based Sparse Approximations   32
       3.2.2. Network Implementation of ℓ_1-Approximation   33
    3.3. Deep ℓ_0 Encoders   34
       3.3.1. Deep ℓ_0-Regularized Encoder   34
       3.3.2. Deep M-Sparse ℓ_0 Encoder   36

3.3.3.   Theoretical Properties   37
3.4.   Task-Driven Optimization   37
3.5.   Experiment   38
    3.5.1.   Implementation   38
    3.5.2.   Simulation on ℓ_0 Sparse Approximation   38
    3.5.3.   Applications on Classification   40
    3.5.4.   Applications on Clustering   42
3.6.   Conclusions and Discussions on Theoretical Properties   43
References   44

4.   Single Image Super-Resolution: From Sparse Coding to Deep Learning   47
Ding Liu, Thomas S. Huang

4.1.   Robust Single Image Super-Resolution via Deep Networks with Sparse Prior   47
    4.1.1.   Introduction   47
    4.1.2.   Related Work   49
    4.1.3.   Sparse Coding Based Network for Image SR   50
    4.1.4.   Network Cascade for Scalable SR   54
    4.1.5.   Robust SR for Real Scenarios   57
    4.1.6.   Implementation Details   60
    4.1.7.   Experiments   61
    4.1.8.   Subjective Evaluation   69
    4.1.9.   Conclusion and Future Work   72
4.2.   Learning a Mixture of Deep Networks for Single Image Super-Resolution   73
    4.2.1.   Introduction   73
    4.2.2.   The Proposed Method   74
    4.2.3.   Implementation Details   76
    4.2.4.   Experimental Results   77
    4.2.5.   Conclusion and Future Work   81
References   83

5.   From Bi-Level Sparse Clustering to Deep Clustering   87
Zhangyang Wang

5.1.   A Joint Optimization Framework of Sparse Coding and Discriminative Clustering   87
    5.1.1.   Introduction   87
    5.1.2.   Model Formulation   88
    5.1.3.   Clustering-Oriented Cost Functions   90
    5.1.4.   Experiments   93
    5.1.5.   Conclusion   98
    5.1.6.   Appendix   99
5.2.   Learning a Task-Specific Deep Architecture for Clustering   101
    5.2.1.   Introduction   101
    5.2.2.   Related Work   102
    5.2.3.   Model Formulation   103
    5.2.4.   A Deeper Look: Hierarchical Clustering by DTAGnet   106

5.2.5.   Experiment Results   107
5.2.6.   Conclusion   117
References   117

6.   Signal Processing   **121**
Zhangyang Wang, Ding Liu, Thomas S. Huang

6.1.   Deeply Optimized Compressive Sensing   121
6.1.1.   Background   121
6.1.2.   An End-to-End Optimization Model of CS   122
6.1.3.   DOCS: Feed-Forward and Jointly Optimized CS   124
6.1.4.   Experiments   126
6.1.5.   Conclusion   129
6.2.   Deep Learning for Speech Denoising   130
6.2.1.   Introduction   130
6.2.2.   Neural Networks for Spectral Denoising   131
6.2.3.   Experimental Results   134
6.2.4.   Conclusion and Future Work   139
References   140

7.   Dimensionality Reduction   **143**
Shuyang Wang, Zhangyang Wang, Yun Fu

7.1.   Marginalized Denoising Dictionary Learning with Locality Constraint   143
7.1.1.   Introduction   143
7.1.2.   Related Works   145
7.1.3.   Marginalized Denoising Dictionary Learning with Locality Constraint   147
7.1.4.   Experiments   157
7.1.5.   Conclusion   164
7.1.6.   Future Works   165
7.2.   Learning a Deep ℓ_∞ Encoder for Hashing   165
7.2.1.   Introduction   166
7.2.2.   ADMM Algorithm   168
7.2.3.   Deep ℓ_∞ Encoder   168
7.2.4.   Deep ℓ_∞ Siamese Network for Hashing   170
7.2.5.   Experiments in Image Hashing   172
7.2.6.   Conclusion   178
References   178

8.   Action Recognition   **183**
Yu Kong, Yun Fu

8.1.   Deeply Learned View-Invariant Features for Cross-View Action Recognition   183
8.1.1.   Introduction   183
8.1.2.   Related Work   185
8.1.3.   Deeply Learned View-Invariant Features   186

	8.1.4. Experiments	191
8.2.	Hybrid Neural Network for Action Recognition from Depth Cameras	198
	8.2.1. Introduction	198
	8.2.2. Related Work	199
	8.2.3. Hybrid Convolutional-Recursive Neural Networks	201
	8.2.4. Experiments	206
8.3.	Summary	209
	References	210

9. Style Recognition and Kinship Understanding   213

Shuhui Jiang, Ming Shao, Caiming Xiong, Yun Fu

9.1.	Style Classification by Deep Learning	213
	9.1.1. Background	213
	9.1.2. Preliminary Knowledge of Stacked Autoencoder (SAE)	217
	9.1.3. Style Centralizing Autoencoder	217
	9.1.4. Consensus Style Centralizing Autoencoder	221
	9.1.5. Experiments	226
9.2.	Visual Kinship Understanding	230
	9.2.1. Background	230
	9.2.2. Related Work	232
	9.2.3. Family Faces	233
	9.2.4. Regularized Parallel Autoencoders	234
	9.2.5. Experimental Results	239
9.3.	Research Challenges and Future Works	246
	References	246

10. Image Dehazing: Improved Techniques   251

Yu Liu, Guanlong Zhao, Boyuan Gong, Yang Li, Ritu Raj, Niraj Goel, Satya Kesav, Sandeep Gottimukkala, Zhangyang Wang, Wenqi Ren, Dacheng Tao

10.1.	Introduction	251
10.2.	Review and Task Description	252
	10.2.1. Haze Modeling and Dehazing Approaches	253
	10.2.2. RESIDE Dataset	253
10.3.	Task 1: Dehazing as Restoration	254
10.4.	Task 2: Dehazing for Detection	257
	10.4.1. Solution Set 1: Enhancing Dehazing and/or Detection Modules in the Cascade	257
	10.4.2. Solution Set 2: Domain-Adaptive Mask-RCNN	257
10.5.	Conclusion	260
	References	261

11. Biomedical Image Analytics: Automated Lung Cancer Diagnosis   263

Steve Kommrusch, Louis-Noël Pouchet

11.1.	Introduction	263

11.2. Related Work   264
11.3. Methodology   265
11.4. Experiments   268
11.5. Conclusion   270
Acknowledgments   271
References   271

Index   *273*

CONTRIBUTORS

Yun Fu

Department of Electrical and Computer Engineering and College of Computer and Information Science (Affiliated), Northeastern University, Boston, MA, United States

Niraj Goel

Department of Computer Science and Engineering, Texas A&M University, College Station, TX, United States

Boyuan Gong

Department of Computer Science and Engineering, Texas A&M University, College Station, TX, United States

Sandeep Gottimukkala

Department of Computer Science and Engineering, Texas A&M University, College Station, TX, United States

Thomas S. Huang

Department of Electrical and Computer Engineering, University of Illinois at Urbana-Champaign, Champaign, IL, United States

Shuhui Jiang

Department of Electrical and Computer Engineering, Northeastern University, Boston, MA, United States

Satya Kesav

Department of Computer Science and Engineering, Texas A&M University, College Station, TX, United States

Steve Kommrusch

Colorado State University, Fort Collins, CO, United States

Yu Kong

B. Thomas Golisano College of Computing and Information Sciences, Rochester Institute of Technology, Rochester, NY, United States

Yang Li

Department of Electrical and Computer Engineering, Texas A&M University, College Station, TX, United States

Ding Liu

Beckman Institute for Advanced Science and Technology, Urbana, IL, United States

Yu Liu

Department of Electrical and Computer Engineering, Texas A&M University, College Station, TX, United States

Louis-Noël Pouchet

Colorado State University, Fort Collins, CO, United States

Ritu Raj

Department of Computer Science and Engineering, Texas A&M University, College Station, TX, United States

Wenqi Ren

Chinese Academy of Sciences, Beijing, China

Ming Shao

Computer and Information Science, University of Massachusetts Dartmouth, Dartmouth, MA, United States

Dacheng Tao

University of Sydney, Sydney, NSW, Australia

Shuyang Wang

Department of Electrical and Computer Engineering, Northeastern University, Boston, MA, United States

Zhangyang Wang

Department of Computer Science and Engineering, Texas A&M University, College Station, TX, United States

Caiming Xiong

Salesforce Research, Palo Alto, CA, United States

Guanlong Zhao

Department of Computer Science and Engineering, Texas A&M University, College Station, TX, United States

ABOUT THE EDITORS

Zhangyang Wang is an Assistant Professor of Computer Science and Engineering (CSE) at the Texas A&M University (TAMU). During 2012–2016, he was a PhD student in the Electrical and Computer Engineering (ECE) Department, at the University of Illinois, Urbana-Champaign (UIUC), working with Professor Thomas S. Huang. Prior to that he obtained the BE degree at the University of Science and Technology of China (USTC) in 2012. Dr. Wang's research has been addressing machine learning, computer vision and multimedia signal processing problems using advanced feature learning and optimization techniques. He has co-authored over 50 papers, and published several books and chapters. He has been granted 3 patents, and has received over 20 research awards and scholarships. He regularly serves as guest editor, area chair, session chair, TPC member, tutorial speaker and workshop organizer at leading conferences and journals. His research has been covered by worldwide media, such as BBC, Fortune, International Business Times, UIUC news and alumni magazine. For more see http://www.atlaswang.com.

Yun Fu received the BEng degree in Information Engineering and the MEng degree in Pattern Recognition and Intelligence Systems from Xi'an Jiaotong University, China, as well as the MS degree in Statistics and the PhD degree in Electrical and Computer Engineering from the University of Illinois at Urbana-Champaign. He is an interdisciplinary faculty member affiliated with College of Engineering and the College of Computer and Information Science at Northeastern University since 2012. His research interests are in machine learning, computational intelligence, Big Data mining, computer vision, pattern recognition, and cyber-physical systems. He has extensive publications in leading journals, books/book chapters and international conferences/workshops. He serves as associate editor, chair, PC member and reviewer of many top journals and international conferences/workshops. He received seven prestigious Young Investigator Awards from NAE, ONR, ARO, IEEE, INNS, UIUC, and Grainger Foundation; seven Best Paper Awards from IEEE, IAPR, SPIE, and SIAM; three major Industrial Research Awards from Google, Samsung, and Adobe. He is currently an Associate Editor of the IEEE Transactions on Neural Networks and Learning Systems (TNNLS). He is a fellow of IAPR and SPIE, a Lifetime Senior Member of ACM, Senior Member of IEEE, Lifetime Member of AAAI, OSA, and Institute of Mathematical Statistics, member of Global Young Academy (GYA), INNS and was a Beckman Graduate Fellow during 2007–2008.

Thomas S. Huang received the ScD degree from the Massachusetts Institute of Technology in 1963. He is currently a Research Professor of Electrical and Computer Engineering and the Swanlund Endowed Chair Professor at the University of Illinois, Urbana-Champaign. He has authored or coauthored 21 books and over 600 papers on network theory, digital filtering, image processing, and computer vision. His current research interests include computer vision, image compression and enhancement, pattern recognition, and multimodal signal processing. He is a member of the United States National Academy of Engineering. He is a fellow of the International Association of Pattern Recognition and the Optical Society of America. He was a recipient of the IS&T and SPIE Imaging Scientist of the Year Award and the IBM Faculty Award in 2006. In 2008, he served as the Honorary Chair of the ACM Conference on Content-Based Image and Video Retrieval and the IEEE Conference on Computer Vision and Pattern Recognition. He has received numerous awards, including the Honda Lifetime Achievement Award, the IEEE Jack Kilby Signal Processing Medal, the King-Sun Fu Prize of the International Association for Pattern Recognition, and the Azriel Rosenfeld Life Time Achievement Award at the International Conference on Computer Vision.

PREFACE

Deep learning has achieved tremendous success in various applications of machine learning, data analytics, and computer vision. It is easy to be parallelized with a low inference complexity and could be jointly tuned in an end-to-end manner. However, generic deep architectures, often referred to as "black-box" methods, largely ignore the problem-specific formulations and domain knowledge. They rely on stacking somewhat ad-hoc modules, which makes it prohibitive to interpret their working mechanisms. Despite a few hypotheses and intuitions, it is widely recognized as difficult to understand why deep models work, and how they can be related to classical machine learning models. On the other hand, sparsity and low rankness are well exploited regularization in classical machine learning. By exploiting the latent low-dimensional subspace structure of high-dimension data, they have also led to great success in many image processing and understanding tasks.

This book provides an overview on the recent research trend on integrated deep learning models with sparse and low rank models. It will be suitable for the audiences who have basic knowledge of deep learning and sparse/low rank models, and place strong emphasis on the concepts and applications, in the hope that it can reach a broader audience. The research advances covered in this book bridge the classical sparse and low rank models that emphasize problem-specific prior and interpretability, with deep network models that allow for larger learning capacity and better utilization of Big Data. The toolkit of deep learning will be shown to be closely tied with the sparse/low rank models and algorithms. Such a viewpoint is expected to motivate a rich variety of theoretical and analytic tools, to guide the architecture design and interpretation of deep models. The theoretical and modeling progress will be complemented with many applications in computer vision, machine learning, signal processing, data mining, and more.

The Authors

ACKNOWLEDGMENTS

We thank Ana Claudia A. Garcia, Tim Pitts, Kamesh Ramajogi, and other Elsevier staff for careful guidance throughout the publishing process. We appreciate the many research collaborations and discussions, that we have had with the students and alumnus in the UIUC Image Formation and Processing (IFP) group. We are also grateful to all the chapter authors for carefully working on their contributed parts.

The Authors

CHAPTER 1

Introduction

Zhangyang Wang*, Ding Liu†
*Department of Computer Science and Engineering, Texas A&M University, College Station, TX, United States
†Beckman Institute for Advanced Science and Technology, Urbana, IL, United States

Contents

1.1.	Basics of Deep Learning	1
1.2.	Basics of Sparsity and Low-Rankness	2
1.3.	Connecting Deep Learning to Sparsity and Low-Rankness	3
1.4.	Organization	4
References		4

1.1. BASICS OF DEEP LEARNING

Machine learning makes computers learn from data without explicitly programming them. However, classical machine learning algorithms often find it challenging to extract semantic features directly from raw data, e.g., due to the well-known "semantic gap" [1], which calls for the assistance from domain experts to hand-craft many well-engineered feature representations, on which the machine learning models operate more effectively. In contrast, the recently popular deep learning relies on multilayer neural networks to derive semantically meaningful representations, by building multiple simple features to represent a sophisticated concept. Deep learning requires less hand-engineered features and expert knowledge. Taking image classification as an example [2], a deep learning-based image classification system represents an object by gradually extracting edges, textures, and structures, from lower to middle-level hidden layers, which becomes more and more associated with the target semantic concept as the model grows deeper. Driven by the emergence of big data and hardware acceleration, the intricacy of data can be extracted with higher and more abstract level representation from raw inputs, gaining more power for deep learning to solve complicated, even traditionally intractable problems. Deep learning has achieved tremendous success in visual object recognition [2–5], face recognition and verification [6,7], object detection [8–11], image restoration and enhancement [12–17], clustering [18], emotion recognition [19], aesthetics and style recognition [20–23], scene understanding [24,25], speech recognition [26], machine translation [27], image synthesis [28], and even playing Go [29] and poker [30].

A basic neural network is composed of a set of perceptrons (artificial neurons), each of which maps inputs to output values with a simple activation function. Among recent deep neural network architectures, convolutional neural networks (CNNs) and recurrent neural networks (RNNs) are the two main streams, differing in their connectivity patterns. CNNs deploy convolution operations on hidden layers for weight sharing and parameter reduction. CNNs can extract local information from grid-like input data, and have mainly shown successes in computer vision and image processing, with many popular instances such as LeNet [31], AlexNet [2], VGG [32], GoogLeNet [33], and ResNet [34]. RNNs are dedicated to processing sequential input data with variable length. RNNs produce an output at each time step. The hidden neuron at each time step is calculated based on input data and hidden neurons at the previous time step. To avoid vanishing/exploding gradients of RNNs in long term dependency, long short-term memory (LSTM) [35] and gated recurrent unit (GRU) [36] with controllable gates are widely used in practical applications. Interested readers are referred to a comprehensive deep learning textbook [37].

1.2. BASICS OF SPARSITY AND LOW-RANKNESS

In signal processing, the classical way to represent a multidimensional signal is to express it as a linear combination of the components in a (chosen in advance and also learned) basis. The goal of linearly transforming a signal with respect to a basis is to have a more predictable pattern in the resultant linear coefficients. With an appropriate basis, such coefficients often exhibit some desired characteristics for signals. One important observation is that, for most natural signals such as image and audio, most of the coefficients are zero or close to zero if the basis is properly selected: the technique is usually termed as *sparse coding*, and the basis is called the dictionary [38]. A sparse prior can have many interpretations in various contexts, such as smoothness, feature selection, etc. Ensured by theories from compressive sensing [39], under certain favorable conditions, the sparse solutions can be reliably obtained using the ℓ_1-norm, instead of the more straightforward but intractable ℓ_0-norm. Beyond the element-wise sparsity model, more elaborate structured sparse models have also been developed [40,41]. The learning of basis (called dictionary) further boosts the power of sparse coding [42–44].

More generally, the sparsity belongs to the well-received principle of *parsimony*, i.e., preferring a simple representation to a more complex one. The sparsity level (number of nonzero elements) is a natural measure of representation complexity of vector-valued features. In the case of matrix-valued features, the matrix rank provides another notion of parsimony, assuming high-dimensional data lies close to a low-dimensional subspace or manifold. Similarly to sparse optimization, a series of works have shown that rank minimization can be achieved through convex optimization [45] or efficient heuristics [46], paving the path to high-dimensional data analysis such as video processing [47–52].

1.3. CONNECTING DEEP LEARNING TO SPARSITY AND LOW-RANKNESS

Beyond their proven success in conventional machine learning algorithms, the sparse and low-rank structures are widely found to be effective for regularizing deep learning, for improving model generalization, training behaviors, data efficiency [53], and/or compactness [54]. For example, adding ℓ_1 (or ℓ_2) decay term limits the weights of the neurons. Another popular tool to avoid overfitting, dropout [2], is a simple regularization approach that improves the generalization of deep networks, by randomly putting hidden neurons to zero in the training stage, which could be viewed as a stochastic form of enforcing sparsity. Besides, the inherent sparse properties of both deep network weights and activations have also been widely observed and utilized for compressing deep models [55] and improving their energy efficiency [56,57]. As for low-rankness, much research has also been devoted to learning low-rank convolutional filters [58] and network compression [59].

Our focus of this book is to explore a deeper structural connection between sparse/low-rank models and deep models. While many examples will be detailed in the remainder of the book, we here briefly state the main idea. We start from the following *regularized regression* form, which represents a large family of feature learning models, such as ridge regression, sparse coding, and low-rank representation

$$\mathbf{a} = \arg\min_{\mathbf{a}} \tfrac{1}{2}\|\mathbf{x} - \Phi(\mathbf{D}, \mathbf{a})\|_F^2 + \Psi(\mathbf{a}). \tag{1.1}$$

Here $\mathbf{x} \in R^n$ denotes the input data, $\mathbf{a} \in R^m$ is the feature to learn, and $\mathbf{D} \in R^{n \times m}$ is the representation basis. Function $\Phi(\mathbf{D}, \mathbf{a})$: $R^{n \times m} \times R^m \to R^n$ defines the form of feature representation. The regularization term $\Psi(\mathbf{a})$: $R^m \to R$ further incorporates the problem-specific prior knowledge. Not surprisingly, many instances of Eq. (1.1) could be solved by a similar class of iterative algorithms

$$\mathbf{z}^{k+1} = \mathcal{N}(\mathcal{L}_1(\mathbf{x}) + \mathcal{L}_2(\mathbf{z}^k)), \tag{1.2}$$

where $\mathbf{z}^k \in R^m$ denotes the intermediate output of the kth iteration, $k = 0, 1, \ldots, \mathcal{L}_1$ and $\mathcal{L}_2$ are linear (or convolutional) operators, while $\mathcal{N}$ is a simple nonlinear operator. Equation (1.2) could be expressed by a recursive system, whose fixed point is expected to be the solution $\mathbf{a}$ of Eq. (1.1). Furthermore, the recursive system could be *unfolded* and *truncated* to k iterations, to construct a $(k+1)$-layer feed-forward network. Without any further tuning, the resulting architecture will output a k-iteration approximation of the exact solution $\mathbf{a}$. We use the sparse coding model [38] as a popular instance of Eq. (1.1), which corresponds to $\Psi(\mathbf{a}) = \lambda\|\mathbf{a}\|_1$, $\Phi(\mathbf{D}, \mathbf{a}) = \mathbf{D}\mathbf{a}$, with $\|\mathbf{D}\|_2 = 1$ by default. Then, the concrete function forms are given as ($\mathbf{u}$ is a vector and u_i is its ith element)

$$\mathcal{L}_1(\mathbf{x}) = \mathbf{D}^T\mathbf{x}, \ \ \mathcal{L}_2(\mathbf{z}^k) = (\mathbf{I} - \mathbf{D}^T\mathbf{D})\mathbf{z}^k, \ \ \mathcal{N}(\mathbf{u})_i = \text{sign}(u_i)(|u_i| - \lambda)_+, \tag{1.3}$$

where $\mathcal{N}$ is an element-wise soft shrinkage function. The unfolded and truncated version of Eq. (1.3) was first proposed in [60], called the learned iterative shrinkage and thresholding algorithm (LISTA). Recent works [61,18,62–64] followed LISTA and developed various models, and many jointly optimized the unfolded model with discriminative tasks [65].

A simple but interesting variant comes out by enforcing $\Psi(\mathbf{a}) = \lambda \|\mathbf{a}\|_1$, $\mathbf{a} \geq 0$, and $\Phi(\mathbf{D}, \mathbf{a}) = \mathbf{Da}$, $\|\mathbf{D}\|_2 = 1$, Eq. (1.2) could be adapted to solve the nonnegative sparse coding problem

$$\mathcal{L}_1(\mathbf{x}) = \mathbf{D}^T\mathbf{x} - \lambda, \ \mathcal{L}_2(\mathbf{z}^k) = (\mathbf{I} - \mathbf{D}^T\mathbf{D})\mathbf{z}^k, \ \mathcal{N}(\mathbf{u})_i = \max(u_i, 0). \qquad (1.4)$$

A by-product of applying nonnegativity is that the original sparsity coefficient λ now occurs in $\mathcal{L}_1$ as the bias term of this layer, rather than appearing in $\mathcal{N}$ as in Eq. (1.3). As a result, $\mathcal{N}$ in Eq. (1.4) now has exactly the same form as the popular neuron of rectified linear unit (ReLU) [2]. We further make an aggressive approximation of Eq. (1.4), by setting $k = 0$ and assuming $\mathbf{z}^0 = 0$, and have

$$\mathbf{z} = \mathcal{N}(\mathbf{D}^T\mathbf{x} - \lambda). \qquad (1.5)$$

Note that even if a nonzero $\mathbf{z}^0$ is assumed, it could be absorbed into the bias term $-\lambda$. Equation (1.5) is exactly a fully-connected layer followed by ReLU neurons, one of the most standard building blocks in existing deep models. Convolutional layers could be derived similarly by looking at a convolutional sparse coding model [66] rather than a linear one. Such a hidden structural resemblance reveals the potential to bridge many sparse and low-rank models with current successful deep models, potentially enhancing the generalization, compactness and interpretability of the latter.

1.4. ORGANIZATION

In the remainder of this book, Chapter 2 will first introduce the bi-level sparse coding model, using the example of hyperspectral image classification. Chapters 3, 4 and 5 will then present three concrete examples (classification, superresolution, and clustering), to show how (bi-level) sparse coding models could be naturally converted to and trained as deep networks. From Chapter 6 to Chapter 9, we will delve into the extensive applications of deep learning aided by sparsity and low-rankness, in signal processing, dimensionality reduction, action recognition, style recognition and kinship understanding, respectively.

REFERENCES

[1] Zhao R, Grosky WI. Narrowing the semantic gap-improved text-based web document retrieval using visual features. IEEE Transactions on Multimedia 2002;4(2):189–200.

[2] Krizhevsky A, Sutskever I, Hinton GE. Imagenet classification with deep convolutional neural networks. In: NIPS; 2012.

[3] Wang Z, Chang S, Yang Y, Liu D, Huang TS. Studying very low resolution recognition using deep networks. In: Proceedings of the IEEE conference on computer vision and pattern recognition; 2016. p. 4792–800.

[4] Liu D, Cheng B, Wang Z, Zhang H, Huang TS. Enhance visual recognition under adverse conditions via deep networks. arXiv preprint arXiv:1712.07732, 2017.

[5] Wu Z, Wang Z, Wang Z, Jin H. Towards privacy-preserving visual recognition via adversarial training: a pilot study. arXiv preprint arXiv:1807.08379, 2018.

[6] Bodla N, Zheng J, Xu H, Chen J, Castillo CD, Chellappa R. Deep heterogeneous feature fusion for template-based face recognition. In: 2017 IEEE winter conference on applications of computer vision, WACV 2017; 2017. p. 586–95.

[7] Ranjan R, Bansal A, Xu H, Sankaranarayanan S, Chen J, Castillo CD, et al. Crystal loss and quality pooling for unconstrained face verification and recognition. CoRR 2018. arXiv:1804.01159 [abs].

[8] Ren S, He K, Girshick R, Sun J. Faster R-CNN: towards real-time object detection with region proposal networks. In: Advances in neural information processing systems; 2015. p. 91–9.

[9] Yu J, Jiang Y, Wang Z, Cao Z, Huang T. Unitbox: an advanced object detection network. In: Proceedings of the 2016 ACM on multimedia conference. ACM; 2016. p. 516–20.

[10] Gao J, Wang Q, Yuan Y. Embedding structured contour and location prior in siamesed fully convolutional networks for road detection. In: Robotics and automation (ICRA), 2017 IEEE international conference on. IEEE; 2017. p. 219–24.

[11] Xu H, Lv X, Wang X, Ren Z, Bodla N, Chellappa R. Deep regionlets for object detection. In: The European conference on computer vision (ECCV); 2018.

[12] Timofte R, Agustsson E, Van Gool L, Yang MH, Zhang L, Lim B, et al. NTIRE 2017 challenge on single image super-resolution: methods and results. In: Computer vision and pattern recognition workshops (CVPRW), 2017 IEEE conference on. IEEE; 2017. p. 1110–21.

[13] Li B, Peng X, Wang Z, Xu J, Feng D. AOD-Net: all-in-one dehazing network. In: Proceedings of the IEEE international conference on computer vision; 2017. p. 4770–8.

[14] Li B, Peng X, Wang Z, Xu J, Feng D. An all-in-one network for dehazing and beyond. arXiv preprint arXiv:1707.06543, 2017.

[15] Li B, Peng X, Wang Z, Xu J, Feng D. End-to-end united video dehazing and detection. arXiv preprint arXiv:1709.03919, 2017.

[16] Liu D, Wen B, Jiao J, Liu X, Wang Z, Huang TS. Connecting image denoising and high-level vision tasks via deep learning. arXiv preprint arXiv:1809.01826, 2018.

[17] Prabhu R, Yu X, Wang Z, Liu D, Jiang A. U-finger: multi-scale dilated convolutional network for fingerprint image denoising and inpainting. arXiv preprint arXiv:1807.10993, 2018.

[18] Wang Z, Chang S, Zhou J, Wang M, Huang TS. Learning a task-specific deep architecture for clustering. SDM 2016.

[19] Cheng B, Wang Z, Zhang Z, Li Z, Liu D, Yang J, et al. Robust emotion recognition from low quality and low bit rate video: a deep learning approach. arXiv preprint arXiv:1709.03126, 2017.

[20] Wang Z, Yang J, Jin H, Shechtman E, Agarwala A, Brandt J, et al. DeepFont: identify your font from an image. In: Proceedings of the 23rd ACM international conference on multimedia. ACM; 2015. p. 451–9.

[21] Wang Z, Yang J, Jin H, Shechtman E, Agarwala A, Brandt J, et al. Real-world font recognition using deep network and domain adaptation. arXiv preprint arXiv:1504.00028, 2015.

[22] Wang Z, Chang S, Dolcos F, Beck D, Liu D, Huang TS. Brain-inspired deep networks for image aesthetics assessment. arXiv preprint arXiv:1601.04155, 2016.

[23] Huang TS, Brandt J, Agarwala A, Shechtman E, Wang Z, Jin H, et al. Deep learning for font recognition and retrieval. In: Applied cloud deep semantic recognition. Auerbach Publications; 2018. p. 109–30.

[24] Farabet C, Couprie C, Najman L, LeCun Y. Learning hierarchical features for scene labeling. IEEE Transactions on Pattern Analysis and Machine Intelligence 2013;35(8):1915–29.

[25] Wang Q, Gao J, Yuan Y. A joint convolutional neural networks and context transfer for street scenes labeling. IEEE Transactions on Intelligent Transportation Systems 2017.

[26] Saon G, Kuo HKJ, Rennie S, Picheny M. The IBM 2015 English conversational telephone speech recognition system. arXiv preprint arXiv:1505.05899, 2015.

[27] Sutskever I, Vinyals O, Le QV. Sequence to sequence learning with neural networks. In: Advances in neural information processing systems; 2014. p. 3104–12.

[28] Goodfellow I, Pouget-Abadie J, Mirza M, Xu B, Warde-Farley D, Ozair S, et al. Generative adversarial nets. In: Advances in neural information processing systems; 2014. p. 2672–80.

[29] Silver D, Huang A, Maddison CJ, Guez A, Sifre L, Van Den Driessche G, et al. Mastering the game of go with deep neural networks and tree search. Nature 2016;529(7587):484–9.

[30] Moravčík M, Schmid M, Burch N, Lisý V, Morrill D, Bard N, et al. DeepStack: expert-level artificial intelligence in no-limit poker. arXiv preprint arXiv:1701.01724, 2017.

[31] LeCun Y, et al. LeNet-5, convolutional neural networks. URL: http://yann.lecun.com/exdb/lenet, 2015.

[32] Simonyan K, Zisserman A. Very deep convolutional networks for large-scale image recognition. arXiv preprint arXiv:1409.1556, 2014.

[33] Szegedy C, Liu W, Jia Y, Sermanet P, Reed S, Anguelov D, et al. Going deeper with convolutions. In: Proceedings of the IEEE conference on computer vision and pattern recognition; 2015. p. 1–9.

[34] He K, Zhang X, Ren S, Sun J. Deep residual learning for image recognition. In: Proceedings of the IEEE conference on computer vision and pattern recognition; 2016. p. 770–8.

[35] Gers FA, Schmidhuber J, Cummins F. Learning to forget: continual prediction with LSTM. Neural Computation 2000;12(10):2451–71.

[36] Chung J, Gulcehre C, Cho K, Bengio Y. Empirical evaluation of gated recurrent neural networks on sequence modeling. arXiv preprint arXiv:1412.3555, 2014.

[37] Goodfellow I, Bengio Y, Courville A. Deep learning. MIT Press; 2016.

[38] Wang Z, Yang J, Zhang H, Wang Z, Yang Y, Liu D, et al. Sparse coding and its applications in computer vision. World Scientific; 2015.

[39] Baraniuk RG. Compressive sensing [lecture notes]. IEEE Signal Processing Magazine 2007;24(4):118–21.

[40] Huang J, Zhang T, Metaxas D. Learning with structured sparsity. Journal of Machine Learning Research Nov. 2011;12:3371–412.

[41] Xu H, Zheng J, Alavi A, Chellappa R. Template regularized sparse coding for face verification. In: 23rd International conference on pattern recognition, ICPR 2016; 2016. p. 1448–54.

[42] Xu H, Zheng J, Alavi A, Chellappa R. Cross-domain visual recognition via domain adaptive dictionary learning. CoRR 2018. arXiv:1804.04687 [abs].

[43] Xu H, Zheng J, Chellappa R. Bridging the domain shift by domain adaptive dictionary learning. In: Proceedings of the British machine vision conference 2015, BMVC 2015; 2015. p. 96.1–96.12.

[44] Xu H, Zheng J, Alavi A, Chellappa R. Learning a structured dictionary for video-based face recognition. In: 2016 IEEE winter conference on applications of computer vision, WACV 2016; 2016. p. 1–9.

[45] Candès EJ, Li X, Ma Y, Wright J. Robust principal component analysis? Journal of the ACM (JACM) 2011;58(3):11.

[46] Wen Z, Yin W, Zhang Y. Solving a low-rank factorization model for matrix completion by a nonlinear successive over-relaxation algorithm. Mathematical Programming Computation 2012:1–29.

[47] Wang Z, Li H, Ling Q, Li W. Robust temporal-spatial decomposition and its applications in video processing. IEEE Transactions on Circuits and Systems for Video Technology 2013;23(3):387–400.

[48] Li H, Lu Z, Wang Z, Ling Q, Li W. Detection of blotch and scratch in video based on video decomposition. IEEE Transactions on Circuits and Systems for Video Technology 2013;23(11):1887–900.

[49] Yu Z, Li H, Wang Z, Hu Z, Chen CW. Multi-level video frame interpolation: exploiting the interaction among different levels. IEEE Transactions on Circuits and Systems for Video Technology 2013;23(7):1235–48.

[50] Yu Z, Wang Z, Hu Z, Li H, Ling Q. Video error concealment via total variation regularized matrix completion. In: Image processing (ICIP), 2012 19th IEEE international conference on. IEEE; 2012. p. 1633–6.

[51] Yu Z, Wang Z, Hu Z, Ling Q, Li H. Video frame interpolation using 3-d total variation regularized completion. In: Image processing (ICIP), 2012 19th IEEE international conference on. IEEE; 2012. p. 857–60.

[52] Wang Z, Li H, Ling Q, Li W. Mixed Gaussian-impulse video noise removal via temporal-spatial decomposition. In: Circuits and systems (ISCAS), 2012 IEEE international symposium on. IEEE; 2012. p. 1851–4.

[53] Zhang X, Wang Z, Liu D, Ling Q. DADA: deep adversarial data augmentation for extremely low data regime classification. arXiv preprint arXiv:1809.00981, 2018.

[54] Wu J, Wang Y, Wu Z, Wang Z, Veeraraghavan A, Lin Y. Deep k-means: re-training and parameter sharing with harder cluster assignments for compressing deep convolutions. arXiv preprint arXiv:1806.09228, 2018.

[55] Han S, Mao H, Dally WJ. Deep compression: compressing deep neural networks with pruning, trained quantization and Huffman coding. arXiv preprint arXiv:1510.00149, 2015.

[56] Liu B, Wang M, Foroosh H, Tappen M, Pensky M. Sparse convolutional neural networks. In: Proceedings of the IEEE conference on computer vision and pattern recognition; 2015. p. 806–14.

[57] Lin Y, Sakr C, Kim Y, Shanbhag N. PredictiveNet: an energy-efficient convolutional neural network via zero prediction. In: Circuits and systems (ISCAS), 2017 IEEE international symposium on. IEEE; 2017. p. 1–4.

[58] Ioannou Y, Robertson D, Shotton J, Cipolla R, Criminisi A. Training CNNs with low-rank filters for efficient image classification. arXiv preprint arXiv:1511.06744, 2015.

[59] Sainath TN, Kingsbury B, Sindhwani V, Arisoy E, Ramabhadran B. Low-rank matrix factorization for deep neural network training with high-dimensional output targets. In: Acoustics, speech and signal processing (ICASSP), 2013 IEEE international conference on. IEEE; 2013. p. 6655–9.

[60] Gregor K, LeCun Y. Learning fast approximations of sparse coding. In: ICML; 2010.

[61] Wang Z, Ling Q, Huang T. Learning deep ℓ_0 encoders. AAAI 2016.

[62] Wang Z, Chang S, Liu D, Ling Q, Huang TS. D3: deep dual-domain based fast restoration of jpeg-compressed images. In: IEEE CVPR; 2016.

[63] Wang Z, Yang Y, Chang S, Ling Q, Huang TS. Learning a deep ℓ_∞ encoder for hashing. 2016.

[64] Liu D, Wang Z, Wen B, Yang J, Han W, Huang TS. Robust single image super-resolution via deep networks with sparse prior. IEEE TIP 2016.

[65] Coates A, Ng AY. The importance of encoding versus training with sparse coding and vector quantization. In: ICML; 2011.

[66] Wohlberg B. Efficient convolutional sparse coding. In: ICASSP. IEEE; 2014.

CHAPTER 2

Bi-Level Sparse Coding: A Hyperspectral Image Classification Example*

Zhangyang Wang
Department of Computer Science and Engineering, Texas A&M University, College Station, TX, United States

Contents

2.1. Introduction	9
2.2. Formulation and Algorithm	12
2.2.1 Notations	12
2.2.2 Joint Feature Extraction and Classification	12
2.2.3 Bi-level Optimization Formulation	15
2.2.4 Algorithm	15
2.3. Experiments	17
2.3.1 Classification Performance on AVIRIS Indiana Pines Data	20
2.3.2 Classification Performance on AVIRIS Salinas Data	23
2.3.3 Classification Performance on University of Pavia Data	23
2.4. Conclusion	26
2.5. Appendix	27
References	28

2.1. INTRODUCTION

The spectral information contained in hyperspectral imagery allows characterization, identification, and classification of land-covers with improved accuracy and robustness. However, several critical issues should be addressed in the classification of hyperspectral data, among which are the following [1,2,38]: (1) small amount of available labeled data; (2) high dimensionality of each spectral sample; (3) spatial variability of spectral signatures; (4) high cost of sample labeling. In particular, the large number of spectral channels and small number of labeled training samples pose the problem of the curse of dimensionality and as a consequence resulting in the risk of overfitting the training data. For these reasons, desirable properties of a hyperspectral image classifier should be its ability to produce accurate land-cover maps when working within a high-dimensional feature space, low-sized training datasets, and high levels of spatial spectral signature variability.

---

* ©2015 IEEE. Reprinted, with permission, from Wang, Zhangyang, Nasrabadi, Nasser, and Huang, Thomas S. "Semi-supervised Hyperspectral Classification using Task-driven Dictionary Learning with Regularization." IEEE Transactions on Geosciences and Remote Sensing (2015).

Many supervised and unsupervised classifiers have been developed to tackle the hyperspectral data classification problem [3]. Classical supervised methods, such as artificial neural networks [4,5] and support vector machines (SVMs) [6–9], were readily revealed to be inefficient when dealing with a high number of spectral bands and lack of labeled data. In [10], SVM was regularized with an unnormalized graph Laplacian, thus leading to the Laplacian SVM (LapSVM) that adopts the manifold assumption for semisupervised classification. Another framework based on neural network was presented in [11]. It consists of adding a flexible embedding regularizer to the loss function used for training neural networks, and leads to improvements in both classification accuracy and scalability on several hyperspectral image classification problems. In recent years, kernel-based methods have often been adopted for hyperspectral image classification [12–15]. They are certainly able to handle efficiently the high-dimensional input feature space and deal with the noisy samples in a robust way [16]. More recently, sparse representation has been increasingly popular for image classification. The sparse representation-based classification (SRC) [17] is mainly based on the observation that despite the high dimensionality of natural signals, signals belonging to the same class usually lie in a low-dimensional subspace. In [18], an SRC-based algorithm for hyperspectral classification was presented, that utilizes the sparsity of the input sample with respect to a given overcomplete training dictionary. It is based on a sparsity model where a test spectral pixel is approximately represented by a few training samples (atoms) among the entire atoms from a dictionary. The weights associated with the atoms are called the sparse code. The class label of the test pixel is then determined by the characteristics of the recovered sparse code. Experimental results show remarkable improvements in discriminative effects. However, the main difficulty with all supervised methods is that the learning process heavily depends on the quality of the training dataset. Even worse, labeled hyperspectral training samples are only available in a very limited number due to the cost of sample labeling. On the other hand, unsupervised methods are not sensitive to the number of labeled samples since they operate on the whole dataset, but the relationships between clusters and class labels are not ensured [19]. Moreover, typically in hyperspectral classification, a preliminary feature selection/extraction step is undertaken to reduce the high input space dimensionality, which is time-consuming, scenario-dependent, and needs prior knowledge.

As a trade-off, semisupervised classification methods become a natural alternative to yield better performance. In semisupervised learning literature, the algorithms are provided with some available supervised information in the form of labeled data in addition to the wealth of unlabeled data. Such a framework has recently attracted a considerable amount of research in remote sensing, such as the Laplacian SVM (LapSVM) [9,10], transductive SVM [20], biased-SVM [21] and graph-based methods [22]. Even though the above mentioned algorithms exhibit good performance in classifying hyperspectral images, most of them are based on the assumption that spectrally similar instances should share the same label. However in practice, we may have very different spectra corresponding to the same material, which sometimes makes the above strict assumption no longer valid. Moreover, in most recent hyperspectral classification approaches [23,24], the spatial information is exploited together with the spectral features, encouraging pixels in the local neighborhood to have similar labels. The spatial smoothness assumption holds well in the homogenous regions of hyperspectral images. However, conventional approaches often fail to capture the spatial variability of spectral signatures, e.g., on the border of regions belonging to different classes.

In this chapter, we introduce a hyperspectral image classification method, tackling the problems imposed by the special characteristics of hyperspectral images, namely, high-input dimension of pixels, low number of labeled samples, and spatial variability of the spectral signatures. To this end, the proposed method has the following characteristics and technical contributions:

- **Semisupervised.** Extending the task-driven dictionary learning formulation in [25] to the semisupervised framework for hyperspectral classification, the huge number of unlabeled samples in the image are exploited together with a limited amount of labeled samples to improve the classification performance in a task-driven setting.

- **Joint optimization of feature extraction and classification.** Almost all prior research on hyperspectral classifier design can be viewed as the combinations of two independent parts, extraction of features and a training procedure for designing the classifier. Although, in some prior work raw spectral pixels are used directly, it is widely recognized that features extracted from the input pixels, such as the sparse code, often promote a more discriminative and robust classification [17]. However, to consider the two stages separately typically leads to a suboptimal performance, because the extracted features are not optimized for the best performance of the following classification step. We jointly optimize the classifier parameters and dictionary atoms. This is different from the classical data-driven feature extraction approach [18] that only tries to reconstruct the training samples well. Our joint task-driven formulation ensures that the learned sparse code features are optimal for the classifier.

- **Incorporation of spatial information.** We incorporate spatial information by adding a spatial Laplacian regularization [9] to the probabilistic outputs of the classifier, i.e., the likelihood of the predicted labels. This is more flexible than the popular "naive" Laplacian smoothness constraint that simply enforces all pixels in a local window to have similar learned features.

A novel formulation of bi-level optimization is designed to meet our requirements [26, 27], which is solved by a stochastic gradient descent algorithm [28]. The proposed method is then evaluated on three popular datasets and we see an impressive improvement in performance on all of them. Even for quite ill-posed classification problems, i.e., very small number of high dimensional labeled samples, the proposed method gains a remarkable and stable improvement in performance over comparable methods.

The rest of this chapter is organized as follows. Section 2.2 manifests a step-by-step construction of our formulation in details, followed by the optimization algorithm to solve it. Section 2.3 discusses the classification results of the proposed method in comparison to several other competitive methods, with a wide range of available labeled samples. It also investigates the influences of both the unlabeled samples and dictionary atoms on the classifier's performance, as well as the discriminability of the obtained dictionary. Section 2.4 includes some concluding remarks and indications for the future work.

2.2. FORMULATION AND ALGORITHM

2.2.1 Notations

Consider a hyperspectral image $\mathbf{X} \in R^{m \times n}$ of n pixels, each of which consists of an m-dimensional spectral vector. Let $\mathbf{X} = [\mathbf{x}_1, \mathbf{x}_2, \ldots, \mathbf{x}_n]$ denote the pixel set in a hyperspectral image, with each spectral pixel $\mathbf{x}_i \in R^{m \times 1}$, $i = 1, 2, \ldots, n$. For all the corresponding labels $\mathbf{y} = [y_1, y_2, \ldots, y_n]$, we assume l labels $[y_1, y_2, \ldots, y_l]$ are known, constituting a labeled training set $\mathbf{X}_l = [\mathbf{x}_1, \mathbf{x}_2, \ldots, \mathbf{x}_l]$, while making $\mathbf{X}_u = [\mathbf{x}_{l+1}, \mathbf{x}_{l+2}, \ldots, \mathbf{x}_n]$ the unlabeled training set with $u = n - l$. We assume that the number of labeled samples is uniformly selected for each class. This means that for a K-class classification, each class has $l_c = \frac{l}{K}$ labeled samples.

Without loss of generality, we let all $y_i \in \{-1, 1\}$ to focus on discussing a binary classification. However, the proposed classifier can be naturally extended to a multiclass case, by either replacing the binary classifier with the multiclass classifier (e.g., soft-max classifier [30]), or adopting the well-known one-versus-one or one-versus-all strategy.

Our goal is to jointly learn a dictionary $\mathbf{D}$ consisting of a set of basis for extracting the sparse code (feature vector), and the classification parameter $\mathbf{w}$ for a binary classifier applied to the extracted feature vector, while guaranteeing them to be optimal to each other.

2.2.2 Joint Feature Extraction and Classification

2.2.2.1 Sparse Coding for Feature Extraction

In [18], the authors suggest that the spectral signatures of pixels belonging to the same class are assumed to approximately lie in a low-dimensional subspace. Pixel can be compactly represented by only a few sparse coefficients (sparse code). We adopt the sparse code as the input features, since extensive literature has examined the outstanding effect of SRC for a more discriminative and robust classification [17].

We assume that all the data samples $\mathbf{X} = [\mathbf{x}_1, \mathbf{x}_2, \ldots, \mathbf{x}_n]$, $\mathbf{x}_i \in R^{m \times 1}$, $i = 1, 2, \ldots, n$, are encoded into their corresponding sparse codes $\mathbf{A} = [\mathbf{a}_1, \mathbf{a}_2, \ldots, \mathbf{a}_n]$, $\mathbf{a}_i \in R^{p \times 1}$,

$i = 1, 2, \ldots, n$, using a learned dictionary $\mathbf{D} = [\mathbf{d}_1, \mathbf{d}_2, \ldots, \mathbf{d}_p]$, where $\mathbf{d}_i \in R^{m \times 1}$, $i = 1, 2, \ldots, p$ are the learned atoms. It should be noted that the initial dictionary is generated by assigning equal number of atoms to each class. This means that for a K-class classification, there are $p_c = \frac{p}{K}$ atoms assigned to each class in a dictionary consisting of p atoms.

The sparse representation is obtained by the following convex optimization:

$$\mathbf{A} = \arg\min_{\mathbf{A}} \tfrac{1}{2} ||\mathbf{X} - \mathbf{DA}||_F^2 + \lambda_1 \sum_i ||\mathbf{a}_i||_1 + \lambda_2 ||\mathbf{A}||_F^2, \qquad (2.1)$$

or rewritten in a separate form for each $\mathbf{x}_i$ as

$$\mathbf{a}_i = \arg\min_{\mathbf{a}_i} \tfrac{1}{2} ||\mathbf{x}_i - \mathbf{D}\mathbf{a}_i||_2^2 + \lambda_1 ||\mathbf{a}_i||_1 + \lambda_2 ||\mathbf{a}_i||_2^2. \qquad (2.2)$$

Note $\lambda_2 > 0$ is necessary for proving the differentiability of the objective function (see [2.1] in Appendix). However, setting $\lambda_2 = 0$ proves to work well in practice [25].

Obviously, the effect of sparse coding (2.1) largely depends on the quality of dictionary $\mathbf{D}$. The authors in [18] suggest to construct the dictionary by directly selecting atoms from the training samples. More sophisticated methods are widely used in SRC literature, discussing on how to learn a more compact and effective dictionary from a given training dataset, e.g., the K-SVD algorithm [31].

We recognize that many structured sparsity constraints (priors) [18,32] can also be considered for dictionary learning. They usually exploit the correlations among the neighboring pixels or their features. For example, the SRC dictionary has an inherent group-structured property since it is composed of several class-wise subdictionaries, i.e., the atoms belonging to the same class are grouped together to form a subdictionary. Therefore, it would be reasonable to enforce each pixel to be compactly represented by groups of atoms instead of individual ones. This could be accomplished by encouraging coefficients of only certain groups to be active, like the group lasso [33]. While the performance may be improved by enforcing structured sparsity priors, the algorithm will be considerably more complicated. Therefore, we do not take into account any structured sparsity prior here, and leave them for our future study.

2.2.2.2 Task-Driven Functions for Classification

Classical loss functions in SRC are often defined by the reconstruction error of data samples [18,39]. The performances of such learned classifiers highly hinge on the quality of the input features, which is only suboptimal without the joint optimization with classifier parameters. In [34], the authors study a straightforward joint representation and classification framework, by adding a penalty term to the classification error in addition to the reconstruction error. The authors of [35,36] propose to enhance the dictionary's representative and discriminative power by integrating both the discriminative sparse-code error and the classification error into a single objective function. The approach jointly learns a single dictionary and a predictive linear classifier. However, being a semisupervised method, the unlabeled data does not contribute much to promoting the discriminative effect in [36], as only the reconstruction error is considered on the unlabeled set except for an "expansion" strategy applied to a small set of highly-confident unlabeled samples.

In order to obtain an optimal classifier with regard to the input feature, we exploit a task-driven formulation which aims to minimize a classification-oriented loss [25]. We incorporate the sparse codes $\mathbf{a}_i$, which are dependent on the atoms of the dictionary $\mathbf{D}$ that are to be learned, into the training of the classifier parameter $\mathbf{w}$. The logistic loss is used in the objective function for the classifier. We recognize that the proposed formulation can be easily extended to other classifiers, e.g., SVM. The loss function for the labeled samples is directly defined by the logistic loss

$$L(\mathbf{A}, \mathbf{w}, \mathbf{x}_i, y_i) = \sum_{i=1}^{l} \log(1 + e^{-y_i \mathbf{w}^T \mathbf{a}_i}). \tag{2.3}$$

For unlabeled samples, the label of each $\mathbf{x_i}$ is unknown. We propose to introduce the predicted confidence probability p_{ij} that sample $\mathbf{x_i}$ has label y_j ($y_j = 1$ or -1), which is naturally set as the likelihood of the logistic regression

$$p_{ij} = p(y_j | \mathbf{w}, \mathbf{a}_i, \mathbf{x}_i) = \frac{1}{1 + e^{-y_j \mathbf{w}^T \mathbf{a}_i}}, \quad y_j = 1 \quad \text{or} \quad -1. \tag{2.4}$$

The loss function for the unlabeled samples then turns into an entropy-like form

$$U(\mathbf{A}, \mathbf{w}, \mathbf{x}_i) = \sum_{i=l+1}^{l+u} \sum_{y_j} p_{ij} L(\mathbf{a}_i, \mathbf{w}, \mathbf{x}_i, y_j), \tag{2.5}$$

which can be viewed as a weighted sum of loss under different classification outputs y_j.

Furthermore, we can similarly define p_{ij} for the labeled sample $\mathbf{x}_i$, that is 1 when y_j is the given correct label y_i and 0 elsewhere. The joint loss functions for all the training samples can thus be written into a unified form

$$T(\mathbf{A}, \mathbf{w}) = \sum_{i=1}^{l+u} \sum_{y_j} p_{ij} L(\mathbf{a}_i, \mathbf{w}, \mathbf{x}_i, y_j). \tag{2.6}$$

A semisupervised task-driven formulation has also been proposed in [25]. However, it is posed as a naive combination of supervised and unsupervised steps. The unlabeled data are only used to minimize the reconstruction loss, without contributing to promoting the discriminative effect. In contrast, our formulation (2.6) clearly distinguishes itself by assigning an adaptive confidence weight (2.4) to each unlabeled sample, and minimizes a classification-oriented loss over both labeled and unlabeled samples. By doing so, unlabeled samples also contribute to improving the discriminability of learned features and classifier, jointly with the labeled samples, rather than only optimized for reconstruction loss.

2.2.2.3 Spatial Laplacian Regularization

We first introduce the weighting matrix $\mathbf{G}$, where G_{ik} characterizes the similarity between a pair of pixels $\mathbf{x}_i$ and $\mathbf{x}_k$. We define G_{ik} in the form of shift-invariant bilateral Gaussian filtering [37] (with controlling parameters σ_d and σ_s)

$$G_{ik} = \exp(-\frac{d(\mathbf{x}_i, \mathbf{x}_k)}{2\sigma_d^2}) \cdot \exp(-\frac{||\mathbf{x}_i - \mathbf{x}_k||_2^2}{2\sigma_s^2}), \tag{2.7}$$

which measures both the spatial Euclidean distance ($d(\mathbf{x}_i, \mathbf{x}_k)$) and the spectral similarity between an arbitrary pair of pixels in a hyperspectral image. Larger G_{ik} represents higher similarity and vice versa. Further, rather than simply enforcing pixels within a local window to share the same label, G_{ik} is defined over the whole image and encourages both spatially neighboring and spectrally similar pixels to have similar classification outputs. It makes our spatial constraints much more flexible and effective. Using the above similarity weights, we define the spatial Laplacian regularization function

$$S(\mathbf{A}, \mathbf{w}) = \sum_{i=1}^{l+u} \sum_{y_j} \sum_{k}^{l+u} G_{ik} ||p_{ij} - p_{kj}||_2^2. \tag{2.8}$$

2.2.3 Bi-level Optimization Formulation

Finally, the objective cost function for the joint minimization formulation can be expressed by the following bi-level optimization (the quadratic term of $\mathbf{w}$ is to avoid overfitting)

$$\begin{aligned} \min_{\mathbf{D}, \mathbf{w}} \quad & T(\mathbf{A}, \mathbf{w}) + S(\mathbf{A}, \mathbf{w}) + \frac{\lambda}{2} ||\mathbf{w}||_2^2 \\ s.t. \quad & \mathbf{A} = \arg\min_{\mathbf{A}} \frac{1}{2} ||\mathbf{X} - \mathbf{DA}||_F^2 + \lambda_1 \sum_i ||\mathbf{a}_i||_1 + \lambda_2 ||\mathbf{A}||_F^2. \end{aligned} \tag{2.9}$$

Bi-level optimization [26] has been investigated in both theory and application sides. In [27], the authors propose a general bi-level sparse coding model for learning dictionaries across coupled signal spaces. Another similar formulation has been studied in [25] for general regression tasks.

In the testing stage, each test sample is first represented by solving (2.2) over the learned $\mathbf{D}$. The resulting sparse coefficients are fed to the trained logistic classifier with the previously learned $\mathbf{w}$. The test sample is classified into the class of the highest output probability (2.4).

2.2.4 Algorithm

Built on the similar methodologies of [25] and [27], we solve (2.9) using a projected first order stochastic gradient descent (SGD) algorithm, whose detailed steps are outlined in Algorithm 2.1. At a high level overview, it consists of an outer stochastic gradient descent loop that incrementally samples the training data. It uses each sample to approximate gradients with respect to the classifier parameter $\mathbf{w}$ and the dictionary $\mathbf{D}$, which are then used to update them. Next, we briefly explain a few key technical points of the Algorithm 2.1.

2.2.4.1 Stochastic Gradient Descent

The stochastic gradient descent (SGD) algorithm [28] is an iterative, "online" approach for optimizing an objective function, based on a sequence of approximate gradients obtained by randomly sampling from the training data set. In the simplest case, SGD estimates the objective function gradient on the basis of a single randomly selected example $\mathbf{x}_t$

$$w_{t+1} = w_t - \rho_t \nabla_w F(\mathbf{x}_t, w_t), \tag{2.10}$$

where F is a loss function, w is a weight being optimized and ρ_t is a step size known as the "learning rate". The stochastic process $\{w_t, t = 1, \dots\}$ depends upon the sequence of randomly selected examples $\mathbf{x}_t$ from the training data. It thus optimizes the empirical cost, which is hoped to be a good proxy for the expected cost.

Following the derivations in [25], we can show that the objective function in (2.9), denoted as $B(\mathbf{A}, \mathbf{w})$ for simplicity, is differentiable on $\mathbf{D} \times \mathbf{w}$, and that

$$\nabla_{\mathbf{w}} B(\mathbf{A}, \mathbf{w}) = \mathbb{E}_{\mathbf{x}, y}[\nabla_{\mathbf{w}} T(\mathbf{A}, \mathbf{w}) + \nabla_{\mathbf{w}} S(\mathbf{A}, \mathbf{w}) + \lambda \mathbf{w}],$$
$$\nabla_{\mathbf{D}} B(\mathbf{A}, \mathbf{w}) = \mathbb{E}_{\mathbf{x}, y}[-\mathbf{D}\boldsymbol{\beta}^* \mathbf{A}^{\mathbf{T}} + (\mathbf{X}_t - \mathbf{D}\mathbf{A})\boldsymbol{\beta}^{*T}], \tag{2.11}$$

where $\boldsymbol{\beta}^*$ is a vector defined by the following property:

$$\boldsymbol{\beta}_{SC}^* = 0, \quad \boldsymbol{\beta}_S^* = (\mathbf{D}_S^T \mathbf{D}_S + \lambda_2 \mathbf{I})^{-1} \nabla_{\mathbf{A}_S}[T(\mathbf{A}, \mathbf{w}) + S(\mathbf{A}, \mathbf{w})], \tag{2.12}$$

and S are the indices of the nonzero coefficients of $\mathbf{A}$. The proof of the above equations is given in the Appendix.

2.2.4.2 Sparse Reconstruction

The most computationally intensive step in Algorithm 2.1 is solving the sparse coding (step 3). We adopt the feature-sign algorithm [39] for efficiently solving the exact solution to the sparse coding problem.

Remark on SGD convergence and sampling strategy. As a typical case in machine learning, we use SGD in a setting where it is not guaranteed to converge in theory, but behaves well in practice, as shown in our experiments. (The convergence proof of SGD [29] for nonconvex problems indeed assumes three times differentiable cost functions.)

SGD algorithms are typically designed to minimize functions whose gradients have the form of an expectation. While an i.i.d. (independent and identically distributed) sampling process is required, it cannot be computed in a batch mode. In our algorithm,

---

Algorithm 2.1: Stochastic gradient descent algorithm for solving (2.9).

Require: $\mathbf{X}, \mathbf{Y}$; $\lambda, \lambda_1, \lambda_2, \sigma_d$ and σ_s; $\mathbf{D}_0$ and $\mathbf{w}_0$ (initial dictionary and classifier parameter); ITER (number of iterations); t_0, ρ (learning rate)

1: FOR t=1 to ITER DO
2: Draw a subset $(\mathbf{X}_t, \mathbf{Y}_t)$ from $(\mathbf{X}, \mathbf{Y})$
3: Sparse coding: compute $\mathbf{A}^*$ using feature-sign algorithm:
$$\mathbf{A}^* = \arg\min_{\mathbf{A}} \tfrac{1}{2}||\mathbf{X}_t - \mathbf{DA}||_2^2 + \lambda_1 \sum_i ||\mathbf{a}_i||_1 + \tfrac{\lambda_2}{2}||\mathbf{A}||_2^2$$
4: Compute the active set S (the nonzero support of $\mathbf{A}$)
5: Compute $\boldsymbol{\beta}^*$: Set $\boldsymbol{\beta}_{SC}^* = 0$ and $\boldsymbol{\beta}_S^* = (\mathbf{D}_S^T\mathbf{D}_S + \lambda_2\mathbf{I})^{-1}\nabla_{\mathbf{A_S}}[T(\mathbf{A}, \mathbf{w}) + S(\mathbf{A}, \mathbf{w})]$
6: Choose the learning rate $\rho_t = \min(\rho, \rho\frac{t_0}{t})$
7: Update $\mathbf{D}$ and $\mathbf{W}$ by a projected gradient step:
$$\mathbf{w} = \textstyle\prod_{\mathbf{w}}[\mathbf{w} - \rho_t(\nabla_{\mathbf{w}}T(\mathbf{A}, \mathbf{w}) + \nabla_{\mathbf{w}}S(\mathbf{A}, \mathbf{w}) + \lambda\mathbf{w})]$$
$$\mathbf{D} = \textstyle\prod_{\mathbf{D}}[\mathbf{D} - \rho_t(\nabla_{\mathbf{D}}(-\mathbf{D}\boldsymbol{\beta}^*\mathbf{A}^T + (\mathbf{X}_t - \mathbf{DA})\boldsymbol{\beta}^{*T}))]$$
where $\prod_{\mathbf{w}}$ and $\prod_{\mathbf{D}}$ are respectively orthogonal projections on the embedding spaces of $\mathbf{w}$ and $\mathbf{D}$.
8: END FOR

Ensure: $\mathbf{D}$ and $\mathbf{w}$

---

instead of sampling one per iteration, we adopt a mini-batch strategy by drawing more samples at a time. Authors in [25] further pointed out that solving multiple elastic-net problems with the same dictionary $\mathbf{D}$ can be accelerated by the precomputation of the matrix $\mathbf{D}^T\mathbf{D}$. In practice, we draw a set of 200 samples in each iteration, which produces steadily good results in all our experiments under universal settings.

Strictly speaking, drawing samples from the distribution of training data should be made i.i.d. (step 2 in Algorithm 2.1). However, this is practically difficult since the distribution itself is typically unknown. As an approximation, samples are instead drawn by iterating over random permutations of the training set [29].

2.3. EXPERIMENTS

In this section, we evaluate the proposed method on three popular datasets, and compare it with some related approaches in the literature, including:

- Laplacian Support Vector Machine (LapSVM) [9,10], which is a semisupervised extension of the SVM and applies the spatial manifold assumption to SVM. The classification is directly executed on raw pixels without any feature extraction, which follows the original setting in [10].
- Semisupervised Classification (SSC) approach [40] that employs a modified clustering assumption.

- Semisupervised hyperspectral image segmentation that adopts Multinomial Logistic Regression with Active Learning (MLR-AL) [41].

Regarding parameter choices of the three methods, we try our best to follow the settings in their original papers. For LapSVM, the regularization parameters γ_1, γ_2 are selected from $[10^{-5}, 10^{5}]$ according to a five-fold cross-validation procedure. In SSC, the width parameter of Gaussian function is tuned using a five-fold cross-validation procedure. The parameter setting in MLR-AL follows that of the original paper [41].

Besides the above mentioned three algorithms, we also include the following algorithms in the comparison, in order to illustrate the merits of both joint optimization and spatial Laplacian regularization on the classifier outputs:

- Non-joint optimization of feature extraction and classification (Non-Joint). It refers to conducting the following two stages of the optimization sequentially:

    1. Feature extraction:
    $$\mathbf{A} = \arg\min_{\mathbf{A}} \tfrac{1}{2}\|\mathbf{X} - \mathbf{DA}\|_F^2 + \lambda_1 \sum_i \|\mathbf{a_i}\|_1 + \lambda_2 \|\mathbf{A}\|_F^2.$$
    2. Learning a classifier:
    $$\min_{\mathbf{w}} \quad T(\mathbf{A}, \mathbf{w}) + \tfrac{\lambda}{2}\|\mathbf{w}\|_2^2. \tag{2.13}$$

    The training of $\mathbf{D}$ is independent of the learning of the classifier parameter $\mathbf{w}$. This is different from the joint optimization of the dictionary and classifier as is done in (2.9) by the task-driven formulation.

- Non-joint optimization of feature extraction and classification, with spatial Laplacian regularization (Non-Joint + Laplacian). It is the same as the Non-Joint method except for adding a spatial Laplacian regularization term $S(\mathbf{A}, \mathbf{w})$ to the second subproblem:

    1. Feature extraction:
    $$\mathbf{A} = \arg\min_{\mathbf{A}} \tfrac{1}{2}\|\mathbf{X} - \mathbf{DA}\|_F^2 + \lambda_1 \sum_i \|\mathbf{a}_i\|_1 + \lambda_2 \|\mathbf{A}\|_F^2.$$
    2. Learning a classifier:
    $$\min_{\mathbf{w}} \quad T(\mathbf{A}, \mathbf{w}) + S(\mathbf{A}, \mathbf{w}) + \tfrac{\lambda}{2}\|\mathbf{w}\|_2^2. \tag{2.14}$$

- The proposed joint method without spatial Laplacian regularization (Joint), which is done by dropping the $S(\mathbf{A}, \mathbf{W})$ term in (2.9)

$$\min_{\mathbf{D},\mathbf{W}} \quad T(\mathbf{A}, \mathbf{w}) + \tfrac{\lambda}{2}\|\mathbf{w}\|_2^2$$
$$s.t. \quad \mathbf{A} = \arg\min_{\mathbf{A}} \tfrac{1}{2}\|\mathbf{X} - \mathbf{DA}\|_F^2 + \lambda_1 \sum_i \|\mathbf{a}_i\|_1 + \lambda_2 \|\mathbf{A}\|_F^2. \tag{2.15}$$

- The proposed joint method with spatial Laplacian regularization (Joint + Laplacian), by minimizing our proposed bi-level formulation (2.9).

Parameter setting. For the proposed method, the regularization parameter λ in (2.9) is fixed to be 10^{-2}, and λ_2 in (2.2) is set to 0 to exploit sparsity. The elastic-net parameter λ_1 in (2.2) is generated by a cross-validation procedure, which is similar to the one in [25]. The values of λ_1 are 0.225, 0.25, and 0.15 for the three experiments in Sections 2.3.1–2.3.3, respectively; σ_d and σ_s in (2.7) are fixed as 3 and 3000, respectively. The learning rate ρ is set as 1, and maximum number ITER is set as 1000 for all. Although, we have verified that these choices of parameters work well in extensive experiments, we recognize that a finer tuning of them may further improve the performance.

In particular, we would like to mention the initializations of $\mathbf{D}$ and $\mathbf{w}$. For the two non-joint methods, $\mathbf{D}$ is initialized by solving the first subproblem (feature extraction) in (2.13) or (2.14). In this subproblem, for each class, we initialize its subdictionary atoms randomly. We then employ several iterations of K-SVD using only the available labeled data for that class, and finally combine all the output class-wise sub-dictionaries into a single initial dictionary $\mathbf{D}$. Next, we solve $\mathbf{A}$ based on $\mathbf{D}$, and continue to feed $\mathbf{A}$ into the second subproblems (learning a classifier) in (2.13) and (2.14) for good initializations of $\mathbf{w}$, for Non-Joint and Non-Joint + Laplacian, respectively. For the two joint methods, we use the results of Non-Joint, and Non-Joint + Laplacian, to initialize $\mathbf{D}$ and $\mathbf{w}$ of Joint and Joint + Laplacian, respectively.

The one-versus-all strategy is adopted for addressing multiclass problems, which means that we train K different binary logistic classifiers with K corresponding dictionaries for a K-class problem. For each test sample, the classifier with the maximum score will provide the class label. When the class number is large, this one-versus-all approach has proven to be more scalable than learning a single large dictionary with a multi-class loss [25], while providing very good results.

For the two joint methods, we assign only five dictionary atoms per class to initialize the dictionary, which means for a K-class problem we have $p_c = 5$ and $p = 5K$ for the total dictionary size. For the two non-joint methods, 50 dictionary atoms ($p_c = 50$) are assigned per class in the first subproblems of (2.13) and (2.14), respectively. The reason why we use the term "atoms per class" are two-fold: (1) we initialize our dictionary by first applying KSVD to each class to obtain a class-wise subdictionary. This helps to improve the class discriminability of the learned dictionary more than just applying KSVD to the whole data. Therefore, we need to specify how many atoms are assigned per class in the initialization stage. Note when Algorithm 2.1 starts, the atoms become all entangled, and further it is impossible to identify how many (and which) atoms are representing a specific class in the final learned dictionary; (2) As each dataset has a different number of classes, and empirically, more classes demand more dictionary atoms to represent. Note, however, if we assign atoms in proportion to the number of samples per class, some minor classes will tend to be severely underrepresented. In all experiments here, we by default use all the unlabeled pixels (denoted as "ALL") from each hyperspectral image for semi-supervised training.

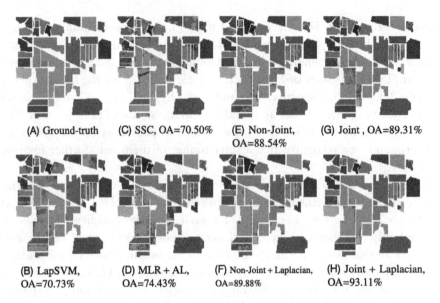

(A) Ground-truth   (C) SSC, OA=70.50%   (E) Non-Joint, OA=88.54%   (G) Joint , OA=89.31%

(B) LapSVM, OA=70.73%   (D) MLR + AL, OA=74.43%   (F) Non-Joint + Laplacian, OA=89.88%   (H) Joint + Laplacian, OA=93.11%

Figure 2.1 Classification maps for the AVIRIS Indian Pines scene using different methods with 10 labeled samples per class.

2.3.1 Classification Performance on AVIRIS Indiana Pines Data

The AVIRIS sensor generates 220 bands across the spectral range from 0.2 to 2.4 μm. In the experiment, the number of bands is reduced to 200 by removing 20 water absorption bands. The AVIRIS Indiana Pines hyperspectral image has the spatial resolution of 20 m and 145×145 pixels. It contains 16 classes, most of which are different types of crops (e.g., corns, soybeans, and wheats). The ground–truth classification map is shown in Fig. 2.1A.

Table 2.1 evaluates the influence of the number of labeled samples per class l_c on the classification of AVIRIS Indiana Pines data, with l_c varying from 2 to 10. The dictionary consists of only $p = 80$ atoms to represent all the 16 classes for the joint methods, and $p = 800$ for the non–joint methods. The bold value in each column indicates the best result among all the seven methods. As can be seen from the table, the classification results improve for all the algorithms with the increase in the number of labeled samples. The last two methods, i.e., "Joint" and "Joint + Laplacian", outperform the other five methods in terms of overall accuracy (OA) significantly. It is also observed that the "Joint + Laplacian" method obtains further improvement over the "Joint" method, showing the advantage of spatial Laplacian regularization. Amazingly, we notice that even when there are as few as three samples per class, the OA of the proposed method ("Joint + Laplacian") is still higher than 80%.

Table 2.1 Overall classification results (%) for the AVIRIS Indiana Pines data with different numbers of labeled samples per class (u = ALL, $\lambda = 10^{-2}, \lambda_1 = 0.225, \lambda_2 = 0, \rho = 1, \sigma_d = 3, \sigma_s = 3000$)

l_c	2	3	4	5	6	7	8	9	10
LapSVM [9]	57.80	61.32	63.1	66.39	68.27	69.00	70.15	70.04	70.73
SSC [40]	44.61	56.98	58.27	60.56	60.79	64.19	66.81	69.40	70.50
MLR+AL [41]	52.34	56.16	59.21	61.47	65.16	69.21	72.14	73.89	74.43
Non-Joint ($p_c = 50$)	63.72	69.21	71.87	76.88	79.04	81.81	85.23	87.77	88.54
Non-Joint + Laplacian ($p_c = 50$)	66.89	72.37	75.33	78.78	81.21	84.98	87.25	88.61	89.88
Joint ($p_c = 5$)	69.81	76.03	80.42	82.91	84.81	85.76	86.95	87.54	89.31
Joint + Laplacian ($p_c = 5$)	**76.55**	**80.63**	**84.28**	**86.33**	**88.27**	**90.68**	**91.87**	**92.53**	**93.11**

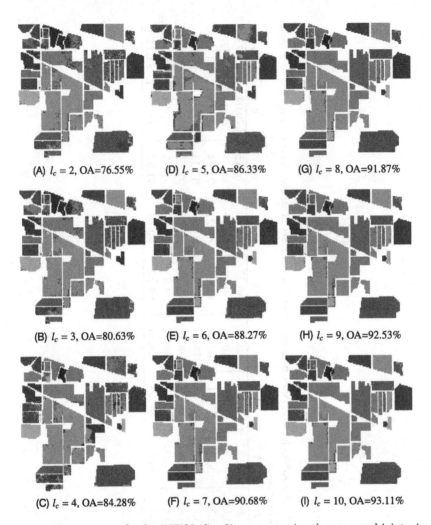

(A) $l_c = 2$, OA=76.55%   (D) $l_c = 5$, OA=86.33%   (G) $l_c = 8$, OA=91.87%

(B) $l_c = 3$, OA=80.63%   (E) $l_c = 6$, OA=88.27%   (H) $l_c = 9$, OA=92.53%

(C) $l_c = 4$, OA=84.28%   (F) $l_c = 7$, OA=90.68%   (I) $l_c = 10$, OA=93.11%

Figure 2.2 Classification maps for the AVIRIS Indian Pines scene using the proposed Joint + Laplacian method with different numbers of labeled samples per class.

Fig. 2.1 demonstrates the classification maps obtained by all the methods when 10 labeled samples are used per class. The proposed method, either with or without spatial regularization, obtains much less misclassifications compared with the other methods. What is more, the homogenous areas in (H) are significantly better preserved than that in (G), which again confirms the effectiveness of the spatial Laplacian regularization on the output of the classifier. Fig. 2.2 visually demonstrates that along with the increase of l_c, the classification results gradually improve, and both the regional and scattered misclassifications are reduced dramatically.

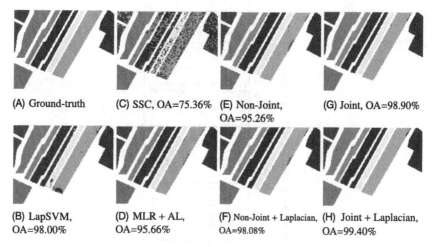

(A) Ground-truth (C) SSC, OA=75.36% (E) Non-Joint, OA=95.26% (G) Joint, OA=98.90%

(B) LapSVM, OA=98.00% (D) MLR + AL, OA=95.66% (F) Non-Joint + Laplacian, OA=98.08% (H) Joint + Laplacian, OA=99.40%

Figure 2.3 Classification maps for the AVIRIS Salinas scene using different methods with 10 labeled samples per class.

2.3.2 Classification Performance on AVIRIS Salinas Data

This dataset is collected over the Valley of Salinas, Southern California, in 1998. This hyperspectral image is of size 217×512, with 16 different classes of objects in total. In our experiment, a nine-class subset is considered, including vegetables, bare soils, and vineyard fields. The ground-truth classification map is shown in Fig. 2.3A. As AVIRIS Salinas is recognized to be easier for classification than AVIRIS Indian Pines, all methods obtain high OAs as listed in Table 2.2, while the "Joint + Laplacian" method marginally stands out. When we turn to Fig. 2.3B–H for the comparison in classification maps, however, the "Joint + Laplacian" method is visually much superior in reducing scattered misclassifications.

2.3.3 Classification Performance on University of Pavia Data

The ROSIS sensor collected this data during a flight campaign over the Pavia district in northern Italy. A total of 103 spectral bands were used for data acquisition in this dataset, comprising 610×610 pixel images with a geometric resolution of 1.3 m. A few samples contain no information and were discarded before the classification. The ground-truth data shows a total of nine distinct classes, and has been portrayed visually in Fig. 2.4A. Similar conclusions can be attained from both Table 2.3 and Fig. 2.4 that once again verify the merits of both the joint optimization framework and spatial regularization.

Table 2.2 Overall classification results (%) for the AVIRIS Salinas data with different numbers of labeled samples per class (u = ALL, $\lambda = 10^{-2}$, $\lambda_1 = 0.25$, $\lambda_2 = 0$, $\rho = 1$, $\sigma_d = 3$, $\sigma_s = 3000$)

l_c	2	3	4	5	6	7	8	9	10
LapSVM [9]	**90.77**	91.53	**92.95**	93.50	94.77	95.08	96.05	97.17	98.00
SSC [40]	59.47	61.84	64.90	67.19	71.04	73.04	72.81	73.51	75.36
MLR+AL [41]	78.98	82.32	84.31	86.27	85.86	89.41	92.27	93.78	95.66
Non-Joint ($p_c = 50$)	85.88	87.21	89.29	90.76	91.42	92.87	93.95	94.78	95.26
Non-Joint + Laplacian ($p_c = 50$)	87.67	89.28	91.54	92.67	93.93	95.28	96.79	97.83	98.08
Joint ($p_c = 5$)	89.71	90.03	91.42	92.12	93.25	94.54	96.05	97.45	98.90
Joint + Laplacian ($p_c = 5$)	90.65	**91.59**	92.28	**93.63**	**95.22**	**96.58**	**97.81**	**98.53**	**99.40**

Table 2.3 Overall classification results (%) for the university of Pavia data with different numbers of labeled samples per class (u = ALL, $\lambda = 10^{-2}$, $\lambda_1 = 0.15$, $\lambda_2 = 0$, $\rho = 1$, $\sigma_d = 3$, $\sigma_s = 3000$)

l_c	2	3	4	5	6	7	8	9	10
LapSVM [9]	64.77	67.83	69.25	71.05	72.97	74.38	76.75	78.17	79.88
SSC [40]	69.54	72.84	74.69	76.21	77.24	78.43	79.81	80.25	80.95
MLR+AL [41]	76.27	78.66	79.30	80.22	81.36	82.41	83.27	84.78	85.53
Non-Joint ($p_c = 50$)	74.21	75.27	76.22	76.83	78.24	79.51	79.67	80.83	81.26
Non-Joint + Laplacian ($p_c = 50$)	**79.23**	80.26	82.58	**84.07**	**86.21**	86.88	87.56	88.23	88.78
Joint ($p_c = 5$)	74.21	76.73	79.24	80.82	82.35	84.54	86.97	87.27	88.08
Joint + Laplacian ($p_c = 5$)	78.56	**80.29**	**82.84**	83.76	85.12	**87.58**	**88.33**	**89.52**	**90.41**

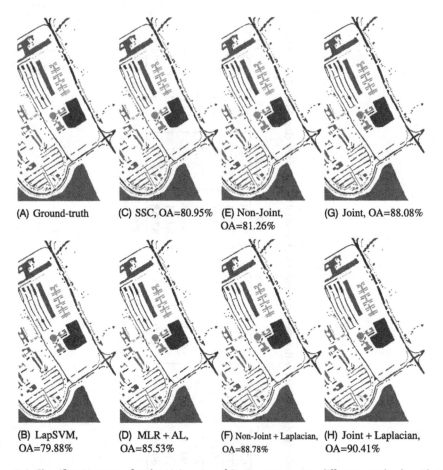

(A) Ground-truth   (C) SSC, OA=80.95%   (E) Non-Joint,   (G) Joint, OA=88.08%
                                          OA=81.26%

(B) LapSVM,   (D) MLR + AL,   (F) Non-Joint + Laplacian,   (H) Joint + Laplacian,
OA=79.88%   OA=85.53%   OA=88.78%   OA=90.41%

Figure 2.4 Classification maps for the University of Pavia scene using different methods with 10 labeled samples per class.

2.4. CONCLUSION

In this chapter, we develop a semisupervised hyperspectral image classification method based on task-driven dictionary learning and spatial Laplacian regularization on the output of the logistic regression classifier. We jointly optimize both the dictionary for feature extraction and the associated classifier parameter, while both the spectral and the spatial information are explored to improve the classification accuracy. Experimental results verify the superior performance of our proposed method on three popular datasets, both quantitatively and qualitatively. A good and stable accuracy is produced in even quite ill-posed problem settings (high dimensional spaces with small number of labeled samples). In the future, we would like to explore the applications of the proposed method to general image classification and segmentation problems.

2.5. APPENDIX

Denote $\mathbf{X} \in \mathcal{X}$, $\mathbf{y} \in \mathcal{Y}$ and $\mathbf{D} \in \mathcal{D}$. Let the objective function $B(\mathbf{A}, \mathbf{w})$ in (2.9) be denoted as B for short. The differentiability of B with respect to $\mathbf{w}$ is easy to show, using only the compactness of $\mathcal{X}$ and $\mathcal{Y}$, as well as the fact that B is twice differentiable.

We will therefore focus on showing that B is differentiable with respect to $\mathbf{D}$, which is more difficult since $\mathbf{A}$, and thus $\mathbf{a}_i$, is not differentiable everywhere. Without loss of generality, we use a vector $\mathbf{a}$ instead of $\mathbf{A}$ for simplifying the derivations hereinafter. In some cases, we may equivalently express $\mathbf{a}$ as $\mathbf{a}(\mathbf{D}, \mathbf{w})$ in order to emphasize the functional dependence.

We recall the following theorem [2.1] that is proved in [42]:

Theorem 2.1 (Regularity of the elastic net solution). *Consider the formulation in (2.1). Assume $\lambda_2 > 0$, and that both $\mathcal{X}$ and $\mathcal{Y}$ are compact. Then,*

- *$\mathbf{a}$ is uniformly Lipschitz on $\mathcal{X} \times \mathcal{D}$.*
- *Let $\mathbf{D} \in \mathcal{D}$, σ be a positive scalar and $\mathbf{s}$ be a vector in $\{-1, 0, 1\}^p$. Define $K_s(\mathbf{D}, \sigma)$ as the set of vectors $\mathbf{x}$ satisfying for all j in $\{1, \dots, p\}$,*

$$
\begin{aligned}
|\mathbf{d}_j^T(\mathbf{x} - \mathbf{Da}) - \lambda_2 \mathbf{a}[j]| \leq \lambda_1 - \sigma \quad &\text{if} \quad \mathbf{s}[j] = 0, \\
\mathbf{s}[j]\mathbf{a}[j] \geq \sigma \quad &\text{if} \quad \mathbf{s}[j] \neq 0.
\end{aligned}
\tag{2.16}
$$

Then there exists $\kappa > 0$ independent of $\mathbf{s}$, $\mathbf{D}$ and σ so that for all $\mathbf{x} \in K_s(\mathbf{D}, \sigma)$, the function $\mathbf{a}$ is twice continuously differentiable on $B_{\kappa\sigma}(\mathbf{x}) \times B_{\kappa\sigma}(\mathbf{D})$, where $B_{\kappa\sigma}(\mathbf{x})$ and $B_{\kappa\sigma}(\mathbf{D})$ denote the open balls of radius $\kappa\sigma$ respectively centered on $\mathbf{x}$ and $\mathbf{D}$.

Built on [2.1] and given a small perturbation $\mathbf{E} \in R^{m \times p}$, it follows that

$$
\begin{aligned}
B(\mathbf{a}(\mathbf{D} + \mathbf{E}), \mathbf{w}) - B(\mathbf{a}(\mathbf{D}), \mathbf{w}) = \\
\nabla_z B_{\mathbf{w}}^T(\mathbf{a}(\mathbf{D} + \mathbf{E}) - \mathbf{a}(\mathbf{D})) + O(\|\mathbf{E}\|_F^2),
\end{aligned}
\tag{2.17}
$$

where the term $O(\|\mathbf{E}\|_F^2)$ is based on the fact that $\mathbf{a}(\mathbf{D}, \mathbf{x})$ is uniformly Lipschitz and $\mathcal{X} \times \mathcal{D}$ is compact. It is then possible to show that

$$
\begin{aligned}
B(\mathbf{a}(\mathbf{D} + \mathbf{E}), \mathbf{w}) - B(\mathbf{a}(\mathbf{D}), \mathbf{w}) = \\
Tr(\mathbf{E}^T g(\mathbf{a}(\mathbf{D} + \mathbf{E}), \mathbf{w})) + O(\|\mathbf{E}\|_F^2)
\end{aligned}
\tag{2.18}
$$

where g has the form given in (2.11). This shows that f is differentiable on $\mathcal{D}$, and its gradient with respect to $\mathbf{D}$ is g.

REFERENCES

[1] Swain PH, Davis SM. Fundamentals of pattern recognition in remote sensing. In: Remote sensing: the quantitative approach. New York: McGraw-Hill; 1978.

[2] Plaza A, Benediktsson J, Boardman J, Brazile J, Bruzzone L, Camps-Valls G, Chanussot J, Fauvel M, Gamba P, Gualtieri A, Marconcini M, Tiltoni J, Trianni G. Recent advances in techniques for hyperspectral image processing. Remote Sensing of Environment Sept. 2009;113(s1):s110–22.

[3] Richards JA, Jia X. Remote sensing digital image analysis: an introduction. Berlin, Germany: Springer-Verlag; 1999.

[4] Bischof H, Leona A. Finding optimal neural networks for land use classification. IEEE Transactions on Geoscience and Remote Sensing Jan. 1998;36(1):337–41.

[5] Yang H, van der Meer F, Bakker W, Tan ZJ. A back-propagation neural network for mineralogical mapping from AVIRIS data. International Journal on Remote Sensing Jan. 1999;20(1):97–110.

[6] Cristianini N, Shawe-Taylor J. An introduction to support vector machines and other kernel-based learning methods. Cambridge University Press; 2000.

[7] Scholkopf B, Smola A. Learning with kernels: support vector machines, regularization, optimization and beyond. Cambridge, MA: MIT Press; 2002.

[8] Bovolo F, Bruzzone L, Carline L. A novel technique for subpixel image classification based on support vector machine. IEEE Transactions on Image Processing Nov. 2010;19(11):2983–99.

[9] Belkin M, Niyogi P, Sindhwani V. Manifold regularization: a geometric framework for learning from labeled and unlabeled examples. Journal of Maching Learning Research Nov. 2006;7:2399–434.

[10] Gomez-Chova L, Camps-Valls G, Munoz-Mari J, Calpe J. Semi-supervised image classification with Laplacian support vector machines. IEEE Geoscience and Remote Sensing Letters Jul. 2008;5(3):336–40.

[11] Ratle F, Camps-Valls G, Weston J. Semi-supervised neural networks for efficient hyperspectral image classification. IEEE Transactions on Geoscience and Remote Sensing May. 2010;48(5):2271–82.

[12] Camps-Valls G, Gomez-Chova L, Munoz-Marı J, Vila-Frances J, Calpe-Maravilla J. Composite kernels for hyperspectral image classification. IEEE Geoscience and Remote Sensing Letters Jan. 2006;3(1):93–7.

[13] Huang C, Davis LS, Townshend JRG. An assessment of support vector machines for land cover classification. International Journal on Remote Sensing Feb. 2002;23(4):725–49.

[14] Camps-Valls G, Gomez-Chova L, Calpe J, Soria E, Martín JD, Alonso L, Moreno J. Robust support vector method for hyperspectral data classification and knowledge discovery. IEEE Transactions on Geoscience and Remote Sensing Jul. 2004;42(7):1530–42.

[15] Camps-Valls G, Bruzzone L. Kernel-based methods for hyperspectral image classification. IEEE Transactions on Geoscience and Remote Sensing Jun. 2005;43(6):1351–62.

[16] Shawe-Taylor J, Cristianini N. Kernel methods for pattern analysis. Cambridge, UK: Cambridge University Press; 2004.

[17] Wright J, Yang A, Ganesh A, Sastry SS, Ma Y. Robust face recognition via sparse representation. IEEE Transactions on Pattern Analysis and Machine Intelligence Feb. 2009;31(2):210–27.

[18] Chen Y, Nasrabadi NM, Tran TD. Hyperspectral image classification using dictionary-based sparse representation. IEEE Transactions on Geoscience and Remote Sensing 2011;49(10):3973–85.

[19] Bidhendi SK, Shirazi AS, Fotoohi N, Ebadzadeh MM. Material classification of hyperspectral images using unsupervised fuzzy clustering methods. In: Proceedings of third international IEEE conference on signal-image technologies and internet-based system (SITIS); 2007. p. 619–23.

[20] Bruzzone L, Chi M, Marconcini M. A novel transductive SVM for semisupervised classification of remote sensing images. IEEE Transactions on Geoscience and Remote Sensing 2006;44(11):3363–73.

[21] Jordi MM, Francesca B, Luis GC, Lorenzo B, Gustavo CV. Semisupervised one-class support vector machines for classification of remote sensing data. IEEE Transactions on Geoscience and Remote Sensing 2010;48(8):3188–97.

[22] Gu YF, Feng K. L_1-graph semisupervised learning for hyperspectral image classification. In: Proceedings of IEEE international conference on geoscience and remote sensing symposium (IGARSS); 2012. p. 1401–4.

[23] Li J, Bioucas-Dias J, Plaza A. Spectral-spatial hyperspectral image segmentation using subspace multinomial logistic regression and Markov random fields. IEEE Transactions on Geoscience and Remote Sensing 2012;50(3):809–23.

[24] Mathur A, Foody GM. Multiclass and binary SVM classification: implications for training and classification users. IEEE Geoscience and Remote Sensing Letters Apr. 2008;5(2):241–5.

[25] Mairal J, Bach F, Ponce J. Task-driven dictionary learning. IEEE Transactions on Pattern Analysis and Machine Intelligence Mar. 2012;34(2):791–804.

[26] Colson B, Marcotte P, Savard G. An overview of bilevel optimization. Annals of Operations Research Sept. 2007;153(1):235–56.

[27] Yang J, Wang Z, Lin Z, Shu X, Huang T. Bilevel sparse coding for coupled feature spaces. In: Proceedings of the IEEE conference on computer vision and pattern recognition (CVPR); 2012. p. 2360–7.

[28] Kushner HJ, Yin GG. Stochastic approximation and recursive algorithms with applications. Second edition. Springer; 2003.

[29] Bottou Léon. Online algorithms and stochastic approximations. In: Online learning and neural networks. Cambridge University Press; 1998.

[30] Duan K, Keerthi S, Chu W, Shevade S, Poo A. Multi-category classification by soft-max combination of binary classifiers. Lecture Notes in Computer Science 2003;2709:125–34.

[31] Aharon M, Elad M, Bruckstein A. K-SVD: an algorithm for designing overcomplete dictionaries for sparse representation. IEEE Transactions on Signal Processing Nov. 2006;54(11):4311–22.

[32] Sun X, Qu Q, Nasrabadi NM, Tran TD. Structured priors for sparse-representation-based hyperspectral image classification. IEEE Geoscience and Remote Sensing Letters 2014;11(4):1235–9.

[33] Simon N, Friedman J, Hastie T, Tibshirani R. A sparse group Lasso. Journal of Computational and Graphical Statistics 2013;22(2):231–45.

[34] Pham D, Venkatesh S. Joint learning and dictionary construction for pattern recognition. In: Proceedings of IEEE conference on computer vision and pattern recognition (CVPR); 2008. p. 1–8.

[35] Jiang Z, Lin Z, Davis L. Learning a discriminative dictionary for sparse coding via label consistent K-SVD. In: Proceedings of IEEE conference on computer vision and pattern recognition (CVPR); 2011. p. 1697–704.

[36] Zhang G, Jiang Z, Davis L. Online semi-supervised discriminative dictionary learning for sparse representation. In: Proceedings of IEEE Asian conference on computer vision (ACCV); 2012. p. 259–73.

[37] Tomasi C, Manduchi R. Bilateral filtering for gray and color images. In: Proceedings of the IEEE international conference on computer vision (ICCV); 1998. p. 839–46.

[38] Wang Qi, Meng Zhaotie, Li Xuelong. Locality adaptive discriminant analysis for spectral-spatial classification of hyperspectral images. IEEE Geoscience and Remote Sensing Letters 2017;14(11):2077–81.

[39] Lee H, Battle A, Raina R, Ng AY. Efficient sparse coding algorithms. In: Advances in neural information processing systems (NIPS); 2007. p. 801–8.

[40] Wang Y, Chen S, Zhou Z. New semi-supervised classification method based on modified clustering assumption. IEEE Transactions on Neural Network May 2012;23(5):689–702.

[41] Li J, Bioucas-Dias J, Plaza A. Semisupervised hyperspectral image segmentation using multinomial logistic regression with active learning. IEEE Transactions on Geoscience and Remote Sensing Nov. 2010;48(11):4085–98.

[42] Marial J, Bach F, Ponce J, Sapiro G. Online dictionary learning for sparse coding. In: Proceeding of international conference on machine learning (ICML); 2009. p. 689–96.

CHAPTER 3

Deep ℓ_0 Encoders: A Model Unfolding Example*

Zhangyang Wang

Department of Computer Science and Engineering, Texas A&M University, College Station, TX, United States

Contents

3.1. Introduction	31
3.2. Related Work	32
3.2.1 ℓ_0- and ℓ_1-Based Sparse Approximations	32
3.2.2 Network Implementation of ℓ_1-Approximation	33
3.3. Deep ℓ_0 Encoders	34
3.3.1 Deep ℓ_0-Regularized Encoder	34
3.3.2 Deep M-Sparse ℓ_0 Encoder	36
3.3.3 Theoretical Properties	37
3.4. Task-Driven Optimization	37
3.5. Experiment	38
3.5.1 Implementation	38
3.5.2 Simulation on ℓ_0 Sparse Approximation	38
3.5.3 Applications on Classification	40
3.5.4 Applications on Clustering	42
3.6. Conclusions and Discussions on Theoretical Properties	43
References	44

3.1. INTRODUCTION

Sparse signal approximation has gained popularity over the last decade. The sparse approximation model suggests that a natural signal could be compactly approximated by only a few atoms out of a properly given dictionary, where the weights associated with the dictionary atoms are called the sparse codes. Proven to be both robust to noise and scalable to high-dimensional data, sparse codes are known as powerful features, and benefit a wide range of signal processing applications, such as source coding [1], denoising [2], source separation [3], pattern classification [4], and clustering [5].

We are particularly interested in the ℓ_0-based sparse approximation problem, which is the fundamental formulation of sparse coding [6]. The nonconvex ℓ_0 problem is

* Reprinted, with permission, from Wang, Zhangyang, Ling, Qing, and Huang, Thomas S. "Learning deep ℓ_0 encoders", AAAI (2016).

intractable and often instead attacked by minimizing surrogate measures, such as the ℓ_1-norm, which leads to more tractable computational methods. However, it has been both theoretically and practically discovered that solving ℓ_0 sparse approximation is still preferable in many cases.

More recently, deep learning has attracted great attentions in many feature learning problems [7]. The advantages of deep learning lie in its composition of multiple nonlinear transformations to yield more abstract and descriptive embedding representations. With the aid of gradient descent, it also scales linearly in time and space with the number of train samples.

It has been noticed that sparse approximation and deep learning bear certain connections [8]. Their similar methodology has been lately exploited in [9], [10], [11], [12]. By turning sparse coding models into deep networks, one may expect faster inference, larger learning capacity, and better scalability. The network formulation also facilitates the integration of task-driven optimization.

In this chapter, we investigate two typical forms of ℓ_0-based sparse approximation problems, the ℓ_0-regularized problem, and the M-sparse problem. Based on solid iterative algorithms [13], we formulate them as feed-forward neural networks [8], called **deep ℓ_0 encoders**, through introducing novel neurons and pooling functions. We study their applications in image classification and clustering; in both cases the models are optimized in a task-driven, end-to-end manner. Impressive performances are observed in numerical experiments.

3.2. RELATED WORK

3.2.1 ℓ_0- and ℓ_1-Based Sparse Approximations

Finding the sparsest, or minimum ℓ_0-norm, representation of a signal given a dictionary of basis atoms is an important problem in many application domains. Consider a data sample $\mathbf{x} \in R^{m \times 1}$ that is encoded into its sparse code $\mathbf{a} \in R^{p \times 1}$ using a learned dictionary $\mathbf{D} = [\mathbf{d}_1, \mathbf{d}_2, \ldots, \mathbf{d}_p]$, where $\mathbf{d}_i \in R^{m \times 1}, i = 1, 2, \ldots, p$ are the learned atoms. The sparse codes are obtained by solving the ℓ_0-**regularized problem** (λ is a constant)

$$\mathbf{a} = \arg\min_{\mathbf{a}} \tfrac{1}{2}||\mathbf{x} - \mathbf{D}\mathbf{a}||_F^2 + \lambda ||\mathbf{a}||_0. \tag{3.1}$$

Alternatively, one could explicitly impose constraints on the number of nonzero coefficients of the solution, by solving the M-**sparse problem**

$$\mathbf{a} = \arg\min_{\mathbf{a}} ||\mathbf{x} - \mathbf{D}\mathbf{a}||_F^2 \qquad s.t. \quad ||\mathbf{a}||_0 \leq M. \tag{3.2}$$

Unfortunately, these optimization problems are often intractable because there is a combinatorial increase in the number of local minima as the number of the candidate basis vectors increases. One potential remedy is to employ a convex surrogate measure, such as the ℓ_1-norm, in place of the ℓ_0-norm that leads to a more tractable optimization problem. For example, (3.1) could be relaxed as

$$\mathbf{a} = \arg\min_{\mathbf{a}} \tfrac{1}{2}||\mathbf{x} - \mathbf{Da}||_F^2 + \lambda||\mathbf{a}||_1. \tag{3.3}$$

It creates an unimodal optimization problem that can be solved via linear programming techniques. The downside is that we have now introduced a mismatch between the ultimate goal and the objective function [14]. Under certain conditions, the minimum ℓ_1-norm solution equals to that of the minimum ℓ_0-norm [6]. But in practice, the ℓ_1-approximation is often used way beyond these conditions, and is thus quite heuristic. As a result, we often get a solution which is not exactly minimizing the original ℓ_0-norm.

That said, ℓ_1-approximation is found to work practically well for many sparse coding problems. Yet in certain applications, we intend to control the exact number of nonzero elements, such as basis selection [14], where ℓ_0-approximation is indispensable. Beyond that, ℓ_0-approximation is desirable for performance concerns in many ways. In compressive sensing literature, empirical evidence [15] suggested that using an iterative reweighted ℓ_1-scheme to approximate the ℓ_0-solution often improved the quality of signal recovery. In image enhancement, it was shown in [16] that ℓ_0 data fidelity was more suitable for reconstructing images corrupted with impulse noise. For the purpose of image smoothing, the authors of [17] utilized ℓ_0 gradient minimization to globally control how many nonzero gradients are used to approximate prominent structures in a structure-sparsity-management manner. Recent work [18] revealed that ℓ_0 sparse subspace clustering can completely characterize the set of minimal union-of-subspace structure, without additional separation conditions required by its ℓ_1 counterpart.

3.2.2 Network Implementation of ℓ_1-Approximation

In [8], a feed-forward neural network, as illustrated in Fig. 3.1, was proposed to efficiently approximate the ℓ_1-based sparse code $\mathbf{a}$ of the input signal $\mathbf{x}$; the sparse code a is obtained by solving (3.3) for a given dictionary $\mathbf{D}$ in advance. The network has a finite number of stages, each of which updates the intermediate sparse code $\mathbf{z}^k$ ($k = 1, 2$) according to

$$\mathbf{z}^{k+1} = s_\theta(\mathbf{Wx} + \mathbf{Sz}^k), \tag{3.4}$$

where s_θ is an element-wise shrinkage function ($\mathbf{u}$ is a vector and $\mathbf{u}_i$ is its ith element, $i = 1, 2, \ldots, p$),

$$[s_\theta(\mathbf{u})]_i = \text{sign}(\mathbf{u}_i)(|\mathbf{u}_i| - \theta_i)_+. \tag{3.5}$$

The parameterized encoder, named learned ISTA (LISTA), is a natural network implementation of the iterative shrinkage and thresholding algorithm (ISTA). LISTA learned

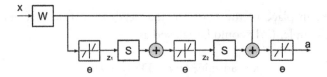

Figure 3.1 A LISTA network [8] with two time-unfolded stages.

all its parameters $\mathbf{W}$, $\mathbf{S}$ and θ from training data using a back–propagation algorithm [19]. In this way, a good approximation of the underlying sparse code can be obtained after a fixed small number of stages.

In [10], the authors leveraged a similar idea on fast trainable regressors and constructed feed-forward network approximations of the learned sparse models. Such a process-centric view was later extended in [11] to develop a principled process of learned deterministic fixed-complexity pursuits, in lieu of iterative proximal gradient descent algorithms, for structured sparse and robust low rank models. Very recently, [9] further summarized the methodology of the problem-level and model-based "deep unfolding", and developed new architectures as inference algorithms for both Markov random fields and nonnegative matrix factorization. Our work shares the same prior ideas, yet studies the unexplored ℓ_0-problems and obtains further insights.

3.3. DEEP ℓ_0 ENCODERS

3.3.1 Deep ℓ_0-Regularized Encoder

To solve the optimization problem in (3.1), an iterative hard-thresholding (IHT) algorithm was derived in [13], namely

$$\mathbf{a}^{k+1} = h_{\lambda^{0.5}}(\mathbf{a}^k + \mathbf{D}^{\mathbf{T}}(\mathbf{x} - \mathbf{D}\mathbf{a}^k)), \qquad (3.6)$$

where $\mathbf{a}^k$ denotes the intermediate result of the kth iteration, and h_θ is an element–wise **hard thresholding** operator given by

$$[h_{\lambda^{0.5}}(\mathbf{u})]_i = \begin{cases} 0 & \text{if} \quad |\mathbf{u}_i| < \lambda^{0.5}, \\ \mathbf{u}_i & \text{if} \quad |\mathbf{u}_i| \geq \lambda^{0.5}. \end{cases} \qquad (3.7)$$

Equation (3.6) could be alternatively rewritten as

$$\mathbf{a}^{k+1} = h_\theta(\mathbf{W}\mathbf{x} + \mathbf{S}\mathbf{a}^k),$$
$$\mathbf{W} = \mathbf{D}^T, \mathbf{S} = \mathbf{I} - \mathbf{D}^T\mathbf{D}, \theta = \lambda^{0.5}, \qquad (3.8)$$

and expressed as the block diagram in Fig. 3.2, which outlines a recurrent network form of solving (3.6).

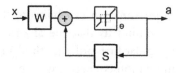

Figure 3.2 The block diagram of solving (3.6).

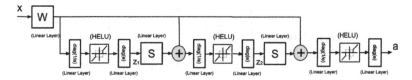

Figure 3.3 Deep ℓ_0-regularized encoder, with two time-unfolded stages.

By time-unfolding and truncating Fig. 3.2 to a fixed number of K iterations ($K = 2$ in this chapter by default),[1] we obtain a feed-forward network structure in Fig. 3.3, where $\mathbf{W}$, $\mathbf{S}$ and θ are shared among both stages, named **deep ℓ_0-regularized encoder**. Furthermore, $\mathbf{W}$, $\mathbf{S}$ and θ are all to be learnt, instead of being directly constructed from any precomputed $\mathbf{D}$. Although the equations in (3.8) do not directly apply any more to solving the deep ℓ_0-regularized encoder, they can usually serve as a high-quality initialization of the latter.

Note that the activation thresholds θ are less straightforward to update. We rewrite (3.5) as $[h_\theta(\mathbf{u})]_i = \theta_i h_1(\mathbf{u}_i/\theta_i)$. It indicates that the original neuron with trainable thresholds can be decomposed into two linear scaling layers, plus a unit-hard-threshold neuron, which is later called Hard thrEsholding Linear Unit (**HELU**) by us. The weights of the two scaling layers are diagonal matrices defined by θ and its element-wise reciprocal, respectively.

Discussion about HELU. While being inspired by LISTA, the differentiating point of deep ℓ_0-regularized encoder lies in the HELU neuron. Compared to classical neuron functions such as logistic, sigmoid, and ReLU [20], as well as the soft shrinkage and thresholding operation (3.5) in LISTA, HELU does not penalize large values, yet enforces strong (in theory infinite) penalty for small values. As such, HELU tends to produce highly sparse solutions.

The neuron form of LISTA (3.5) could be viewed as a double-sided and translated variant of ReLU, which is continuous and piecewise linear. In contrast, HELU is a **discontinuous** function that rarely occurs in existing deep network neurons. As pointed out by [21], HELU has countably many discontinuities and is thus (Borel) measurable,

[1] We test larger K values (3 or 4). In several cases they do bring performance improvements, but add complexity, too.

in which case the universal approximation capability of the network is not compromised. However, experiments remind us that the algorithmic learnability with such discontinuous neurons (using popular first-order methods) is in question, and the training is in general hard. For computation concerns, we replace HELU with the following continuous and piecewise linear function HELU_σ, during network training:

$$[\text{HELU}_\sigma(\mathbf{u})]_i = \begin{cases} 0 & \text{if} \quad |\mathbf{u}_i| \leq 1 - \sigma, \\ \frac{(\mathbf{u}_i - 1 + \sigma)}{\sigma} & \text{if} \quad 1 - \sigma < \mathbf{u}_i < 1, \\ \frac{(\mathbf{u}_i + 1 - \sigma)}{\sigma} & \text{if} -1 < \mathbf{u}_i < \sigma - 1, \\ \mathbf{u}_i & \text{if} \quad |\mathbf{u}_i| \geq 1. \end{cases} \tag{3.9}$$

Obviously, HELU_σ becomes HELU when $\sigma \to 0$. To approximate HELU, we tend to choose very small σ, while avoiding putting the training to be ill-posed. As a practical strategy, we start with a moderate σ (0.2 by default), and divide it by 10 after each epoch. After several epochs, HELU_σ turns very close to the ideal HELU.

In [22], the authors introduced an ideal hard thresholding function for solving sparse coding, whose formulation was close to HELU. Note that [22] approximates the ideal function with a sigmoid function, which has connections with our HELU_σ approximation. In [23], a similar truncated linear ReLU was utilized in the networks.

3.3.2 Deep M-Sparse ℓ_0 Encoder

Both the ℓ_0-regularized problem in (3.1) and deep ℓ_0-regularized encoder have no explicit control on the sparsity level of the solution. One would therefore turn to the M-sparse problem in (3.2), and derive the following iterative algorithm [13]:

$$\mathbf{a}^{k+1} = h_M(\mathbf{a}^k + \mathbf{D}^\mathbf{T}(\mathbf{x} - \mathbf{D}\mathbf{a}^k)). \tag{3.10}$$

Equation (3.10) resembles (3.6), except that h_M is now a nonlinear operator retaining the M coefficients with the **top M-largest absolute values**. Following the same methodology as in the previous section, the iterative form could be time–unfolded and truncated to the **deep M-sparse encoder**, as in Fig. 3.4. To deal with the h_M operation, we refer to the popular concepts of pooling and unpooling [24] in deep networks, and introduce the pairs of $\max_M$ pooling and unpooling, as in Fig. 3.4.

Discussion about $\max_M$ pooling/unpooling. Pooling is popular in convolutional networks to obtain translation-invariant features [7]. It is yet less common in other forms of deep networks [25]. The unpooling operation was introduced in [24] to insert the pooled values back to the appropriate locations of feature maps for reconstruction purposes.

In our proposed deep M-sparse encoder, the pooling and unpooling operation pair is used to construct a projection from R^m to its subset $S := \{s \in R^m \mid ||s||_0 \leq M\}$. The

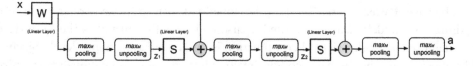

Figure 3.4 Deep *M*-sparse encoder, with two time-unfolded stages.

max$_M$ pooling and unpooling functions are intuitively defined as

$$[\mathbf{p}_M, \mathbf{idx}_M] = \text{max}_M.\text{pooling}(\mathbf{u}),$$
$$\mathbf{u}_M = \text{max}_M.\text{unpooling}(\mathbf{p}_M, \mathbf{idx}_M).$$
(3.11)

For each input $\mathbf{u}$, the *pooled map* $\mathbf{p}_M$ records the top *M*-largest values (irrespective of sign), and the *switch* $\mathbf{idx}_M$ records their locations. The corresponding unpooling operation takes the elements in $\mathbf{p}_M$ and places them in $\mathbf{u}_M$ at the locations specified by $\mathbf{idx}_M$, the remaining elements being set to zero. The resulting $\mathbf{u}_M$ is of the same dimension as $\mathbf{u}$, but has exactly no more than *M* nonzero elements. In back propagation, each position in $\mathbf{idx}_M$ is propagated with the entire error signal.

3.3.3 Theoretical Properties

It is showed in [13] that the iterative algorithms in both (3.6) and (3.10) are guaranteed not to increase the cost functions. Under mild conditions, their targeted fixed points are local minima of the original problems. As the next step after the time truncation, the deep encoder models are to be solved by the stochastic gradient descent (SGD) algorithm, which converges to stationary points under a few stricter assumptions than those satisfied in this chapter [26].[2] However, the entanglement of the iterative algorithms and the SGD algorithm makes the overall convergence analysis really difficult.

One must emphasize that in each step, the back propagation procedure requires only operations of order O(p) [8]. The training algorithm takes O(Cnp) time (C is the constant absorbing epochs, stage numbers, etc.). The testing process is purely feed-forward and is therefore dramatically faster than traditional inference methods by solving (3.1) or (3.2). SGD is also easy to be parallelized.

3.4. TASK-DRIVEN OPTIMIZATION

It is often desirable to jointly optimize the learned sparse code features and the targeted task so that they mutually reinforce each other. The authors of [27] associated label information with each dictionary item by adding discriminable regularization terms to

---

[2] As a typical case, we use SGD in a setting where it is not guaranteed to converge in theory, but behaves well in practice.

the objective. Recent work [28], [29] developed task-driven sparse coding via bi-level optimization models, where (ℓ_1-based) sparse coding is formulated as the lower-level constraint while a task-oriented cost function is minimized as its upper-level objective. The above approaches in sparse coding are complicated and computationally expensive. It is much more convenient to implement end-to-end task-driven training in deep architectures, by concatenating the proposed deep encoders with certain task-driven loss functions.

In this chapter, we mainly discuss two tasks: classification and clustering, while being aware of other immediate extensions, such as semisupervised learning. Assuming K classes (or clusters), and $\boldsymbol{\omega} = [\boldsymbol{\omega}_1, \ldots, \boldsymbol{\omega}^k]$ as the set of parameters of the loss function, where $\boldsymbol{\omega}_i$ corresponds to the jth class (cluster), $j = 1, 2, \ldots, K$. For the **classification** case, one natural choice is the well-known softmax loss function. For the **clustering** case, since the true cluster label of each $\mathbf{x}$ is unknown, we define the predicted confidence probability p_j that sample $\mathbf{x}$ belongs to cluster j as the likelihood of softmax regression

$$p_j = p(j|\boldsymbol{\omega}, \boldsymbol{a}) = \frac{e^{-\boldsymbol{\omega}_j^T \mathbf{a}}}{\sum_{l=1}^{K} e^{-\boldsymbol{\omega}_l^T \mathbf{a}}}. \tag{3.12}$$

The predicted cluster label of $\mathbf{a}$ is the cluster j where it achieves the largest p_j.

3.5. EXPERIMENT

3.5.1 Implementation

Two proposed deep ℓ_0 encoders are implemented with the CUDA ConvNet package [7]. We use a constant learning rate of 0.01 with no momentum, and a batch size of 128. In practice, given that the model is well initialized, the training takes approximately 1 hour on the MNIST dataset, on a workstation with 12 Intel Xeon 2.67 GHz CPUs and 1 GTX680 GPU. It is also observed that the training efficiency of our model scales approximately linearly with the size of data.

While many neural networks train well with random initializations without pretraining, given that the training data is sufficient, it has been discovered that poorly initialized networks can hamper the effectiveness of first-order methods (e.g., SGD) [30]. For the proposed models, it is, however, much easier to initialize the model in the right regime, benefiting from the analytical relationships between sparse coding and network hyperparameters in (3.8).

3.5.2 Simulation on ℓ_0 Sparse Approximation

We first compare the performance of different methods on ℓ_0 sparse code approximation. The first 60,000 samples of the MNIST dataset are used for training and the last 10,000 for testing. Each patch is resized to 16×16 and then preprocessed to remove its mean and normalize its variance. The patches with small standard deviations are discarded. A sparsity coefficient $\lambda = 0.5$ is used in (3.1), and the sparsity level $M = 32$ is fixed in (3.2). The sparse code dimension (dictionary size) p is to be varied.

Our prediction task resembles the setup in [8]: first learning a dictionary from training data, followed by solving sparse approximation (3.3) with respect to the dictionary, and finally training the network as a regressor from input samples to the solved sparse codes. The only major difference here lies in that unlike the ℓ_1-based problems, the nonconvex ℓ_0-based minimization could only reach a (nonunique) local minimum. To improve stability, we first solve the ℓ_1-problems to obtain a good initialization for ℓ_0-problems, and then run the iterative algorithms (3.6) or (3.10) until convergence. The obtained sparse codes are called "optimal codes" hereinafter and used in both training and testing evaluation (as "ground-truth"). One must keep in mind that we are not seeking to produce approximate sparse code for all possible input vectors, but only for *input vectors drawn from the same distribution as our training samples*.

We compare the proposed deep ℓ_0 encoders with the iterative algorithms under different number of iterations. In addition, we include a *baseline encoder* into comparison, which is a fully-connected feed-forward network, consisting of three hidden layers of dimension p with ReLu neurons. The baseline encoder thus has the same parameter capacity as deep ℓ_0 encoders.[3] We apply dropout to the baseline encoders, with the probabilities of retaining the units being 0.9, 0.9, and 0.5. The proposed encoders do not apply dropout.

The deep ℓ_0 encoders and the baseline encoder are first trained, and all are then evaluated on the testing set. We calculate the total prediction errors, i.e., the normalized squared errors between the optimal codes and the predicted codes, as in Tables 3.1 and 3.2. For the M-sparse case, we also compare their recovery of nonzero supports in Table 3.3, by counting the mismatched nonzero element locations between optimal and predicted codes (averaged on all samples). Immediate conclusions from the numerical results are as follows:

- The proposed deep encoders have outstanding generalization performance, thanks to the effective regularization brought by their architectures, which are derived from specific problem formulations (i.e., (3.1) and (3.2)) as priors. The "general-architecture" baseline encoders, which have the same parameter complexity, appear to overfit the training set and generalize much worse.

- While the deep encoders only unfold two stages, they outperform their iterative counterparts even when the latter have passed 10 iterations. Meanwhile, the former enjoy much faster inference as being feed-forward.

- The deep ℓ_0-regularized encoder obtains a particularly low prediction error. It is interpretable that while the iterative algorithm has to work with a fixed λ, the deep

---

[3] Except for the "diag(θ)" layers in Fig. 3.3, each of which contains only p free parameters.

Table 3.1 Prediction error (%) comparison of all methods on solving the ℓ_0-regularized problem (3.1)

p	128	256	512
Iterative (2 iterations)	17.52	18.73	22.40
Iterative (5 iterations)	8.14	6.75	9.37
Iterative (10 iterations)	3.55	4.33	4.08
Baseline Encoder	8.94	8.76	10.17
Deep ℓ_0–Regularized Encoder	0.92	0.91	0.81

Table 3.2 Prediction error (%) comparison of all methods on solving the M-sparse problem (3.2)

p	128	256	512
Iterative (2 iterations)	17.23	19.27	19.31
Iterative (5 iterations)	10.84	12.52	12.40
Iterative (10 iterations)	5.67	5.44	5.20
Baseline Encoder	14.04	16.76	12.86
Deep M–Sparse Encoder	2.94	2.87	3.29

Table 3.3 Averaged nonzero support error comparison of all methods on solving the M-sparse problem (3.2)

p	128	256	512
Iterative (2 iterations)	10.8	13.4	13.2
Iterative (5 iterations)	6.1	8.0	8.8
Iterative (10 iterations)	4.6	5.6	5.3
Deep M–Sparse Encoder	2.2	2.7	2.7

ℓ_0-regularized encoder is capable of "fine-tuning" this hyperparameter automatically (after diag(θ) is initialized from λ), by exploring the training data structure.

- The deep M-sparse encoder is able to find the nonzero support with high accuracy.

3.5.3 Applications on Classification

Since the task-driven models are trained from end to end, **no precomputation of a is needed**. For classification, we evaluate our methods on the MNIST dataset, and the AVIRIS Indiana Pines hyperspectral image dataset (see [31] for details). We compare our two proposed deep encoders with two competitive sparse coding-based methods: (1) task-driven sparse coding (TDSC) in [28], with the original setting followed and all parameters carefully tuned; (2) a pre-trained LISTA followed by supervised tuning with

Table 3.4 Classification error rate (%) comparison of all methods on the MNIST dataset

p	128	256	512
TDSC	0.71	0.55	0.53
Tuned LISTA	0.74	0.62	0.57
Deep ℓ_0-Regularized	0.72	0.58	0.52
Deep M-Sparse ($M = 10$)	0.72	0.57	0.53
Deep M-Sparse ($M = 20$)	0.69	0.54	0.51
Deep M-Sparse ($M = 30$)	0.73	0.57	0.52

Table 3.5 Classification error rate (%) comparison of all methods on the AVIRIS Indiana Pines dataset

p	128	256	512
TDSC	15.55	15.27	15.21
Tuned LISTA	16.12	16.05	15.97
Deep ℓ_0-Regularized	15.20	15.07	15.01
Deep M-Sparse ($M = 10$)	13.77	13.56	13.52
Deep M-Sparse ($M = 20$)	14.67	14.23	14.07
Deep M-Sparse ($M = 30$)	15.14	15.02	15.00

softmax loss. Note that for the deep M-sparse encoder, M is not known in advance and has to be tuned. To our surprise, the fine-tuning of M is likely to improve the performances significantly, which is analyzed next. The overall error rates are compared in Tables 3.4 and 3.5.

In general, the proposed deep ℓ_0 encoders provide superior results to the deep ℓ_1-based method (tuned LISTA). TDSC also generates competitive results, but at the cost of the high complexity for inference, i.e., solving conventional sparse coding. It is of particular interest to us that when supplied with specific M values, the deep M-sparse encoder can generate remarkably improved results.[4] Especially in Table 3.5, when $M = 10$, the error rate is around 1.5% lower than that of $M = 30$. Note that in the AVIRIS Indiana Pines dataset, the training data volume is much smaller than that of MNIST. In this way, we conjecture that it might not be sufficiently effective to make the training process depend fully on data; instead, crafting a stronger sparsity prior by smaller M could help learn more discriminative features.[5] Such a behavior provides us with an important hint to **impose suitable structural priors to deep networks**.

---

[4] To get a good estimate of M, one might first try to perform (unsupervised) sparse coding on a subset of samples.

[5] Interestingly, there are a total of 16 classes in the AVIRIS Indiana Pines dataset. When $p = 128$, each class has on average 8 "atoms" for class-specific representation. Therefore $M = 10$ approximately coincides with the sparse representation classification (SRC) principle [31] of forcing sparse codes to be compactly focused on one class of atoms.

Table 3.6 Clustering error rate (%) comparison of all methods on the COIL 20 dataset

p	128	256	512
[29]	17.75	17.14	17.15
[33]	14.47	14.17	14.08
Deep ℓ_0-Regularized	14.52	14.27	14.06
Deep M-Sparse ($M = 10$)	14.59	14.25	14.03
Deep M-Sparse ($M = 20$)	14.84	14.33	14.15
Deep M-Sparse ($M = 30$)	14.77	14.37	14.12

Table 3.7 Clustering error rate (%) comparison of all methods on the CMU PIE dataset

p	128	256	512
[29]	17.50	17.26	17.20
[33]	16.14	15.58	15.09
Deep ℓ_0-Regularized	16.08	15.72	15.41
Deep M-Sparse ($M = 10$)	16.77	16.46	16.02
Deep M-Sparse ($M = 20$)	16.44	16.23	16.05
Deep M-Sparse ($M = 30$)	16.46	16.17	16.01

3.5.4 Applications on Clustering

For clustering, we evaluate our methods on the COIL 20 and the CMU PIE datasets [32]. Two state-of-the-art methods to compare are the jointly optimized sparse coding and clustering method proposed in [29], as well as the graph-regularized deep clustering method in [33].[6] The overall error rates are compared in Tables 3.6 and 3.7.

Note that the method in [33] incorporated Laplacian regularization as an additional prior while the others did not. It is thus no wonder that this method often performs better than others. Even without any graph information utilized, the proposed deep encoders are able to obtain very close performances, and outperform [33] in certain cases. On the COIL 20 dataset, the lowest error rate is reached by the deep M-sparse ($M = 10$) Encoder, when $p = 512$, followed by the deep ℓ_0-regularized encoder.

On the CMU PIE dataset, the deep ℓ_0-regularized encoder leads to competitive accuracy with [33], and outperforms all deep M-sparse encoders with noticeable margins, which is different from other cases. Previous work discovered that sparse approximations over CMU PIE had significant errors [34], which is also verified by us. Therefore, hardcoding exact sparsity could even hamper the model performance.

---

[6] Both papers train their model under both soft-max and max-margin type losses. To ensure fair comparison, we adopt the former, with the same form of loss function as ours.

Remark. From the experiments, we gain additional insights in designing deep architectures:

- If one expects the model to explore the data structure by itself, and provided that there is sufficient training data, then the deep ℓ_0-regularized encoder (and its peers) might be preferred as its all parameters, including the desired sparsity, are fully learnable from the data.
- If one has certain correct prior knowledge of the data structure, including but not limited to the exact sparsity level, one should choose deep M-sparse encoder, or other models of its type that are designed to maximally enforce that prior. The methodology could be especially useful when the training data is less than sufficient.

We hope the above insights could be of reference to many other deep learning models.

3.6. CONCLUSIONS AND DISCUSSIONS ON THEORETICAL PROPERTIES

We propose deep ℓ_0 encoders to solve the ℓ_0 sparse approximation problem. Rooted in solid iterative algorithms, the deep ℓ_0-regularized encoder and deep M-sparse encoder are developed, each designed to solve one typical formulation, accompanied with the introduction of the novel HELU neuron and $\max_M$ pooling/unpooling. When applied to specific tasks of classification and clustering, the models are optimized in an end-to-end manner. The latest deep learning tools enable us to solve considered problems in a highly effective and efficient fashion. They not only provide us with impressive performance in numerical experiments, but also enlighten with important insights into designing deep models.

While many recent works followed the idea of constructing feed-forward networks by unfolding and truncating iterative algorithms, as fast trainable regressors to approximate the solutions of sparse coding models, progress has been slow towards understanding the efficient approximation from a theoretical perspective. [35] investigated the convergence property of our proposed Deep ℓ_0 Encoders. The authors argued that they can use data to train a transformation of dictionary that can improve its restricted isometry property (RIP) constant, when the original dictionary is highly correlated, causing IHT to fail easily. They moreover showed it beneficial to allow the weights to decouple across layers. However, the analysis in [35] cannot be straightforwardly extended to ISTA although IHT is linearly convergent [36] under rather strong assumptions.

Beside, [37] attempted to explain the mechanism of LISTA by re-factorizing the Gram matrix of dictionary, which tries to nearly diagonalize the Gram matrix with a basis that produces a small perturbation of the ℓ_1 ball. They the re-parameterized LISTA into a new factorized architecture that achieved similar acceleration gain to LISTA. Using an "indirect" proof, [37] was able to show that LISTA can converge faster than ISTA, but still sublinearly. In [38], a similar learning-based model inspired by another iterative algorithm solve LASSO, approximated message passing (AMP), was studied. The idea was advanced in [39] to substitute the AMP proximal operator (soft-thresholding) with a learnable Gaussian denoiser. However, their main theoretical tool, named "state evolution", is not directly applicable to analyzing LISTA.

REFERENCES

[1] Donoho DL, Vetterli M, DeVore RA, Daubechies I. Data compression and harmonic analysis. Information Theory, IEEE Transactions on 1998;44(6):2435–76.

[2] Donoho DL. De-noising by soft-thresholding. Information Theory, IEEE Transactions on 1995;41(3):613–27.

[3] Davies M, Mitianoudis N. Simple mixture model for sparse overcomplete ICA. IEE Proceedings-Vision, Image and Signal Processing 2004;151(1):35–43.

[4] Wright J, Yang AY, Ganesh A, Sastry SS, Ma Y. Robust face recognition via sparse representation. TPAMI 2009;31(2):210–27.

[5] Cheng B, Yang J, Yan S, Fu Y, Huang TS. Learning with l1 graph for image analysis. TIP 2010;19(4).

[6] Donoho DL, Elad M. Optimally sparse representation in general (nonorthogonal) dictionaries via l1 minimization. Proceedings of the National Academy of Sciences 2003;100(5):2197–202.

[7] Krizhevsky A, Sutskever I, Hinton GE. ImageNet classification with deep convolutional neural networks. In: NIPS; 2012. p. 1097–105.

[8] Gregor K, LeCun Y. Learning fast approximations of sparse coding. In: ICML; 2010. p. 399–406.

[9] Hershey JR, Roux JL, Weninger F. Deep unfolding: model-based inspiration of novel deep architectures. arXiv preprint arXiv:1409.2574, 2014.

[10] Sprechmann P, Litman R, Yakar TB, Bronstein AM, Sapiro G. Supervised sparse analysis and synthesis operators. In: NIPS; 2013. p. 908–16.

[11] Sprechmann P, Bronstein A, Sapiro G. Learning efficient sparse and low rank models. In: TPAMI; 2015.

[12] Sun K, Wang Z, Liu D, Liu R. L_p-norm constrained coding with Frank–Wolfe network. arXiv preprint arXiv:1802.10252, 2018.

[13] Blumensath T, Davies ME. Iterative thresholding for sparse approximations. Journal of Fourier Analysis and Applications 2008;14(5–6):629–54.

[14] Wipf DP, Rao BD. l0-Norm minimization for basis selection. In: NIPS; 2004. p. 1513–20.

[15] Candes EJ, Wakin MB, Boyd SP. Enhancing sparsity by reweighted l1 minimization. Journal of Fourier Analysis and Applications 2008;14(5–6):877–905.

[16] Yuan G, Ghanem B. L0tv: a new method for image restoration in the presence of impulse noise. 2015.

[17] Xu L, Lu C, Xu Y, Jia J. Image smoothing via l0 gradient minimization. TOG, vol. 30. ACM; 2011. p. 174.

[18] Wang Y, Wang YX, Singh A. Clustering consistent sparse subspace clustering. arXiv preprint arXiv:1504.01046, 2015.

[19] LeCun YA, Bottou L, Orr GB, Müller KR. Efficient backprop. In: Neural networks: tricks of the trade. Springer; 2012. p. 9–48.

[20] Mhaskar HN, Micchelli CA. How to choose an activation function. In: NIPS; 1994. p. 319–26.

[21] Hornik K, Stinchcombe M, White H. Multilayer feedforward networks are universal approximators. Neural Networks 1989;2(5):359–66.

[22] Rozell CJ, Johnson DH, Baraniuk RG, Olshausen BA. Sparse coding via thresholding and local competition in neural circuits. Neural Computation 2008;20(10):2526–63.

[23] Konda K, Memisevic R, Krueger D. Zero-bias autoencoders and the benefits of co-adapting features. arXiv preprint arXiv:1402.3337, 2014.

[24] Zeiler MD, Taylor GW, Fergus R. Adaptive deconvolutional networks for mid and high level feature learning. In: ICCV. IEEE; 2011. p. 2018–25.

[25] Gulcehre C, Cho K, Pascanu R, Bengio Y. Learned-norm pooling for deep feedforward and recurrent neural networks. In: Machine learning and knowledge discovery in databases. Springer; 2014. p. 530–46.

[26] Bottou L. Large-scale machine learning with stochastic gradient descent. In: Proceedings of COMPSTAT'2010. Springer; 2010. p. 177–86.

[27] Jiang Z, Lin Z, Davis LS. Learning a discriminative dictionary for sparse coding via label consistent K-SVD. In: CVPR. IEEE; 2011. p. 1697–704.

[28] Mairal J, Bach F, Ponce J. Task-driven dictionary learning. TPAMI 2012;34(4):791–804.

[29] Wang Z, Yang Y, Chang S, Li J, Fong S, Huang TS. A joint optimization framework of sparse coding and discriminative clustering. In: IJCAI; 2015.

[30] Sutskever I, Martens J, Dahl G, Hinton G. On the importance of initialization and momentum in deep learning. In: ICML; 2013. p. 1139–47.

[31] Wang Z, Nasrabadi NM, Huang TS. Semisupervised hyperspectral classification using task-driven dictionary learning with Laplacian regularization. TGRS 2015;53(3):1161–73.

[32] Sim T, Baker S, Bsat M. The CMU pose, illumination, and expression (PIE) database. In: Automatic face and gesture recognition, 2002. Proceedings. Fifth IEEE international conference on. IEEE; 2002. p. 46–51.

[33] Wang Z, Chang S, Zhou J, Wang M, Huang TS. Learning a task-specific deep architecture for clustering. arXiv preprint arXiv:1509.00151, 2015.

[34] Yang J, Yu K, Huang T. Supervised translation-invariant sparse coding. In: Computer vision and pattern recognition (CVPR), 2010 IEEE conference on. IEEE; 2010. p. 3517–24.

[35] Xin B, Wang Y, Gao W, Wipf D, Wang B. Maximal sparsity with deep networks? In: Advances in neural information processing systems; 2016. p. 4340–8.

[36] Blumensath T, Davies ME. Iterative hard thresholding for compressed sensing. Applied and Computational Harmonic Analysis 2009;27(3):265–74.

[37] Moreau T, Bruna J. Understanding trainable sparse coding with matrix factorization. In: ICLR; 2017.

[38] Borgerding M, Schniter P, Rangan S. AMP-inspired deep networks for sparse linear inverse problems. IEEE Transactions on Signal Processing 2017;65(16):4293–308.

[39] Metzler CA, Mousavi A, Baraniuk RG. Learned D-AMP: principled neural network based compressive image recovery. In: Advances in neural information processing systems; 2017. p. 1770–81.

CHAPTER 4

Single Image Super-Resolution: From Sparse Coding to Deep Learning

Ding Liu*, Thomas S. Huang[†]

*Beckman Institute for Advanced Science and Technology, Urbana, IL, United States
[†]Department of Electrical and Computer Engineering, University of Illinois at Urbana-Champaign, Champaign, IL, United States

Contents

4.1.	Robust Single Image Super-Resolution via Deep Networks with Sparse Prior	47
	4.1.1 Introduction	47
	4.1.2 Related Work	49
	4.1.3 Sparse Coding Based Network for Image SR	50
	4.1.4 Network Cascade for Scalable SR	54
	4.1.5 Robust SR for Real Scenarios	57
	4.1.6 Implementation Details	60
	4.1.7 Experiments	61
	4.1.8 Subjective Evaluation	69
	4.1.9 Conclusion and Future Work	72
4.2.	Learning a Mixture of Deep Networks for Single Image Super-Resolution	73
	4.2.1 Introduction	73
	4.2.2 The Proposed Method	74
	4.2.3 Implementation Details	76
	4.2.4 Experimental Results	77
	4.2.5 Conclusion and Future Work	81
References		83

4.1. ROBUST SINGLE IMAGE SUPER-RESOLUTION VIA DEEP NETWORKS WITH SPARSE PRIOR[1]

4.1.1 Introduction

Single image super-resolution (SR) aims at obtaining a high-resolution (HR) image from a low-resolution (LR) input image by inferring all the missing high frequency contents. With the known variables in LR images greatly outnumbered by the unknowns in HR images, SR is a highly ill-posed problem and the current techniques are

---

[1] ©2017 IEEE. Reprinted, with permission, from Liu, Ding, Wang, Zhaowen, Wen, Bihan, Yang, Jianchao, Han, Wei, and Huang, Thomas S. "Robust single image super-resolution via deep networks with sparse prior." IEEE Transactions on Image Processing 25.7 (2016): 3194–3207.

Deep Learning Through Sparse and Low-Rank Modeling
DOI: 10.1016/B978-0-12-813659-1.00004-4

far from being satisfactory for many real applications [1,2], such as surveillance, medical imaging and consumer photo editing [3].

To regularize the solution of SR, people have exploited various priors of natural images. Analytical priors, such as bicubic interpolation, work well for smooth regions; while image models based on statistics of edges [4] and gradients [5] can recover sharper structures. Sparse priors are utilized in the patch-based methods [6–8]. HR patch candidates are recovered from similar examples in the LR image itself at different locations and across different scales [9,10].

More recently, inspired by the success achieved by deep learning [11] in other computer vision tasks, people began to use neural networks with deep architecture for image SR. Multiple layers of collaborative auto-encoders are stacked together in [12,13] for robust matching of self-similar patches. Deep convolutional neural networks (CNN) [14] and deconvolutional networks [15] are designed that directly learn the nonlinear mapping from the LR space to HR space in a way similar to coupled sparse coding [7]. As these deep networks allow end-to-end training of all the model components between LR input and HR output, significant improvements have been observed over their shadow counterparts.

The networks in [12,14] are built with generic architectures, which means all their knowledge about SR is learned from training data. On the other hand, people's domain expertise for the SR problem, such as natural image prior and image degradation model, is largely ignored in deep learning based approaches. It is then worthwhile to investigate whether domain expertise can be used to design better deep model architectures, or whether deep learning can be leveraged to improve the quality of handcrafted models.

In this section, we extend the conventional sparse coding model [6] using several key ideas from deep learning, and show that domain expertise is complementary to large learning capacity in further improving SR performance. First, based on the learned iterative shrinkage and thresholding algorithm (LISTA) [16], we implement a feed-forward neural network in which each layer strictly correspond to one step in the processing flow of sparse coding based image SR. In this way, the sparse representation prior is effectively encoded in our network structure; at the same time, all the components of sparse coding can be trained jointly through back-propagation. This simple model, which is named sparse coding based network (SCN), achieves notable improvement over the generic CNN model [14] in terms of both recovery accuracy and human perception, and yet has a compact model size. Moreover, with the correct understanding of each layer's physical meaning, we have a more principled way to initialize the parameters of SCN, which helps to improve optimization speed and quality.

A single network is only able to perform image SR by a particular scaling factor. In [14], different networks are trained for different scaling factors. In this section, we propose a cascade of multiple SCNs to achieve SR for arbitrary factors. This approach, motivated by the self-similarity based SR approach [9], not only increases the scaling flexibility of our model, but also reduces artifacts for large scaling factors. Moreover, inspired by the multi-pass scheme of image denoising [17], we demonstrate that the SR results can be further enhanced by cascading multiple SCNs for SR of a fixed scaling factor. A cascade of SCNs (CSCN) can also benefit from the end-to-end training of deep network with a specially designed multi-scale cost function.

In practical SR scenarios, the real LR measurements usually suffer from various types of corruption, such as noise and blurring. Sometimes the degradation process is even too complicated or unclear. We propose several schemes using our SCN to robustly handle such practical SR cases. When the degradation mechanism is unknown, we fine-tune the generic SCN with the requirement of only a small amount of real training data and manage to adapt our model to the new scenario. When the forward model for LR generation is clear, we propose an iterative SR scheme incorporating SCN with additional regularization based on priors from the degradation mechanism.

Subjective assessment is important to the SR technology because commercial products equipped with such technology are usually evaluated subjectively by the end users. In order to thoroughly compare our model with other prevailing SR methods, we conduct a systematic subjective evaluation of these methods, in which the assessment results are statistically analyzed and one score is given for each method.

In the following, we first review the literature in Sect. 4.1.2 and introduce the SCN model in Sect. 4.1.3. Then the cascade scheme of SCN models is detailed in Sect. 4.1.4. The method for robustly handling images with additional degradation such as noise and blurring is discussed in Sect. 4.1.5. Implementation details are provided in Sect. 4.1.6. Extensive experimental results are reported in Sect. 4.1.7, and the subjective evaluation is described in Sect. 4.1.8. Finally, conclusion and future work are presented in Sect. 4.1.9.

4.1.2 Related Work

Single image SR is the task of recovering an HR image from only one LR observation. A comprehensive review can be found in [18]. Generally, existing methods can be classified into three categories: interpolation based [19], image statistics based [4,20], and example based methods [6,21].

Interpolation based methods include bilinear, bicubic and Lanczos filtering [19], which usually run very fast because of the low algorithm complexity. However, the simplicity of these methods leads to the failure of modeling the complex mapping between the LR feature space and the corresponding HR feature space, generating overly-smoothed unsatisfactory regions.

Image statistics based methods utilize the statistical edge information to reconstruct HR images [4,20]. They rely on the priors of edge statistics in images while facing the shortcoming of losing high-frequency detail information, especially in the case of large upscaling factors.

The current most popular and successful approaches are built on example based learning techniques, which aim to learn the correspondence between the LR feature space and HR feature space through a large number of representative exemplar pairs. The pioneer work in this area includes [22].

Given the origin of exemplar pairs, these methods can be further categorized into three classes: *self-example based* [9,10], *external-example based* methods [6,21] and the joint of them [23]. Self-example based methods only exploit the single input LR image as references, and extract exemplar pairs merely from the LR image across different scales to predict the HR image. Such methods usually work well on the images containing repetitive patterns or textures but lack the richness of image structures outside the input image and thus fail to generate satisfactory prediction for images of other classes. Huang et al. [24] extend this idea by building self-dictionaries for handling geometric transformations.

External-example based methods first utilize the exemplar pairs extracted from a large external dataset, in order to learn the universal image characteristics between the LR feature space and HR feature space, and then apply the learned mapping for SR. Usually, representative patches from external datasets are compactly embodied in pre-trained dictionaries. One representative approach is the sparse coding based method [6,7]. For example, in [7] two coupled dictionaries are trained for the LR feature space and HR patch feature space, respectively, such that the LR patch over LR dictionary and its corresponding HR patch over HR dictionary share the same sparse representation. Although it is able to capture the universal LR–HR correspondence from external datasets and recover fine details and sharpened edges, it suffers from the high computational cost when solving complicated nonlinear optimization problems.

Timofte et al. [21,25] propose a neighboring embedding approach for SR, and formulate the problem as a least squares optimization with l_2 norm regularization, which drastically reduces the computation complexity compared with [6,7]. Neighboring embedding approaches approximate HR patches as a weighted average of similar training patches in a low dimensional manifold.

Random forest is built for SR without dictionary learning in [26,27]. Such an approach achieves fast inference time but usually suffers from the huge model size.

4.1.3 Sparse Coding Based Network for Image SR

In this section, we first introduce the background of sparse coding for image SR and its network implementation. Then we illustrate the design of our proposed sparse coding based network and its advantage over previous models.

4.1.3.1 Image SR Using Sparse Coding

The sparse representation based SR method [6] models the transform from each local patch $y \in \mathbb{R}^{m_y}$ in the bicubic-upscaled LR image to the corresponding patch $x \in \mathbb{R}^{m_x}$ in the HR image. The dimension m_y is not necessarily the same as m_x when image features other than raw pixel are used to represent patch y. It is assumed that the LR (HR) patch y (x) can be represented with respect to an overcomplete dictionary D_y (D_x) using some sparse linear coefficients α_y (α_x) $\in \mathbb{R}^n$, which are known as sparse code. Since the degradation process from x to y is nearly linear, the patch pair can share the same sparse code $\alpha_y = \alpha_x = \alpha$ if the dictionaries D_y and D_x are defined properly. Therefore, for an input LR patch y, the HR patch can be recovered as

$$x = D_x\alpha, \;\; \text{s.t. } \alpha = \arg\min_z \|y - D_y z\|_2^2 + \lambda\|z\|_1, \tag{4.1}$$

where $\|\cdot\|_1$ denotes the ℓ_1 norm, which is convex and sparsity-inducing, and λ is a regularization coefficient.

In order to learn the dictionary pair (D_y, D_x), the goal is to minimize the recovery error of x and y, and thus the loss function L in [7] is defined as

$$L = \frac{1}{2}\left(\gamma\|x - D_x z\|_2^2 + (1-\gamma)\|y - D_y z\|_2^2\right), \tag{4.2}$$

where γ ($0 < \gamma \leq 1$) balances the two reconstruction errors. Then the optimal dictionary pair $\{D_x^*, D_y^*\}$ can be found by minimizing the empirical expectation of (4.2) over all the training LR/HR pairs,

$$\min_{D_x, D_y} \frac{1}{N}\sum_{i=1}^{N} L(D_x, D_y, x_i, y_i)$$
$$\text{s.t. } z_i = \arg\min_\alpha \|y_i - D_y\alpha\|_2^2 + \lambda\|\alpha\|_1, \quad i = 1, 2, \ldots, N, \tag{4.3}$$
$$\|D_x(\cdot, k)\|_2 \leq 1, \;\; \|D_y(\cdot, k)\|_2 \leq 1, \quad k = 1, 2, \ldots, K.$$

Since the objective function in (4.2) is highly nonconvex, the dictionary pair (D_y, D_x) is usually learned alternatively while keeping one of them fixed [7]. The authors of [28,23] also incorporated patch-level self-similarity with dictionary pair learning.

4.1.3.2 Network Implementation of Sparse Coding

There is an intimate connection between sparse coding and neural network, which has been well studied in [29,16]. A feed-forward neural network as illustrated in Fig. 4.1 is proposed in [16] to efficiently approximate the sparse code α of input signal y as it would be obtained by solving (4.1) for a given dictionary D_y. The network has a finite number of recurrent stages, each of which updates the intermediate sparse code according to

$$z_{k+1} = h_\theta(Wy + Sz_k), \tag{4.4}$$

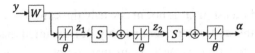

Figure 4.1 An LISTA network [16] with 2 time-unfolded recurrent stages, whose output α is an approximation of the sparse code of input signal y. The linear weights W, S and the shrinkage thresholds θ are learned from data.

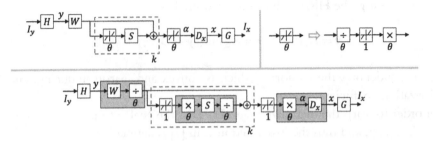

Figure 4.2 (Top left) The proposed SCN model with a patch extraction layer H, an LISTA subnetwork for sparse coding (with k recurrent stages denoted by the dashed box), an HR patch recovery layer D_x, and a patch combination layer G. (Top right) A neuron with an adjustable threshold decomposed into two linear scaling layers and a unit-threshold neuron. (Bottom) The SCN reorganized with unit-threshold neurons and adjacent linear layers merged together in the gray boxes.

where h_θ is an element-wise shrinkage function defined as $[h_\theta(a)]_i = \text{sign}(a_i)(|a_i| - \theta_i)_+$ with positive thresholds θ.

Different from the iterative shrinkage and thresholding algorithm (ISTA) [30,31] which finds an analytical relationship between network parameters (weights W, S and thresholds θ) and sparse coding parameters (D_y and λ), the authors of [16] learn all the network parameters from training data using a back–propagation algorithm called learned ISTA (LISTA). In this way, a good approximation of the underlying sparse code can be obtained within a fixed number of recurrent stages.

4.1.3.3 Network Architecture of SCN

Given the fact that sparse coding can be effectively implemented with an LISTA network, it is straightforward to build a multi-layer neural network that mimics the processing flow of the sparse coding based SR method [6]. Similar to most patch–based SR methods, our sparse coding based network (SCN) takes the bicubic-upscaled LR image I_y as input, and outputs the full HR image I_x. Fig. 4.2 shows the main network structure, and each layer is described in the following.

The input image I_y first goes through a convolutional layer H which extracts features for each LR patch. There are m_y filters of spatial size $s_y \times s_y$ in this layer, so that our input patch size is $s_y \times s_y$ and its feature representation y has m_y dimensions.

Each LR patch y is then fed into an LISTA network with k recurrent stages to obtain its sparse code $\alpha \in \mathbb{R}^n$. Each stage of LISTA consists of two linear layers parameterized by $W \in \mathbb{R}^{n \times m_y}$ and $S \in \mathbb{R}^{n \times n}$, and a nonlinear neuron layer with activation function h_θ. The activation thresholds $\theta \in \mathbb{R}^n$ are also to be updated during training, which complicates the learning algorithm. To restrict all the tunable parameters in our linear layers, we do a simple trick to rewrite the activation function as

$$[h_\theta(a)]_i = \text{sign}(a_i)\theta_i(|a_i|/\theta_i - 1)_+ = \theta_i h_1(a_i/\theta_i). \tag{4.5}$$

Equation (4.5) indicates that the original neuron with an adjustable threshold can be decomposed into two linear scaling layers and a unit-threshold neuron, as shown in Fig. 4.2(top-right). The weights of the two scaling layers are diagonal matrices defined by θ and its element-wise reciprocal, respectively.

The sparse code α is then multiplied with HR dictionary $D_x \in \mathbb{R}^{m_x \times n}$ in the next linear layer, reconstructing HR patch x of size $s_x \times s_x = m_x$.

In the final layer G, all the recovered patches are put back to the corresponding positions in the HR image I_x. This is realized via a convolutional filter of m_x channels with spatial size $s_g \times s_g$. The size s_g is determined as the number of neighboring patches that overlap with the same pixel in each spatial direction. The filter will assign appropriate weights to the overlapped recoveries from different patches and take their weighted average as the final prediction in I_x.

As illustrated in Fig. 4.2(bottom), after some simple reorganizations of the layer connections, the network described above has some adjacent linear layers which can be merged into a single layer. This helps to reduce the computation load and redundant parameters in the network. The layers H and G are not merged because we apply additional nonlinear normalization operations on patches y and x, which will be detailed in Sect. 4.1.6.

Thus, there are a total of 5 trainable layers in our network: 2 convolutional layers H and G, and 3 linear layers shown as gray boxes in Fig. 4.2. The k recurrent layers share the same weights and are therefore conceptually regarded as one. Note that all the linear layers are actually implemented as convolutional layers applied on each patch with filter spatial size of 1×1, a structure similar to the network in network [32]. Also note that all these layers have only weights but no biases (zero biases).

Mean squared error (MSE) is employed as the cost function to train the network, and our optimization objective can be expressed as

$$\min_\Theta \sum_i \| SCN(I_y^{(i)}; \Theta) - I_x^{(i)} \|_2^2, \tag{4.6}$$

where $I_y^{(i)}$ and $I_x^{(i)}$ form the ith pair of LR/HR training data, and $SCN(I_y; \Theta)$ denotes the HR image for I_y predicted using the SCN model with parameter set Θ. All the parameters are optimized through the standard back-propagation algorithm. Although it is possible to use other cost terms that are more correlated with human visual perception than MSE, our experimental results show that simply minimizing MSE leads to improvement in subjective quality.

4.1.3.4 Advantages over Previous Models

The construction of our SCN follows exactly each step in the sparse coding based SR method [6]. If the network parameters are set according to the dictionaries learned in [6], it can reproduce almost the same results. However, after training, SCN learns a more complex regression function and can no longer be converted to an equivalent sparse coding model. The advantage of SCN comes from its ability to jointly optimize all the layer parameters from end to end; while in [6] some variables are manually designed and some are optimized individually by fixing all the others.

Technically, our network is also a CNN and it has similar layers as the CNN model proposed in [14] for patch extraction and reconstruction. The key difference is that we have an LISTA subnetwork specifically designed to enforce sparse representation prior; while in [14] a generic rectified linear unit (ReLU) [33] is used for nonlinear mapping. Since SCN is designed based on our domain knowledge in sparse coding, we are able to obtain a better interpretation of the filter responses and have a better way to initialize the filter parameters in training. We will see in the experiments that all these contribute to better SR results, faster training speed and smaller model size than a vanilla CNN.

4.1.4 Network Cascade for Scalable SR

In this section, we investigate two different network cascade techniques in order to fully exploit our SCN model in SR applications.

4.1.4.1 Network Cascade for SR of a Fixed Scaling Factor

First, we observe that the SR results can be further improved by cascading multiple SCNs trained for the same objective in (4.6), which is inspired by the multi-pass scheme in [17]. The only difference for training these SCNs is to replace the bicubic interpolated input by its latest HR estimate, while the target output remains the same.

The first SCN plays as a function approximator to model the nonlinear mapping from the bicubic upscaled image to the ground-truth image. The following SCN plays as another function approximator, with the starting point changed to a better estimate: the output of its previous SCN.

In other words, the cascade of SCNs as a whole can be considered as a new deeper network having more powerful learning capability, which is able to better approximate the mapping between the LR inputs to the HR counterparts, and these SCNs can be trained jointly to pursue even better SR performance.

4.1.4.2 Network Cascade for Scalable SR

Like most SR models learned from external training examples, the SCN discussed previously can only upscale images by a fixed factor. A separate model needs to be trained for each scaling factor to achieve the best performance, which limits the flexibility and scalability in practical use. One way to overcome this difficulty is to repeatedly enlarge the image by a fixed scale until the resulting HR image reaches a desired size. This practice is commonly adopted in the self-similarity based methods [9,10,12], but is not so popular in other cases for the fear of error accumulation during repetitive upscaling.

In our case, however, it is observed that a cascade of SCNs trained for small scaling factors can generate even better SR results than a single SCN trained for a large scaling factor, especially when the target scaling factor is large (greater than 2). This is illustrated by the example in Fig. 4.3. Here an input image is magnified 4 times in two ways: with a single SCN×4 model through the processing flow (A) → (B) → (D); and with a cascade of two SCN×2 models through (A) → (C) → (E). It can be seen that the input to the second cascaded SCN×2 in (C) is already sharper and contains less artifacts than the bicubic×4 input to the single SCN×4 in (B), which naturally leads to the better final result in (E) than that in (D).

To get a better understanding of the above observation, we can draw a loose analogy between the SR process and a communication system. Bicubic interpolation is like a noisy channel through which an image is "transmitted" from the LR domain to HR domain. And our SCN model (or any SR algorithm) behaves as a receiver which recovers clean signals from noisy observations. A cascade of SCNs is then like a set of relay stations that enhance signal-to-noise ratio before the signal becomes too weak for further transmission. Therefore, cascading will work only when each SCN can restore enough useful information to compensate for the new artifacts it introduces as well as the magnified artifacts from previous stages.

4.1.4.3 Training Cascade of Networks

Taking into account the two aforementioned cascade techniques, we can consider the cascade of all SCNs as a deeper network (CSCN), in which the final output of the consecutive SCNs of the same ground truth is connected to the input of the next SCN with bicubic interpolation in the between. To construct the cascade, besides stacking several SCNs trained individually with respect to (4.6), we can also optimize all of them jointly as shown in Fig. 4.4. Without loss of generality, we assume that each stage in Sect. 4.1.4.2 has the same scaling factor s. Let $\hat{I}_{j,k}$ ($j > 0, k > 0$) denote the output image of the jth SCN in the kth stage upscaled by a total of $\times s^k$ times. In the same stage, each output of SCNs is compared with the associated ground-truth image I_k according to

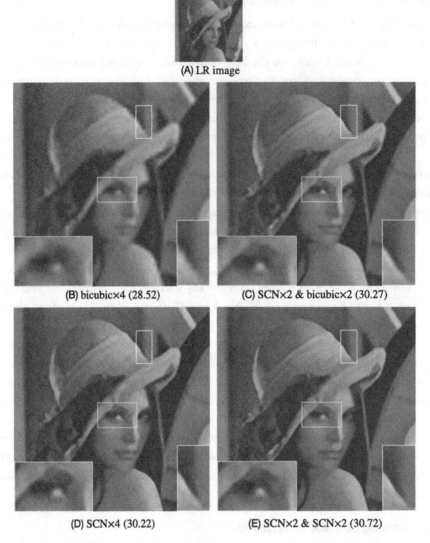

(A) LR image

(B) bicubic×4 (28.52)

(C) SCN×2 & bicubic×2 (30.27)

(D) SCN×4 (30.22)

(E) SCN×2 & SCN×2 (30.72)

Figure 4.3 SR results for the "Lena" image upscaled 4 times. (A) → (B) → (D) represents the processing flow with a single SCN×4 model. (A) → (C) → (E) represents the processing flow with two cascaded SCN×2 models. PSNR is given in parentheses.

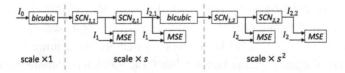

Figure 4.4 Training cascade of SCNs with multi-scale objectives.

the MSE cost, leading to a multi-scale objective function:

$$\min_{\{\boldsymbol{\Theta}_{j,k}\}} \sum_i \sum_j \sum_k \left\| SCN(\hat{\boldsymbol{I}}_{j-1,k}^{(i)}; \boldsymbol{\Theta}_{j,k}) - \boldsymbol{I}_k^{(i)} \right\|_2^2, \tag{4.7}$$

where i denotes the data index, and j, k denote the SCN index. For simplicity of notation, $\hat{\boldsymbol{I}}_{0,k}$ specially denotes the bicubic interpolated image of the final output in the $(k-1)$th stage upscaled by a total of $\times s^{k-1}$ times. This multi-scale objective function makes full use of the supervision information in all scales, sharing a similar idea as heterogeneous networks [34]. All the layer parameters $\{\boldsymbol{\Theta}_{j,k}\}$ in (4.7) could be optimized from end to end by back-propagation. The SCNs share the same training objective can be trained simultaneously, taking advantage of the merit of deep learning. For the SCNs with different training objectives, we use a greedy algorithm here to train them sequentially from the beginning of the cascade so that we do not need to care about the gradient of bicubic layers. Applying back-propagation through a bicubic layer or its trainable surrogate will be considered in a future work.

4.1.5 Robust SR for Real Scenarios

Most of recent SR works generate the LR images for both training and testing by downscaling HR images using bicubic interpolation [6,21]. However, this assumption of the forward model may not always hold in practice. For example, real LR measurements are usually blurred, or corrupted with noise. Sometimes, the LR generation mechanism may be complicated, or even unknown. We now investigate a practical SR problem, and propose two approaches to handle such non-ideal LR measurements, using the generic SCN. In the case that the underlying mechanism of the real LR generation is unclear or complicated, we propose a data-driven approach by fine-tuning the learned generic SCN with a limited number of real LR measurements, as well as their corresponding HR counterparts. On the other hand, if the real training samples are unavailable but the LR generation mechanism is clear, we formulate this inverse problem as the regularized HR image reconstruction problem which can be solved using iterative methods. The proposed methods demonstrate the robustness of our SCN model to different SR scenarios. In the following, we elaborate the details of these two approaches.

4.1.5.1 Data-Driven SR by Fine-Tuning

Deep learning models can be efficiently transferred from one task to another by reusing the intermediate representation in the original neural network [35]. This method has been proven successful on a number of high-level vision tasks, even if there is a limited amount of training data in the new task [36].

The success of super-resolution algorithms usually highly depends on the accuracy of the model of the imaging process. When the underlying mechanism of the generation of LR images is not clear, we can take advantage of the aforementioned merit of deep learning models by learning our model in a data-driven manner, to adapt it for a particular task. Specifically, we start training from the generic SCN model while using very limited amount of training data from a new SR scenario, and manage to adapt it to the new SR scenario and obtain promising results. In this way, it is demonstrated that the SCN has strong capability of learning complex mappings between the non-ideal LR measurements and their HR counterparts, as well as the high flexibility of adapting to various SR tasks.

4.1.5.2 Iterative SR with Regularization

The second approach considers the case that the mechanism of generating the real LR images is relatively simple and clear, indicating the training data is always available if we synthesize LR images with the known degradation process. We propose an iterative SR scheme which incorporates the generic SCN model with additional regularization based on task-related priors (e.g., the known kernel for deblurring, or the data sparsity for denoising). In this section, we specifically discuss handling blurred and noisy LR measurements in details as examples, though the iterative SR methods can be generalized to other practical imaging models.

Blurry Image Upscaling

The real LR images can be generated with various types of blurring. Directly applying the generic SCN model is obviously not optimal. Instead, with the known blurring kernel, we propose to estimate the regularized version of the HR image $\hat{I}_x$ based on the directly upscaled image $\tilde{I}_x$ by the learned SCN as follows:

$$\hat{I}_x = \arg\min_{I} \|I - \tilde{I}_x\|_2, \text{ s.t. } D \cdot B \cdot I = I_y^0 \tag{4.8}$$

where I_y^0 is the original blurred LR input, and the operators B and D are blurring and subsampling, respectively. Similar to the previous work [6], we use back-projection to iteratively estimate the regularized HR input on which our model can perform better. Specifically, given the regularized estimate $\hat{I}_x^{i-1}$ at iteration $i-1$, we estimate a less blurred LR image I_y^{i-1} by downsampling $\hat{I}_x^i$ using bicubic interpolation. The upscaled $\tilde{I}_x^i$ by learned SCN serves the regularizer for the ith iteration as follows:

$$\hat{I}_x^i = \arg\min_{I} \|I - \tilde{I}_x^i\|_2^2 + \|D \cdot B \cdot I - I_y^0\|_2^2. \tag{4.9}$$

Here we use a penalty method to form an unconstrained problem. The upscaled HR image $\tilde{I}_x^i$ can be computed as $SCN(I_y^{i-1}, \Theta)$. The same process is repeated until convergence. We have applied the proposed iterative scheme to LR images generated from Gaussian blurring and subsampling as an example. The empirical performance is illustrated in Sect. 4.1.7.

Noisy Image Upscaling

Noise is a ubiquitous cause of corruption in image acquisition. State-of-the-art image denoising methods usually adopt priors such as patch similarity [37], patch sparsity [38,17], or both [39], as a regularizer in image restoration. In this section, we propose a regularized noisy image upscaling scheme, for specifically handling noisy LR images, in order to obtain improved SR quality. Though any denoising algorithm can be used in our proposed scheme, here we apply spatial similarity combined with transform domain image patch group-sparsity as our regularizer [39], to form the regularized iterative SR problem as an example.

Similar to the method in Sect. 4.1.5.2, we iteratively estimate the less noisy HR image from the denoised LR image. Given the denoised LR estimate $\hat{I}_y^{i-1}$ at iteration $i-1$, we directly upscale it, using the learned generic SCN, to obtain the HR image $\hat{I}_x^{i-1}$. It is then downsampled using bicubic interpolation, to generate the LR image $\tilde{I}_y^i$, which is used in the fidelity term in the ith iteration of LR image denoising. The same process is repeated until convergence. The iterative LR image denoising problem is formulated as follows:

$$\left\{\hat{I}_y^i, \{\hat{\alpha}_i\}\right\} = \arg\min_{I,\{\alpha_i\}} \|I - \tilde{I}_y^i\|_2^2$$
$$+ \sum_{j=1}^{N} \left\{\|W_{3D}G_j I - \alpha_j\|_2^2 + \tau\|\alpha_j\|_0\right\} \tag{4.10}$$

where the operator G_j generates the 3D vectorized tensor, which groups the jth overlapping patch from the LR image I, together with the spatially similar patches within its neighborhood by block matching [39]. The codes $\{\alpha_j\}$ of the patch groups in the domain of 3D sparsifying transform W_{3D} are sparse, which is enforced by the l_0 norm penalty [40]. The weight τ controls the sparsity level, which normally depends on the remaining noise level in $\tilde{I}_y^i$ [41,40].

In (4.10), we use the patch group sparsity as our denoising regularizer. The 3D sparsifying transform W_{3D} can be one of commonly used analytical transforms, such as discrete cosine transform (DCT) or wavelets. The state-of-the-art BM3D denoising algorithm [39] is based on such an approach, but further improved by more sophisticated engineering stages. In order to achieve the best practical SR quality, we demonstrate the empirical performance comparison using BM3D as the regularizer in Sect. 4.1.7. Additionally, our proposed iterative method is a general practical SR framework, which is not dedicated to SCN. One can conveniently extend it to other SR methods, which generate $\tilde{I}_y^i$ in the ith iteration. A performance comparison of these methods is given in Sect. 4.1.7.

4.1.6 Implementation Details

We determine the number of nodes in each layer of our SCN mainly according to the corresponding settings used in sparse coding [7]. Unless otherwise stated, we use input LR patch size $s_y = 9$, LR feature dimension $m_y = 100$, dictionary size $n = 128$, output HR patch size $s_x = 5$, and patch aggregation filter size $s_g = 5$. All the convolution layers have a stride of 1. Each LR patch y is normalized by its mean and variance, and the same mean and variance are used to restore the final HR patch x. We crop 56×56 regions from each image to obtain fixed-sized input samples to the network, which produces outputs of size 44×44.

To reduce the number of parameters, we implement the LR patch extraction layer H as the combination of two layers: the first layer has 4 trainable filters, each of which is shifted to 25 fixed positions by the second layer. Similarly, the patch combination layer G is also split into a fixed layer which aligns pixels in overlapping patches and a trainable layer whose weights are used to combine overlapping pixels. In this way, the number of parameters in these two layers are reduced by more than an order, and there is no observable loss in performance.

We employ a standard stochastic gradient descent algorithm to train our networks with mini-batch size of 64. Based on the understanding of each layer's role in sparse coding, we use Harr-like gradient filters to initialize layer H, and use uniform weights to initialize layer G. All the remaining three linear layers are related to the dictionary pair (D_x, D_y) in sparse coding. To initialize them, we first randomly set D_x and D_y with Gaussian noise, and then find the corresponding layer weights as in ISTA [30]:

$$
\begin{aligned}
w_1 &= C \cdot D_y^T, \\
w_2 &= I - D_y^T D_y, \\
w_3 &= \frac{1}{CL} D_x,
\end{aligned}
\tag{4.11}
$$

where w_1, w_2 and w_3 denote the weights of the three subsequent layers after layer H; L is the upper bound on the largest eigenvalue of $D_y^T D_y$, and C is the threshold value before normalization. We empirically set $L = C = 5$.

The proposed models are all trained using the CUDA ConvNet package [11] on a workstation with 12 Intel Xeon 2.67 GHz CPUs and 1 GTX680 GPU. Training an SCN usually takes less than one day. Note that this package is customized for classification networks, and its efficiency can be further optimized for our SCN model.

In testing, to make the entire image covered by output samples, we crop input samples with overlap and extend the boundary of original image by reflection. Note we shave the image border in the same way as [14] for objective evaluations to ensure fair comparison. Only the luminance channel is processed with our method, and bicubic interpolation is applied to the chrominance channels, as their high frequency components

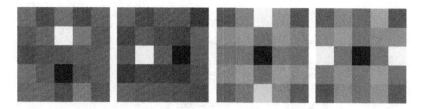

Figure 4.5 The four learned filters in the first layer H.

are less noticeable to human eyes. To achieve arbitrary scaling factors using CSCN, we upscale an image by a factor of 2 repeatedly until it is at least as large as the desired size. Then a bicubic interpolation is used to downscale it to the target resolution if necessary.

When reporting our best results in Sect. 4.1.7.2, we also use the multi-view testing strategy commonly employed in image classification. For patch-based image SR, multi-view testing is implicitly used when predictions from multiple overlapping patches are averaged. Here, besides sampling overlapping patches, we also add more views by flipping and transposing the patch. Such strategy is found to improve SR performance for general algorithms at the sheer cost of computation.

4.1.7 Experiments

We evaluate and compare the performance of our models using the same data and protocols as in [21], which are commonly adopted in SR literature. All our models are learned from a training set with 91 images, and tested on Set5 [42], Set14 [43] and BSD100 [44] which contain 5, 14 and 100 images, respectively. We have also trained on other different larger data sets, and observe marginal performance change (around 0.1 dB). The original images are downsized by bicubic interpolation to generate LR–HR image pairs for both training and evaluation. The training data are augmented with translation, rotation and scaling.

4.1.7.1 Algorithm Analysis

We first visualize the four filters learned in the first layer H in Fig. 4.5. The filter patterns do not change much from the initial first- and second-order gradient operators. Some additional small coefficients are introduced in a highly structured form that capture richer high frequency details.

The performance of several networks during training is measured on Set5 in Fig. 4.6. Our SCN improves significantly over sparse coding (SC) [7], as it leverages data more effectively with end-to-end training. The SCN initialized according to (4.11) can converge faster and better than the same model with random initialization, which indicates that the understanding of SCN based on sparse coding can help its optimization. We also train a CNN model [14] of the same size as SCN, but find its convergence speed

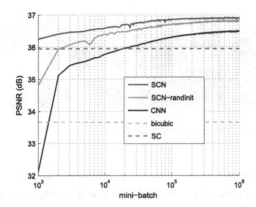

Figure 4.6 The PSNR change for ×2 SR on Set5 during training using different methods: SCN; SCN with random initialization; CNN. The horizontal dashed lines show the benchmarks of bicubic interpolation and sparse coding (SC).

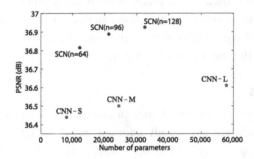

Figure 4.7 PSNR for ×2 SR on Set5 using SCN and CNN with various network sizes.

Table 4.1 Time consumption for SCN to upscale the "baby" image from 256×256 to 512×512 using different dictionary size n

n	64	96	128	256	512
time (s)	0.159	0.192	0.230	0.445	1.214

much slower. It is reported in [14] that training a CNN takes 8×10^8 back-propagations (equivalent to 12.5×10^6 mini-batches here). To achieve the same performance as CNN, our SCN requires less than 1% back-propagations.

The network size of SCN is mainly determined by the dictionary size n. Besides the default value $n = 128$, we have tried other sizes and plot their performance versus the number of network parameters in Fig. 4.7. The PSNR of SCN does not drop too much as n decreases from 128 to 64, but the model size and computation time can be reduced significantly, as shown in Table 4.1. Fig. 4.7 also shows the performance of

Table 4.2 PSNR of different network cascading schemes on Set5, evaluated for different scaling factors in each column

scaling factor	×1.5	×2	×3	×4
SCN×1.5	40.14	36.41	30.33	29.02
SCN×2	**40.15**	**36.93**	32.99	30.70
SCN×3	39.88	36.76	32.87	30.63
SCN×4	39.69	36.54	32.76	30.55
CSCN	**40.15**	**36.93**	**33.10**	**30.86**

CNN with various sizes. Our smallest SCN can achieve higher PSNR than the largest model (CNN-L) in [45] while only using about 20% of parameters.

Different numbers of recurrent stages k have been tested for SCN, and we find increasing k from 1 to 3 only improves performance by less than 0.1 dB. As a tradeoff between speed and accuracy, we use $k = 1$ throughout the section.

In Table 4.2, different network structures with cascade for scalable SR in Sect. 4.1.4.2 (in each row) are compared at different scaling factors (in each column). SCN×a denotes the model trained with fixed scaling factor a without any cascade technique. For a fixed a, we use SCN×a as a basic module and apply it one or more times to super-resolve images for different upscaling factors, which is shown in each row of Table 4.2. It is observed that SCN×2 can perform as well as the scale-specific model for small scaling factor (1.5), and much better for large scaling factors (3 and 4). Note that the cascade of SCN×1.5 does not lead to good results since artifacts quickly get amplified through many repetitive upscalings. Therefore, we use SCN×2 as the default building block for CSCN, and drop the notation ×2 when there is no ambiguity. The last row in Table 4.2 shows that a CSCN trained using the multi-scale objective in (4.7) can further improve the SR results for scaling factors 3 and 4, as the second SCN in the cascade is trained to be robust to the artifacts generated by the first one.

As shown in [45], the amount of training data plays an important role in the field of deep learning. In order to evaluate the effect of various amount of data on training CSCN, we change the training set from a relatively small set of 91 images (Set91) [21] to two other sets: the 199 out of 200 training images[2] in BSD500 dataset (BSD200) [44], and a subset of 7500 images from the ILSVRC2013 dataset [71]. A model of exactly the same architecture without any cascade is trained on each data set, and another 100 images from the ILSVRC2013 dataset are included as an additional test set. From Table 4.3, we can observe that the CSCN trained on BSD200 consistently outperforms its counterpart trained on Set91 by around 0.1 dB on all test data. However, the perfor-

---

[2] Since one out of 200 training images coincides with one image in Set5, we exclude it from our training set.

Table 4.3 Effect of various training sets on the PSNR of ×2 upscaling with single view SCN

Training Set	Test Set			
	Set5	Set14	BSD100	ILSVRC (100)
Set91	36.93	32.56	31.40	32.13
BSD200	**36.97**	**32.69**	**31.55**	32.27
ILSVRC (7.5k)	36.84	32.67	31.51	**32.31**

mance of the model trained on ILSVRC2013 is slightly different from the one trained on BSD200, which shows the saturation of the performance as the amount of training data increases. The inferior quality of images in ILSVRC2013 may be a hurdle to further improve the performance. Therefore, our method is robust to training data and can benefit marginally from a larger set of training images.

4.1.7.2 Comparison with State-of-the-Art

We compare the proposed CSCN with other recent SR methods on all the images in Set5, Set14 and BSD100 for different scaling factors. Table 4.4 shows the PSNR and structural similarity (SSIM) [46] for adjusted anchored neighborhood regression (A+) [25], CNN [14], CNN trained with larger model size and much more data (CNN-L) [45], the proposed CSCN, and CSCN with our multi-view testing (CSCN-MV). We do not list other methods [7,21,43,47,24] whose performance is worse than A+ or CNN-L.

It can be seen from Table 4.4 that CSCN performs consistently better than all previous methods in both PSNR and SSIM, and with multi-view testing the results can be further improved. CNN-L improves over CNN by increasing model parameters and training data. However, it is still not as good as CSCN which is trained with a much smaller size and on a much smaller data set. Clearly, the better model structure of CSCN makes it less dependent on model capacity and training data in improving performance. Our models are generally more advantageous for large scaling factors due to the cascade structure. A larger performance gain is observed on Set5 than the other two test sets because Set5 has more similar statistics as the training set.

The visual qualities of the SR results generated by sparse coding (SC) [7], CNN and CSCN are compared in Fig. 4.8. Our approach produces image patterns with shaper boundaries and richer textures, and is free of the ringing artifacts observable in the other two methods.

Fig. 4.9 shows the SR results on the "chip" image compared among more methods including the self-example based method (SE) [10] and the deep network cascade (DNC) [12]. SE and DNC can generate very sharp edges on this image, but also introduce artifacts and blurs on corners and fine structures due to the lack of self-similar

Table 4.4 PSNR (SSIM) comparison on three test data sets among different methods. Red indicates the best and blue indicates the second best performance. The performance gain of our best model over all the others' best is shown in the last row. (For interpretation of the colors in the tables, the reader is referred to the web version of this chapter)

Data Set	Set5			Set14			BSD100		
Upscaling	×2	×3	×4	×2	×3	×4	×2	×3	×4
A+ [25]	36.55 (0.9544)	32.59 (0.9088)	30.29 (0.8603)	32.28 (0.9056)	29.13 (0.8188)	27.33 (0.7491)	30.78 (0.8773)	28.18 (0.7808)	26.77 (0.7085)
CNN [14]	36.34 (0.9521)	32.39 (0.9033)	30.09 (0.8530)	32.18 (0.9039)	29.00 (0.8145)	27.20 (0.7413)	31.11 (0.8835)	28.20 (0.7794)	26.70 (0.7018)
CNN-L [45]	36.66 (0.9542)	32.75 (0.9090)	30.49 (0.8628)	32.45 (0.9067)	29.30 (0.8215)	27.50 (0.7513)	31.36 (0.8879)	28.41 (0.7863)	26.90 (0.7103)
CSCN	37.00 (0.9557)	33.18 (0.9153)	30.94 (0.8755)	32.65 (0.9081)	29.41 (0.8234)	27.71 (0.7592)	31.46 (0.8891)	28.52 (0.7883)	27.06 (0.7167)
CSCN-MV	37.21 (0.9571)	33.34 (0.9173)	31.14 (0.8789)	32.80 (0.9101)	29.57 (0.8263)	27.81 (0.7619)	31.60 (0.8915)	28.60 (0.7905)	27.14 (0.7191)
Our Improvement	0.55 (0.0029)	0.59 (0.0083)	0.65 (0.0161)	0.35 (0.0034)	0.27 (0.0048)	0.31 (0.0106)	0.24 (0.0036)	0.19 (0.0042)	0.24 (0.0088)

Figure 4.8 SR results given by SC [7] (first row), CNN [14] (second row) and our CSCN (third row). Images from left to right: the "monarch" image upscaled by ×3; the "zebra" image upscaled by ×3; the "comic" image upscaled by ×3.

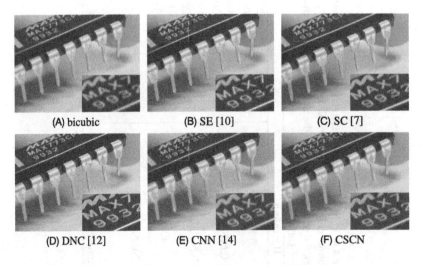

Figure 4.9 The "chip" image upscaled by ×4 using different methods.

patches. On the contrary, the CSCN method recovers all the structures of the characters without any distortion.

4.1.7.3 Robustness to Real SR Scenarios

We evaluate the performance of the proposed practical SR methods in Sect. 4.1.5, by providing the empirical results of several experiments for the two aforementioned approaches.

Data-Driven SR by Fine-Tuning

The proposed method in Sect. 4.1.5.1 is data-driven, and thus the generic SCN can be easily adapted for a particular task, with a small amount of training samples. We demonstrate the performance of this method in the application of enlarging low-DPI scanned document images with heavy noise. We first obtain several pairs of LR and HR images by scanning a document under two settings of 150 and 300 DPI. Then we fine-tune our generic CSCN model using only one pair of scanned images for a few iterations. Fig. 4.11 illustrates the visualization of the upscaled image from the 150 DPI scanned image. As shown by the SR results in Fig. 4.11, the CSCN before adaptation is very sensitive to LR measurement corruption, so the enlarged texts in (B) are much more corrupted than they are in the nearest neighbor upscaled image (A). However, the adapted CSCN model removes almost all the artifacts and can restore clear texts in (C), which is promising for practical applications such as quality enhancement of online scanned books and restoration of legacy documents.

Regularized Iterative SR

We now show experimental results of practical SR for blurred and noisy LR images, using the proposed regularized iterative methods in Sect. 4.1.5.2. We first compare the SR performance on blurry images using the proposed method in Sect. 4.1.5.2 with several other recent methods [50,48,49], using the same test images and settings. All these methods are designed for blurry LR input, while our model is trained on sharp LR input. As shown in Table 4.5, our model achieves much better results than the competitors. Note the speed of our model is also much faster than the conventional sparse coding based methods.

To test the performance of upscaling noisy LR images, we simulate additive Gaussian noise for the LR input images at 4 different noise levels ($\sigma = 5, 10, 15, 20$) as the noisy input images. We compare the practical SR results in Set5 obtained from the following algorithms: directly using SCN, our proposed iterative SCN method using BM3D as denoising regularizer (iterative BM3D-SCN), and fine-tuning SCN with additional noisy training pairs. Note that knowing the underlying corruption model of real LR image (e.g., noise distribution or blurring kernel), one can always synthesize real training pairs for fine-tuning the generic SCN. In other words, once the iterative SR method

Table 4.5 PSNR of ×3 upscaling on LR images with different blurring kernels

Kernel	Gaussian $\sigma = 1.0$			Gaussian $\sigma = 1.6$		
Method	CSR [48]	NLM [49]	SCN	CSR [48]	GSC [50]	SCN
Butterfly	27.87	26.93	**28.70**	28.19	25.48	**29.03**
Parrots	30.17	29.93	**30.75**	30.68	29.20	**30.83**
Parthenon	26.89	–	**27.06**	27.23	26.44	**27.40**
Bike	24.41	24.38	**24.81**	24.72	23.78	**25.11**
Flower	29.14	28.86	**29.50**	29.54	28.30	**29.78**
Girl	**33.59**	33.44	33.57	**33.68**	33.13	33.65
Hat	31.09	30.81	**31.32**	31.33	30.29	**31.62**
Leaves	26.99	26.47	**27.45**	27.60	24.78	**27.87**
Plants	33.92	33.27	**34.35**	34.00	32.33	**34.53**
Raccoon	**29.09**	–	28.99	**29.29**	28.81	29.16
Average	29.32	29.26	**29.65**	29.63	28.25	**29.90**

Table 4.6 PSNR values for ×2 upscaling noisy LR images in Set5 by directly using SCN (Direct SCN), directly using CNN-L (Direct CNN-L), SCN after fine-tuning on new noisy training data (Fine-tuning SCN), the iterative method of BM3D & SCN (Iterative BM3D-SCN), and the iterative method of BM3D & CNN-L (Iterative BM3D-CNN-L)

σ	5	10	15	20
Direct SCN	30.23	25.11	21.81	19.45
Direct CNN-L	30.47	25.32	21.91	19.46
Fine-tuning SCN	33.03	31.00	29.46	28.44
Iterative BM3D-SCN	**33.51**	**31.22**	**29.65**	**28.61**
Iterative BM3D-CNN-L	33.42	31.16	29.62	28.59

is feasible, one can always apply our proposed data–driven method for SR alternatively. However, the converse is not true. Therefore, the knowledge of the corruption model of real measurements can be considered as a stronger assumption, compared to providing real training image pairs. Correspondingly, the SR performances of these two methods are evaluated when both can be applied. We also provide the results of methods directly using another generic SR model: CNN-L [45], and the similar iterative SR method involving CNN-L (iterative BM3D-CNN-L).

The practical SR results are listed in Table 4.6. We observed the improved PSNR using our proposed regularized iterative SR method over all noise levels. The proposed iterative BM3D-SCN achieves much higher PSNR than the method of directly using SCN. The performance gap (in terms of SR PSNR) between iterative BM3D-SCN and direct SCN becomes larger, as the noise level increases. Similar observation can be made when comparing iterative BM3D-CNN-L and direct CNN-L. Compared to the method of fine-tuning SCN, the iterative BM3D-SCN method demonstrates better

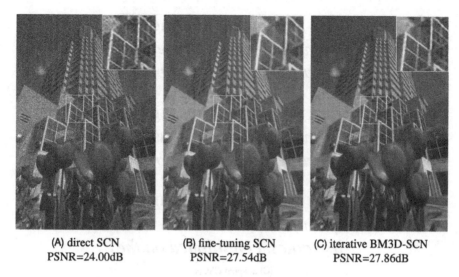

(A) direct SCN
PSNR=24.00dB

(B) fine-tuning SCN
PSNR=27.54dB

(C) iterative BM3D-SCN
PSNR=27.86dB

Figure 4.10 The "building" image corrupted by additive Gaussian noise of $\sigma = 10$ and then upscaled by ×2 using different methods.

empirical performance, with 0.3 dB improvement on average. The iterative BM3D-CNN-L method provides comparable results as the iterative BM3D-SCN method, which demonstrates that our proposed regularized iterative SCN scheme can be easily extended for other SR methods, and is able to effectively handle noisy LR measurements.

An example of upscaling noisy LR images using the aforementioned methods is demonstrated in Fig. 4.10. Both fine-tuning SCN and iterative BM3D-SCN are able to significantly suppress the additive noise, while many artifacts induced by noise are observed in the SR result of direct SCN. It is notable that the fine-tuning SCN method performs better recovering the texture and the iterative BM3D-SCN method is preferable in smooth regions.

4.1.8 Subjective Evaluation

Subjective perception is an important metric to evaluate SR techniques for commercial use, other than the quantitative evaluation. In order to more thoroughly compare various SR methods and quantify the subjective perception, we utilize an online platform for subjective evaluation of SR results from several methods [23], including bicubic, SC [7], SE [10], self-example regression (SER) [51], CNN [14] and CSCN. Each participant is invited to conduct several pair-wise comparisons of SR results from different methods. The SR methods of displayed SR images in each pair are randomly selected. Ground-truth HR images are also included when they are available as references. For

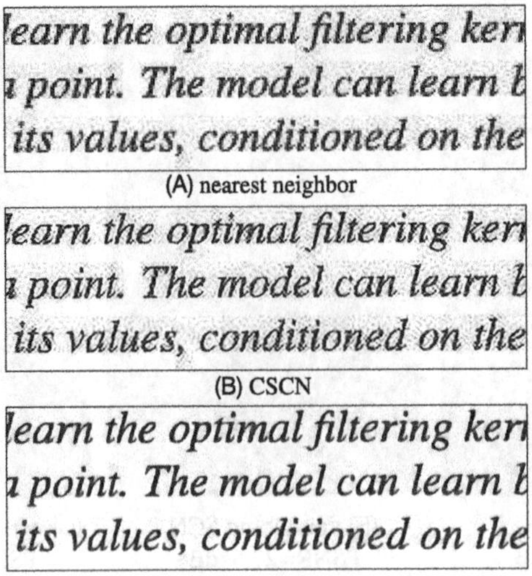

learn the optimal filtering ker

ɪ point. The model can learn t

its values, conditioned on the

(A) nearest neighbor

learn the optimal filtering ker

ɪ point. The model can learn t

its values, conditioned on the

(B) CSCN

learn the optimal filtering ker

ɪ point. The model can learn t

its values, conditioned on the

(C) adapted CSCN

Figure 4.11 Low-DPI scanned document upscaled by ×4 using different methods.

image pair: Clip-1-1 (3/6)

Left is better   Right is better

Figure 4.12 The user interface of a web-based image quality evaluation, where two images are displayed side by side and local details can be magnified by moving mouse over the corresponding region.

each pair, the participant needs to select the better one in terms of perceptual quality. A snapshot of our evaluation webpage[3] is shown in Fig. 4.12.

Specifically, there are SR results over 6 images with different scaling factors: "kid"×4, "chip"×4, "statue"×4, "lion"×3, "temple"×3 and "train"×3. The images are shown in Fig. 4.13. All the visual comparison results are then summarized into a

---

[3] www.ifp.illinois.edu/~wang308/survey.

Figure 4.13 The 6 images used in subjective evaluation.

7×7 winning matrix for 7 methods (including ground truth). A Bradley–Terry [52] model is calculated based to these results and the subjective score is estimated for each method according to this model. In the Bradley–Terry model, the probability that an object X is favored over Y is assumed to be

$$p(X \succ Y) = \frac{e^{s_X}}{e^{s_X} + e^{s_Y}} = \frac{1}{1 + e^{s_Y - s_X}}, \tag{4.12}$$

where s_X and s_Y are the subjective scores for X and Y. The scores s for all the objects can be jointly estimated by maximizing the log-likelihood of the pairwise comparison observations:

$$\max_s \sum_{i,j} w_{ij} \log \left(\frac{1}{1 + e^{s_j - s_i}} \right), \tag{4.13}$$

where w_{ij} is the (i,j)th element in the winning matrix $\boldsymbol{W}$, meaning the number of times when method i is favored over method j. We use the Newton–Raphson method to solve Eq. (4.13) and set the score for ground truth method as 1 to avoid the scale ambiguity.

Now we describe the detailed experiment results. We have a total of 270 participants giving 720 pairwise comparisons over six images with different scaling factors, which are shown in Fig. 4.13. Not every participant completed all the comparisons but their partial responses are still useful.

Fig. 4.14 shows the estimated scores for the six SR methods in our evaluation, with the score for ground truth method normalized to 1. As expected, all the SR methods have much lower scores than ground-truth, showing the great challenge in SR problem.

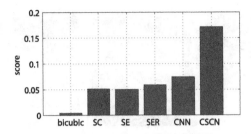

Figure 4.14 Subjective SR quality scores for different methods including bicubic, SC [7], SE [10], SER [51], CNN [14] and the proposed CSCN. The score for ground-truth result is 1.

The bicubic interpolation is significantly worse than other SR methods. The proposed CSCN method outperforms other previous state-of-the-art methods by a large margin, demonstrating its superior visual quality. It should be noted that the visual difference between some image pairs is very subtle. Nevertheless, the human subjects are able to perceive such difference when seeing the two images side by side, and therefore make consistent ratings. The CNN model becomes less competitive in the subjective evaluation than it is in PSNR comparison. This indicates that the visually appealing image appearance produced by CSCN should be attributed to the regularization from sparse representation, which cannot be easily learned by merely minimizing reconstruction error as in CNN.

4.1.9 Conclusion and Future Work

We propose a new approach for image SR by combining the strengths of sparse representation and deep network, and make considerable improvement over existing deep and shallow SR models both quantitatively and qualitatively. Besides producing outstanding SR results, the domain knowledge in the form of sparse coding can also benefit training speed and model compactness. Furthermore, we investigate the cascade of network for both fixed and incremental scaling factors so as to enhance SR performance. In addition, the robustness to real SR scenarios is discussed for handling non-ideal LR measurements. More generally, our observation is in line with other recent extensions made to CNN with better domain knowledge for different tasks.

In a future work, we will apply the SCN model to other problems where sparse coding can be useful. The interaction between deep networks for low- and high-level vision tasks, such as in [53], will also be explored. Another interesting direction to explore is video super-resolution [54], which is the task of inferring a high-resolution video sequence from a low-resolution one. This problem has drawn growing attention in both the research community and industry recently. From the research perspective, this problem is challenging because video signals vary in both temporal and spatial dimensions. In the meantime, with the prevalence of high-definition (HD) display such as HDTV in the market, there is an increasing need for converting low quality video sequences to high-definition so that they can be played on the HD displays in a visually pleasant manner.

There are two types of relation that are utilized for video SR: the intra-frame spatial relation and the inter-frame temporal relation. Neural network based models have successfully demonstrated the strong capability of modeling the spatial relation. Compared with the intra-frame spatial relation, the inter-frame temporal relation is more important for video SR, as researches of vision systems suggest that the human vision system is more sensitive to motion [55]. Thus it is essential for video SR algorithm to capture and model the effect of motion information on visual perception. Sparse priors have been shown useful for video SR [56]. We will try employing the sparse coding domain knowledge in deep network models for utilizing the temporal relation among consecutive LR video frames in the future.

4.2. LEARNING A MIXTURE OF DEEP NETWORKS FOR SINGLE IMAGE SUPER-RESOLUTION[4]

4.2.1 Introduction

The main difficulty of single image SR resides in the loss of much information in the degradation process. Since the known variables from the LR image are usually greatly outnumbered by that from the HR image, this problem is a highly ill-posed problem.

A large number of single image SR methods have been proposed in the literature, including interpolation based method [57], edge model based method [4] and example based method [58,9,6,21,45,24]. Since the former two methods usually suffer the sharp drop in restoration performance with large upscaling factors, the example based method has drawn great attention from the community recently. It usually learns the mapping from LR images to HR images in a patch-by-patch manner, with the help of sparse representation [6,23], random forest [26] and so on. The neighbor embedding method [58,21] and neural network based method [45] are two representatives of this category.

Neighbor embedding is proposed in [58,42] which estimates HR patches as a weighted average of local neighbors with the same weights as in LR feature space, based on the assumption that LR/HR patch pairs share similar local geometry in low-dimensional nonlinear manifolds. The coding coefficients are first acquired by representing each LR patch as a weighted average of local neighbors, and then the HR counterpart is estimated by the multiplication of the coding coefficients with the corresponding training HR patches. Anchored neighborhood regression (ANR) is utilized in [21] to improve the neighbor embedding methods, which partitions the feature space into a number of clusters using the learned dictionary atoms as a set of anchor points. A regressor is then learned for each cluster of patches. This approach has demonstrated superiority over the counterpart of global regression in [21]. Other variants of learning a mixture of SR regressors can be found in [25,59,60].

Recently, neural network based models have demonstrated the strong capability for single image SR [12,45,61], due to its large model capacity and the end-to-end learning strategy to get rid of hand-crafted features.

In this section, we propose a method to combine the merits of the neighborhood embedding methods and the neural network based methods via learning a mixture of neural networks for single image SR. The entire image signal space can be partitioned into several subspaces, and we dedicate one SR module to the image signals in each subspace, the synergy of which allows for a better capture of the complex relation between the LR image signal and its HR counterpart than the generic model. In order to take advantage of the end-to-end learning strategy of neural network based methods, we choose neural networks as the SR inference modules and incorporate these modules into one unified network, and design a branch in the network to predict the pixel-level weights for HR estimates from each SR module before they are adaptively aggregated to form the final HR image.

A systematic analysis of different network architectures is conducted with the focus on the relation between SR performance and various network architectures via extensive experiments, where the benefit of utilizing a mixture of SR models is demonstrated. Our proposed approach is contrasted with other current popular approaches on a large number of test images, and achieves state-of-the-arts performance consistently along with more flexibility of model design choices.

The section is organized as follows. The proposed method is introduced and explained in detail in Sect. 4.2.2. Implementation details are provided in Sect. 4.2.3. Section 4.2.4 describes our experimental results, in which we analyze thoroughly different network architectures and compare the performance of our method with other current SR methods both quantitatively and qualitatively. Finally, in Sect. 4.2.5, we conclude the section and discuss the future work. The more detailed version of this work can be found in [62].

4.2.2 The Proposed Method

First we give an overview of our method. The LR image serves as the input to our method. There are several **SR inference modules** $\{B_i\}_{i=1}^{N}$ in our method. Each of them, B_i, is dedicated to inferring a certain class of image patches, and applied on the LR input image to predict an HR estimate. We also devise an **adaptive weight module**, T, to adaptively combine at the pixel-level the HR estimates from SR inference modules. When we select neural networks as the SR inference modules, all the components can be incorporated into a unified neural network and jointly learned.

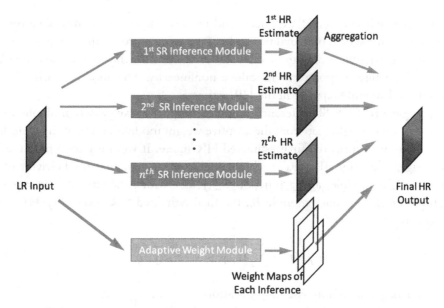

Figure 4.15 An overview of our proposed method. It consists of a number of SR inference modules and an adaptive weight module. Each SR inference module is dedicated to inferring a certain class of image local patterns, and is independently applied on the LR image to predict one HR estimate. These estimates are adaptively combined using pixel-wise aggregation weights from the adaptive weight module in order to form the final HR image.

The final estimated HR image is adaptively aggregated from the estimates of all SR inference modules. By the multi-branch design of our network, the super-resolution performance is improved comparing with its single branch counterpart, which will be shown in Sect. 4.2.4. The overview of our method is shown in Fig. 4.15.

Now we will introduce the network architecture in detail.

SR Inference Module. Taking the LR image as input, each SR inference module is designed to better capture the complex relation between a certain class of LR image signals and its HR counterpart, while predicting an HR estimate. For the sake of inference accuracy, we choose as the SR inference module a recent sparse coding based network (SCN) in [61], which implicitly incorporates the sparse prior into neural networks via employing the learned iterative shrinkage and thresholding algorithm (LISTA), and closely mimics the sparse coding based image SR method [7]. The architecture of SCN is shown in Fig. 4.2. Note that the design of the SR inference module is not limited to SCN, and all other neural network based SR models, e.g., SRCNN [45], can work as the SR inference module as well. The output of B_i serves as an estimate to the final HR frame.

Adaptive Weight Module. The goal of this module is to model the selectivity of the HR estimates from every SR inference module. We propose assigning pixel-wise aggregation weights of each HR estimate, and again the design of this module is open to any operation in the field of neural networks. Taking into account the computation cost and efficiency, we utilize only three convolutional layers for this module, and ReLU is applied on the filter responses to introduce nonlinearity. This module finally outputs the pixel-level weight maps for all the HR estimates.

Aggregation. Each SR inference module's output is pixel-wisely multiplied with its corresponding weight map from the adaptive weight module, and then these products are summed up to form the final estimated HR frame. If we use y to denote the LR input image, a function $W(y; \theta_w)$ with parameters θ_w to represent the behavior of the adaptive weight module, and a function $F_{B_i}(y; \theta_{B_i})$ with parameters θ_{B_i} to represent the output of SR inference module B_i, the final estimated HR image $F(y; \Theta)$ can be expressed as

$$F(y; \Theta) = \sum_{i=1}^{N} W_i(y; \theta_w) \odot F_{B_i}(y; \theta_{B_i}), \qquad (4.14)$$

where $\odot$ denotes the point-wise multiplication.

In training, our model tries to minimize the loss between the target HR frame and the predicted output, as

$$\min_{\Theta} \sum_{j} \| F(y_j; \Theta) - x_j \|_2^2, \qquad (4.15)$$

where $F(y; \Theta)$ represents the output of our model, x_j is the jth HR image and y_j is the corresponding LR image; Θ is the set of all parameters in our model.

If we plug Eq. (4.14) into Eq. (4.15), then the cost function can be expanded as:

$$\min_{\theta_w, \{\theta_{B_i}\}_{i=1}^{N}} \sum_{j} \| \sum_{i=1}^{N} W_i(y_j; \theta_w) \odot F_{B_i}(y_j; \theta_{B_i}) - x_j \|_2^2. \qquad (4.16)$$

4.2.3 Implementation Details

We conduct experiments following the protocols in [21]. Different learning based methods use different training data in the literature. We choose 91 images proposed in [6] to be consistent with [25,26,61]. These training data are augmented with translation, rotation and scaling, providing approximately 8 million training samples of 56×56 pixels.

Our model is tested on three benchmark data sets, which are Set5 [42], Set14 [43] and BSD100 [44]. The ground-truth images are downscaled by bicubic interpolation to generate LR–HR image pairs for both training and testing.

Following the convention in [21,61], we convert each color image into the YCbCr colorspace and only process the luminance channel with our model; bicubic interpolation is applied to the chrominance channels, because the visual system of human is more sensitive to details in intensity than in color.

Each SR inference module adopts the network architecture of SCN, while the filters of all three convolutional layers in the adaptive weight module have the spatial size of 5×5 and the numbers of filters of three layers are set to be 32, 16 and N, which is the number of SR inference modules.

Our network is trained on a machine with 12 Intel Xeon 2.67 GHz CPUs and one Nvidia TITAN X GPU. For the adaptive weight module, we employ a constant learning rate of 10^{-5} and initialize the weights from Gaussian distribution, while we stick to the learning rate and the initialization method in [61] for the SR inference modules. The standard gradient descent algorithm is employed to train our network with a batch size of 64 and the momentum of 0.9.

We train our model for the upscaling factor of 2. For larger upscaling factors, we adopt the model cascade technique in [61] to apply $\times 2$ models several times until the resulting image reaches at least as large as the desired size. The resulting image is down-sized via bicubic interpolation to the target resolution if necessary.

4.2.4 Experimental Results

In this section, we first analyze the architecture of our proposed model and then compare it with several other recent SR methods. Finally, we provide a runtime analysis of our approach and other competing methods.

4.2.4.1 Network Architecture Analysis

In this section we investigate the relation between various numbers of SR inference modules and SR performance. For the sake of our analysis, we increase the number of inference modules as we decrease the module capacity of each of them, so that the total model capacity is approximately consistent and thus the comparison is fair. Since the chosen SR inference module, SCN [61], closely mimics the sparse coding based SR method, we can reduce the module capacity of each inference module by decreasing the embedded dictionary size n (i.e., the number of filters in SCN) for sparse representation. We compare the following cases:

- one inference module with $n = 128$, which is equivalent to the structure of SCN in [61], denoted as SCN $(n = 128)$. Note that there is no need to include the adaptive weight module in this case.
- two inference modules with $n = 64$, denoted as $MSCN$-2 $(n = 64)$.
- four inference modules with $n = 32$, denoted as $MSCN$-4 $(n = 32)$.

The average Peak Signal-to-Noise Ratio (PSNR) and structural similarity (SSIM) [46] are measured to quantitatively compare the SR performance of these models over Set5, Set14 and BSD100 for various upscaling factors ($\times 2$, $\times 3$, $\times 4$), and the results are displayed in Table 4.7.

It can be observed that $MSCN$-2 $(n = 64)$ usually outperforms the original SCN network, i.e., SCN $(n = 128)$, and $MSCN$-4 $(n = 32)$ can achieve the best SR performance

Table 4.7 PSNR (in dB) and SSIM comparisons on Set5, Set14 and BSD100 for ×2, ×3 and ×4 upscaling factors among various network architectures. Red indicates the best and blue indicates the second best performance

Benchmark		SCN ($n = 128$)	MSCN-2 ($n = 64$)	MSCN-4 ($n = 32$)
Set5	×2	36.93 / 0.9552	37.00 / 0.9558	36.99 / 0.9559
	×3	33.10 / 0.9136	33.15 / 0.9133	33.13 / 0.9130
	×4	30.86 / 0.8710	30.92 / 0.8709	30.93 / 0.8712
Set14	×2	32.56 / 0.9069	32.70 / 0.9074	32.72 / 0.9076
	×3	29.41 / 0.8235	29.53 / 0.8253	29.56 / 0.8256
	×4	27.64 / 0.7578	27.76 / 0.7601	27.79 / 0.7607
BSD100	×2	31.40 / 0.8884	31.54 / 0.8913	31.56 / 0.8914
	×3	28.50 / 0.7885	28.56 / 0.7920	28.59 / 0.7926
	×4	27.03 / 0.7161	27.10 / 0.7207	27.13 / 0.7216

by improving the performance marginally over *MSCN-2* ($n = 64$). This demonstrates the effectiveness of our approach, namely that each SR inference model is able to super-resolve its own class of image signals better than one single generic inference model.

In order to further analyze the adaptive weight module, we select several input images, namely, *butterfly*, *zebra*, *barbara*, and visualize the four weight maps for every SR inference module in the network. Moreover, we record the index of the maximum weight across all weight maps at every pixel and generate a *max label map*. These results are displayed in Fig. 4.16.

From these visualizations it can be seen that weight map 4 shows high responses in many uniform regions, and thus mainly contributes to the low frequency regions of HR predictions. On the contrary, weight maps 1, 2 and 3 have large responses in regions with various edges and textures, and restore the high frequency details of HR predictions. These weight maps reveal that these sub-networks work in a complementary manner for constructing the final HR predictions. In the *max label map*, similar structures and patterns of images usually share the same label, indicating that such similar textures and patterns are favored to be super-resolved by the same inference model.

4.2.4.2 Comparison with State-of-the-Art

We conduct experiments on all the images in Set5, Set14 and BSD100 for different upscaling factors (×2, ×3, and ×4), to quantitatively and qualitatively compare our own approach with a number of state-of-the-art image SR methods. Table 4.8 shows the PSNR and SSIM for adjusted anchored neighborhood regression (A+) [25], SRCNN [45], RFL [26], SelfEx [24] and our proposed model, *MSCN-4* ($n = 128$), that consists of four SCN modules with $n = 128$. The single generic SCN without multi-view test-

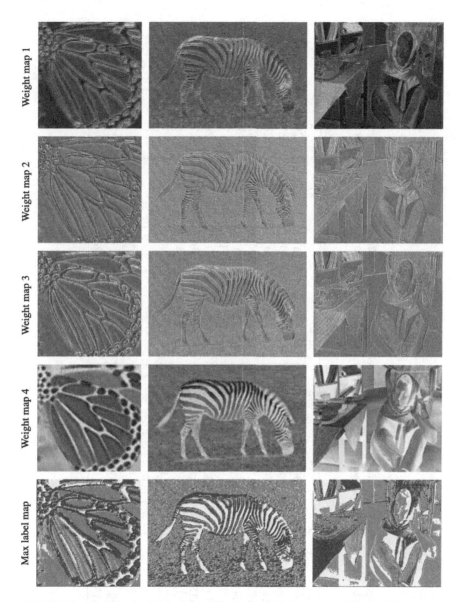

Figure 4.16 Weight maps for the HR estimate from every SR inference module in *MSCN-4* are given in the first four rows. The map (*max label map*) which records the index of the maximum weight across all weight maps at every pixel is shown in the last row. Images from left to right: the *butterfly* image upscaled by ×2; the *zebra* image upscaled by ×2; the *barbara* image upscaled by ×2.

ing in [61], i.e., SCN ($n = 128$), is also included for comparison as the baseline. Note that all the methods use the same 91 images [6] for training except SRCNN [45], which uses 395,909 images from ImageNet as training data.

Table 4.8 PSNR (SSIM) comparison on three test data sets for various upscaling factors among different methods. The best performance is indicated in red and the second best performance is shown in blue. The performance gain of our best model over all the other models' best is shown in the last row

Data Set	Set5			Set14			BSD100		
Upscaling	×2	×3	×4	×2	×3	×4	×2	×3	×4
A+ [25]	36.55 (0.9544)	32.59 (0.9088)	30.29 (0.8603)	32.28 (0.9056)	29.13 (0.8188)	27.33 (0.7491)	31.21 (0.8863)	28.29 (0.7835)	26.82 (0.7087)
SRCNN [45]	36.66 (0.9542)	32.75 (0.9090)	30.49 (0.8628)	32.45 (0.9067)	29.30 (0.8215)	27.50 (0.7513)	31.36 (0.8879)	28.41 (0.7863)	26.90 (0.7103)
RFL [26]	36.54 (0.9537)	32.43 (0.9057)	30.14 (0.8548)	32.26 (0.9040)	29.05 (0.8164)	27.24 (0.7451)	31.16 (0.8840)	28.22 (0.7806)	26.75 (0.7054)
SelfEx [24]	36.49 (0.9537)	32.58 (0.9093)	30.31 (0.8619)	32.22 (0.9034)	29.16 (0.8196)	27.40 (0.7518)	31.18 (0.8855)	28.29 (0.7840)	26.84 (0.7106)
SCN [61]	36.93 (0.9552)	33.10 (0.9144)	30.86 (0.8732)	32.56 (0.9074)	29.41 (0.8238)	27.64 (0.7578)	31.40 (0.8884)	28.50 (0.7885)	27.03 (0.7161)
MSCN-4	37.16 (0.9565)	33.33 (0.9155)	31.08 (0.8740)	32.85 (0.9084)	29.65 (0.8272)	27.87 (0.7624)	31.65 (0.8928)	28.66 (0.7941)	27.19 (0.7229)
Our Improvement	0.23 (0.0013)	0.23 (0.0011)	0.22 (0.0008)	0.29 (0.0010)	0.24 (0.0034)	0.23 (0.0046)	0.25 (0.0044)	0.16 (0.0056)	0.16 (0.0068)

It can be observed that our proposed model achieves the best SR performance consistently over three data sets for various upscaling factors. It outperforms SCN ($n = 128$) which obtains the second best results by about 0.2 dB across all the data sets, owing to the power of multiple inference modules.

We compare the visual quality of SR results among various methods in Fig. 4.17. The region inside the bounding box is zoomed in and shown for the sake of visual comparison. Our proposed model $MSCN$-4 ($n = 128$) is able to recover sharper edges and generate less artifacts in the SR inferences.

4.2.4.3 Runtime Analysis

The inference time is an important factor of SR algorithms other than the SR performance. The relation between the SR performance and the inference time of our approach is analyzed in this section. Specifically, we measure the average inference time of different network structures in our method for upscaling factor ×2 on Set14. The inference time costs versus the PSNR values are displayed in Fig. 4.18, where several other current SR methods [24,26,45,25] are included as reference (the inference time of SRCNN is from the public slower implementation of CPU). We can see that, generally, the more modules our network has, the more inference time is needed and the better SR results are achieved. By adjusting the number of SR inference modules in our network structure, we can achieve a tradeoff between SR performance and computation complexity. However, our slowest network still has the superiority in term of inference time, compared with other previous SR methods.

4.2.5 Conclusion and Future Work

In this section, we propose to jointly learn a mixture of deep networks for single image super-resolution, each of which serves as an SR inference module to handle a certain class of image signals. An adaptive weight module is designed to predict pixel-level aggregation weights of the HR estimates. Various network architectures are analyzed in terms of the SR performance and the inference time, which validates the effectiveness of our proposed model design. Extensive experiments manifest that our proposed model is able to achieve outstanding SR performance along with more flexibility of design.

Recent SR approaches increase the network depth in order to boost SR accuracy [63–67]. Kim et al. [63] proposed a very deep CNN with residual architecture to achieve outstanding SR performance, which utilizes broader contextual information with larger model capacity. Another network was designed by Kim et al. [64], which has recursive architectures with skip connection for image SR to boost performance while only exploiting a small number of model parameters. Tai et al. [65] discovered that many residual SR learning algorithms are based on either global residual learning or local residual learning, which are insufficient for very deep models. Instead, they proposed a model that applies both global and local learning while remaining parameter

Figure 4.17 Visual comparisons of SR results among different methods. From left to right: the *ppt3* image upscaled by ×3; the *102061* image upscaled by ×3; the *butterfly* image upscaled by ×4.

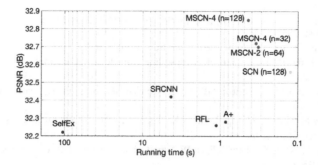

Figure 4.18 The average PSNR and the average inference time for upscaling factor ×2 on Set14 are compared among different network structures of our method and other SR methods. SRCNN uses the public slower implementation of CPU.

efficient via recursive learning. More recently, Tong et al. [67] proposed making use of Densely Connected Networks (DenseNet) [68] instead of ResNet [69] as the building block for image SR. Besides developing deeper networks, we show that increasing the number of parallel branches inside the network can achieve the same goal.

In the future, this approach of image super-resolution will be explored to facilitate other high-level vision tasks [53]. While the visual recognition research has made tremendous progress in recent years, most models are trained, applied, and evaluated on high-quality (HQ) visual data, such as the LFW [70] and ImageNet [71] benchmarks. However, in many emerging applications such as autonomous driving, intelligent video surveillance and robotics, the performances of visual sensing and analytics can be seriously endangered by different corruptions in complex unconstrained scenarios, such as limited resolution. Therefore, image super-resolution may provide one solution to feature enhancement for improving the performance of high-level vision tasks [72].

REFERENCES

[1] Baker S, Kanade T. Limits on super-resolution and how to break them. IEEE TPAMI 2002;24(9):1167–83.

[2] Lin Z, Shum HY. Fundamental limits of reconstruction-based superresolution algorithms under local translation. IEEE TPAMI 2004;26(1):83–97.

[3] Park SC, Park MK, Kang MG. Super-resolution image reconstruction: a technical overview. IEEE Signal Processing Magazine 2003;20(3):21–36.

[4] Fattal R. Image upsampling via imposed edge statistics. In: ACM transactions on graphics (TOG), vol. 26. ACM; 2007. p. 95.

[5] Aly HA, Dubois E. Image up-sampling using total-variation regularization with a new observation model. IEEE TIP 2005;14(10):1647–59.

[6] Yang J, Wright J, Huang TS, Ma Y. Image super-resolution via sparse representation. IEEE TIP 2010;19(11):2861–73.

[7] Yang J, Wang Z, Lin Z, Cohen S, Huang T. Coupled dictionary training for image super-resolution. IEEE TIP 2012;21(8):3467–78.

[8] Gao X, Zhang K, Tao D, Li X. Image super-resolution with sparse neighbor embedding. IEEE TIP 2012;21(7):3194–205.

[9] Glasner D, Bagon S, Irani M. Super-resolution from a single image. In: ICCV. IEEE; 2009. p. 349–56.

[10] Freedman G, Fattal R. Image and video upscaling from local self-examples. ACM Transactions on Graphics 2011;30(2):12.

[11] Krizhevsky A, Sutskever I, Hinton GE. ImageNet classification with deep convolutional neural networks. In: NIPS; 2012. p. 1097–105.

[12] Cui Z, Chang H, Shan S, Zhong B, Chen X. Deep network cascade for image super-resolution. In: ECCV. Springer; 2014. p. 49–64.

[13] Wang Z, Yang Y, Wang Z, Chang S, Han W, Yang J, et al. Self-tuned deep super resolution. In: CVPR workshops; 2015. p. 1–8.

[14] Dong C, Loy CC, He K, Tang X. Learning a deep convolutional network for image super-resolution. In: ECCV. Springer; 2014. p. 184–99.

[15] Osendorfer C, Soyer H, van der Smagt P. Image super-resolution with fast approximate convolutional sparse coding. In: Neural information processing. Springer; 2014. p. 250–7.

[16] Gregor K, LeCun Y. Learning fast approximations of sparse coding. In: ICML; 2010. p. 399–406.

[17] Wen B, Ravishankar S, Bresler Y. Structured overcomplete sparsifying transform learning with convergence guarantees and applications. IJCV 2015;114(2):137–67.

[18] Yang CY, Ma C, Yang MH. Single-image super-resolution: a benchmark. In: ECCV; 2014. p. 372–86.

[19] Duchon CE. Lanczos filtering in one and two dimensions. Journal of Applied Meteorology 1979;18(8):1016–22.

[20] Sun J, Xu Z, Shum HY. Gradient profile prior and its applications in image super-resolution and enhancement. IEEE Transactions on Image Processing 2011;20(6):1529–42.

[21] Timofte R, De V, Gool LV. Anchored neighborhood regression for fast example-based super-resolution. In: ICCV. IEEE; 2013. p. 1920–7.

[22] Freeman WT, Jones TR, Pasztor EC. Example-based super-resolution. IEEE Computer Graphics and Applications 2002;22(2):56–65.

[23] Wang Z, Yang Y, Wang Z, Chang S, Huang TS. Learning super-resolution jointly from external and internal examples. IEEE TIP 2015;24(11):4359–71.

[24] Huang JB, Singh A, Ahuja N. Single image super-resolution from transformed self-exemplars. In: CVPR. IEEE; 2015. p. 5197–206.

[25] Timofte R, De Smet V, Van Gool L. A+: adjusted anchored neighborhood regression for fast super-resolution. In: ACCV. Springer; 2014. p. 111–26.

[26] Schulter S, Leistner C, Bischof H. Fast and accurate image upscaling with super-resolution forests. In: CVPR; 2015. p. 3791–9.

[27] Salvador J, Pérez-Pellitero E. Naive Bayes super-resolution forest. In: Proceedings of the IEEE international conference on computer vision; 2015. p. 325–33.

[28] Wang Z, Wang Z, Chang S, Yang J, Huang T. A joint perspective towards image super-resolution: unifying external- and self-examples. In: Applications of computer vision (WACV), 2014 IEEE winter conference on. IEEE; 2014. p. 596–603.

[29] Kavukcuoglu K, Ranzato M, LeCun Y. Fast inference in sparse coding algorithms with applications to object recognition. arXiv preprint arXiv:1010.3467, 2010.

[30] Daubechies I, Defrise M, De Mol C. An iterative thresholding algorithm for linear inverse problems with a sparsity constraint. Communications on Pure and Applied Mathematics 2004;57(11):1413–57.

[31] Rozell CJ, Johnson DH, Baraniuk RG, Olshausen BA. Sparse coding via thresholding and local competition in neural circuits. Neural Computation 2008;20(10):2526–63.

[32] Lin M, Chen Q, Yan S. Network in network. arXiv preprint arXiv:1312.4400, 2013.

[33] Nair V, Hinton GE. Rectified linear units improve restricted Boltzmann machines. In: ICML; 2010. p. 807–14.

[34] Chang S, Han W, Tang J, Qi GJ, Aggarwal CC, Huang TS. Heterogeneous network embedding via deep architectures. In: ACM SIGKDD. ACM; 2015.

[35] Le QV. Building high-level features using large scale unsupervised learning. In: ICASSP; 2013. p. 8595–8.

[36] Oquab M, Bottou L, Laptev I, Sivic J. Learning and transferring mid-level image representations using convolutional neural networks. In: CVPR; 2014. p. 1717–24.

[37] Buades A, Coll B, Morel JM. A non-local algorithm for image denoising. In: CVPR; 2005.

[38] Aharon M, Elad M, Bruckstein A. K-SVD: an algorithm for designing overcomplete dictionaries for sparse representation. IEEE TSP 2006;54(11):4311–22.

[39] Dabov K, Foi A, Katkovnik V, Egiazarian K. Image denoising by sparse 3D transform-domain collaborative filtering. IEEE TIP 2007;16(8):2080–95.

[40] Wen B, Ravishankar S, Bresler Y. Video denoising by online 3d sparsifying transform learning. In: ICIP; 2015.

[41] Ravishankar S, Wen B, Bresler Y. Online sparsifying transform learning – part i: algorithms. IEEE Journal of Selected Topics in Signal Process 2015;9(4):625–36.

[42] Bevilacqua M, Roumy A, Guillemot C, Morel MLA. Low-complexity single-image super-resolution based on nonnegative neighbor embedding. In: BMVC. BMVA Press; 2012.

[43] Zeyde R, Elad M, Protter M. On single image scale-up using sparse-representations. In: Curves and surfaces. Springer; 2012. p. 711–30.

[44] Martin D, Fowlkes C, Tal D, Malik J. A database of human segmented natural images and its application to evaluating segmentation algorithms and measuring ecological statistics. In: ICCV, vol. 2. IEEE; 2001. p. 416–23.

[45] Dong C, Loy CC, He K, Tang X. Image super-resolution using deep convolutional networks. TPAMI 2015.

[46] Wang Z, Bovik AC, Sheikh HR, Simoncelli EP. Image quality assessment: from error visibility to structural similarity. IEEE TIP 2004;13(4):600–12.

[47] Kim KI, Kwon Y. Single-image super-resolution using sparse regression and natural image prior. IEEE TPAMI 2010;32(6):1127–33.

[48] Dong W, Zhang L, Shi G. Centralized sparse representation for image restoration. In: ICCV; 2011. p. 1259–66.

[49] Zhang K, Gao X, Tao D, Li X. Single image super-resolution with non-local means and steering kernel regression. IEEE TIP 2012;21(11):4544–56.

[50] Lu X, Yuan H, Yan P, Yuan Y, Li X. Geometry constrained sparse coding for single image super-resolution. In: CVPR; 2012. p. 1648–55.

[51] Yang J, Lin Z, Cohen S. Fast image super-resolution based on in-place example regression. In: CVPR; 2013. p. 1059–66.

[52] Bradley RA, Terry ME. Rank analysis of incomplete block designs: I. The method of paired comparisons. Biometrika 1952:324–45.

[53] Wang Z, Chang S, Yang Y, Liu D, Huang TS. Studying very low resolution recognition using deep networks. In: CVPR. IEEE; 2016. p. 4792–800.

[54] Liu D, Wang Z, Fan Y, Liu X, Wang Z, Chang S, et al. Learning temporal dynamics for video super-resolution: a deep learning approach. IEEE Transactions on Image Processing 2018;27(7):3432–45.

[55] Dorr M, Martinetz T, Gegenfurtner KR, Barth E. Variability of eye movements when viewing dynamic natural scenes. Journal of Vision 2010;10(10):28.

[56] Dai Q, Yoo S, Kappeler A, Katsaggelos AK. Sparse representation-based multiple frame video super-resolution. IEEE Transactions on Image Processing 2017;26(2):765–81.

[57] Morse BS, Schwartzwald D. Image magnification using level-set reconstruction. In: CVPR 2001, vol. 1. IEEE; 2001.

[58] Chang H, Yeung DY, Xiong Y. Super-resolution through neighbor embedding. In: CVPR, vol. 1. IEEE; 2004. p. 275–82.

[59] Dai D, Timofte R, Van Gool L. Jointly optimized regressors for image super-resolution. In: Euro-graphics, vol. 7; 2015. p. 8.

[60] Timofte R, Rasmus R, Van Gool L. Seven ways to improve example-based single image super reso-lution. In: CVPR. IEEE; 2016.

[61] Wang Z, Liu D, Yang J, Han W, Huang T. Deep networks for image super-resolution with sparse prior. In: ICCV; 2015. p. 370–8.

[62] Liu D, Wang Z, Nasrabadi N, Huang T. Learning a mixture of deep networks for single image super-resolution. In: ACCV. Springer; 2016. p. 145–56.

[63] Kim J, Lee JK, Lee KM. Accurate image super-resolution using very deep convolutional networks. In: CVPR. IEEE; 2016.

[64] Kim J, Lee JK, Lee KM. Deeply-recursive convolutional network for image super-resolution. In: CVPR; 2016.

[65] Tai Y, Yang J, Liu X. Image super-resolution via deep recursive residual network. In: CVPR; 2017.

[66] Fan Y, Shi H, Yu J, Liu D, Han W, Yu H, et al. Balanced two-stage residual networks for image super-resolution. In: Computer vision and pattern recognition workshops (CVPRW), 2017 IEEE conference on. IEEE; 2017. p. 1157–64.

[67] Tong T, Li G, Liu X, Gao Q. Image super-resolution using dense skip connections. In: ICCV; 2017.

[68] Huang G, Liu Z, Weinberger KQ, van der Maaten L. Densely connected convolutional networks. In: Proceedings of the IEEE conference on computer vision and pattern recognition; 2017.

[69] He K, Zhang X, Ren S, Sun J. Deep residual learning for image recognition. In: Proceedings of the IEEE conference on computer vision and pattern recognition; 2016. p. 770–8.

[70] Huang GB, Mattar M, Berg T, Learned-Miller E. Labeled faces in the wild: a database for studying face recognition in unconstrained environments. In: Workshop on faces in real-life images: detection, alignment, and recognition; 2008.

[71] Deng J, Dong W, Socher R, Li LJ, Li K, Fei-Fei L. ImageNet: a large-scale hierarchical image database. In: Computer vision and pattern recognition, 2009. CVPR 2009. IEEE conference on. IEEE; 2009. p. 248–55.

[72] Liu D, Cheng B, Wang Z, Zhang H, Huang TS. Enhance visual recognition under adverse conditions via deep networks. arXiv preprint arXiv:1712.07732, 2017.

CHAPTER 5

From Bi-Level Sparse Clustering to Deep Clustering

Zhangyang Wang

Department of Computer Science and Engineering, Texas A&M University, College Station, TX, United States

Contents

5.1.	A Joint Optimization Framework of Sparse Coding and Discriminative Clustering	87
	5.1.1 Introduction	87
	5.1.2 Model Formulation	88
	5.1.3 Clustering-Oriented Cost Functions	90
	5.1.4 Experiments	93
	5.1.5 Conclusion	98
	5.1.6 Appendix	99
5.2.	Learning a Task-Specific Deep Architecture for Clustering	101
	5.2.1 Introduction	101
	5.2.2 Related Work	102
	5.2.3 Model Formulation	103
	5.2.4 A Deeper Look: Hierarchical Clustering by DTAGnet	106
	5.2.5 Experiment Results	107
	5.2.6 Conclusion	117
References		117

5.1. A JOINT OPTIMIZATION FRAMEWORK OF SPARSE CODING AND DISCRIMINATIVE CLUSTERING[1]

5.1.1 Introduction

Clustering aims to divide data into groups of similar objects (clusters), and plays an important role in many real world data mining applications. To learn the hidden patterns of the dataset in an unsupervised way, existing clustering algorithms can be described as either generative or discriminative in nature. Generative clustering algorithms model categories in terms of their geometric properties in feature spaces, or as statistical processes of data. Examples include K-means [1] and Gaussian mixture model (GMM) clustering [2], which assume a parametric form of the underlying category distributions.

---

[1] Reprinted, with permission, from Wang, Zhangyang, Yang, Yingzhen, Chang, Shiyu, Li, Jinyan, Fong, Simon, and Huang, Thomas S. "A joint optimization framework of sparse coding and discriminative clustering", IJCAI (2015).

Rather than modeling categories explicitly, discriminative clustering techniques search for the boundaries or distinctions between categories. With fewer assumptions being made, these methods are powerful and flexible in practice. For example, maximum-margin clustering [3–5] aims to find the hyperplane that can separate the data from different classes with a maximum margin. Information-theoretic clustering [6,7] minimizes the conditional entropy of all samples. Many recent discriminative clustering methods have achieved very satisfactory performances [5].

Moreover, many clustering methods extract discriminative features from input data, prior to clustering. The principal component analysis (PCA) feature is a common choice but not necessarily discriminative [8]. Kernel-based clustering methods [9] were explored to find implicit feature representations of input data. In [10], the features are selected for optimizing the discriminativity of the used partitioning algorithm, by solving a linear discriminant analysis (LDA) problem. More recently, sparse codes have been shown to be robust to noise and capable of handling high-dimensional data [11]. Furthermore, ℓ_1-graph [12] builds the graph by reconstructing each data point sparsely and locally with other data. A spectral clustering [13] is followed based on the constructed graph matrix. In [14,15], dictionary learning is combined with the clustering process, which uses Lloyd's-type algorithms that iteratively reassign data to clusters and then optimize the dictionary associated with each cluster. In [8], the authors learned the sparse codes that explicitly preserve the local data manifold structures. Their results indicate that encoding geometrical information will significantly enhance the learning performance. A joint optimization of clustering and manifold structure were further considered in [16]. However, the clustering step in [8,16] is not correlated with the above mentioned discriminative clustering methods.

In this section, we propose to jointly optimize feature extraction and discriminative clustering, in which way they mutually reinforce each other. We focus on sparse codes as the extracted features, and develop our loss functions based on two representative discriminative clustering methods, the entropy-minimization [6] and maximum-margin [3] clustering, respectively. A task-driven bi-level optimization model [17,18] is then built upon the proposed framework. The sparse coding step is formulated as the lower-level constraint, where a graph regularization is enforced to preserve the local manifold structure [8]. The clustering-oriented cost functions are considered as the upper-level objectives to be minimized. Stochastic gradient descent algorithms are developed to solve both bi-level models. Experiments on several popular real datasets verify the noticeable performance improvement led by such a joint optimization framework.

5.1.2 Model Formulation
5.1.2.1 Sparse Coding with Graph Regularization

Sparse codes have proved to be an effective feature for clustering. In [12], the authors suggested that the contribution of one sample to the reconstruction of another sample was a good indicator of similarity between these two samples. Therefore, the reconstruction coefficients (sparse codes) can be used to constitute the similarity graph for spectral clustering. The ℓ_1-graph performs sparse representation for each data point separately without considering the geometric information and manifold structure of the entire data. Further research shows that the graph regularized sparse representations produce superior results in various clustering and classification tasks [8,19]. In this section, we adopt the graph regularized sparse codes as the features for clustering.

We assume that all the data samples $X = [x_1, x_2, \ldots, x_n], x_i \in R^{m \times 1}, i = 1, 2, \ldots, n$, are encoded into their corresponding sparse codes $A = [a_1, a_2, \ldots, a_n], a_i \in R^{p \times 1}, i = 1, 2, \ldots, n$, using a learned dictionary $D = [d_1, d_2, \ldots, d_p]$, where $d_i \in R^{m \times 1}, i = 1, 2, \ldots, p$ are the learned atoms. Moreover, given a pairwise similarity matrix W, the sparse representations that capture the geometric structure of the data according to the manifold assumption should minimize the following objective: $\frac{1}{2} \sum_{i=1}^{n} \sum_{j=1}^{n} W_{ij} ||a_i - a_j||_2^2 = \text{Tr}(ALA^T)$, where L is the graph Laplacian matrix constructed from W. In this section, W is chosen as the Gaussian kernel, $W_{ij} = \exp(-\frac{||x_i - x_j||_2^2}{\delta^2})$, where δ is the controlling parameter selected by cross-validation.

The graph regularized sparse codes are obtained by solving the following convex optimization:

$$A = \arg\min_A \frac{1}{2} ||X - DA||_F^2 + \lambda \sum_i ||a_i||_1 + \alpha \, \text{Tr}(ALA^T) + \lambda_2 ||A||_F^2. \tag{5.1}$$

Note that $\lambda_2 > 0$ is necessary for proving the differentiability of the objective function (see [5.2] in the Appendix). However, setting $\lambda_2 = 0$ proves to work well in practice, and thus the term $\lambda_2 ||A||_F^2$ will be omitted by default hereinafter (except for the differentiability proof).

Obviously, the effect of sparse codes A largely depends on the quality of dictionary D. Dictionary learning methods, such as K-SVD algorithm [20], are widely used in sparse coding literature. In regard to clustering, the authors of [12,19] constructed the dictionary by directly selecting atoms from data samples. Zheng et al. [8] learned the dictionary that can reconstruct input data well. However, it does not necessarily lead to discriminative features for clustering. In contrast, we will optimize D together with the clustering task.

5.1.2.2 Bi-level Optimization Formulation

The objective cost function for the joint framework can be expressed by the following bi-level optimization:

$$\begin{aligned} &\min_{D, w} \quad C(A, w) \\ &s.t. \quad A = \arg\min_A \frac{1}{2} ||X - DA||_F^2 + \lambda \sum_i ||a_i||_1 + \alpha \text{Tr}(ALA^T), \end{aligned} \tag{5.2}$$

where $C(A, w)$ is a cost function evaluating the loss of clustering. It can be formulated differently based on various clustering principles, two of which will be discussed and solved in Sect. 5.1.3.

Bi-level optimization [21] has been investigated in both theory and application. In [21], the authors proposed a general bi-level sparse coding model for learning dictionaries across coupled signal spaces. Another similar formulation has been studied in [17] for general regression tasks.

5.1.3 Clustering-Oriented Cost Functions

Assuming K clusters, let $w = [w_1, \ldots, w_K]$ be the set of parameters of the loss function, where w_i corresponds to the ith cluster, $i = 1, 2, \ldots, K$. We introduce two forms of loss functions, each of which is derived from a representative discriminative clustering method.

5.1.3.1 Entropy-Minimization Loss

Maximization of the mutual information with respect to parameters of the encoder model effectively defines a discriminative unsupervised optimization framework. The model is parameterized similarly to a conditionally trained classifier, but the cluster allocations are unknown [7]. In [22,6], the authors adopted an information-theoretic framework as an implementation of the low–density separation assumption by minimizing the conditional entropy. By substituting the logistic posterior probability into the minimum conditional entropy principle, the authors got the logistics clustering algorithm, which is equivalent to finding a labeling strategy so that the total entropy of data clustering is minimized.

Since the true cluster label of each x_i is unknown, we introduce the predicted confidence probability p_{ij} that sample x_i belongs to cluster j, $i = 1, 2, \ldots, N, j = 1, 2, \ldots, K$, which is set as the likelihood of the multinomial logistic (softmax) regression

$$p_{ij} = p(j|w, a_i) = \frac{1}{1 + e^{-jw^T a_i}}. \tag{5.3}$$

The loss function for all data could be defined accordingly in a entropy-like form

$$C(A, w) = -\sum_{i=1}^{n} \sum_{j=1}^{K} p_{ij} \log p_{ij}. \tag{5.4}$$

The predicted cluster label of a_i is the cluster j where it achieves the largest likelihood probability p_{ij}. The logistics regression can deal with multiclass problems more easily compared with the support vector machine (SVM). The next important thing we need to study is the differentiability of (5.2).

Theorem 5.1. *The objective $C(A, w)$ defined in (5.4) is differentiable on $D \times w$.*

Proof. Denote $\boldsymbol{X} \in \mathcal{X}$, and $\boldsymbol{D} \in \mathcal{D}$. Also let the objective function $C(\boldsymbol{A}, \boldsymbol{w})$ in (5.4) be denoted as C for short. The differentiability of C with respect to $\boldsymbol{w}$ is easy to show, using only the compactness of $\mathcal{X}$, as well as the fact that C is twice differentiable.

We will therefore focus on showing that C is differentiable with respect to $\boldsymbol{D}$, which is more difficult since $\boldsymbol{A}$, and thus $\boldsymbol{a}_i$, is not differentiable everywhere. Without loss of generality, we use a vector $\boldsymbol{a}$ instead of $\boldsymbol{A}$ for simplifying the derivations hereinafter. In some cases, we may equivalently express $\boldsymbol{a}$ as $\boldsymbol{a}(\boldsymbol{D}, \boldsymbol{w})$ in order to emphasize the functional dependence. Based on [5.2] in Appendix, and given a small perturbation $\boldsymbol{E} \in R^{m \times p}$, it follows that

$$C(\boldsymbol{a}(\boldsymbol{D} + \boldsymbol{E}), \boldsymbol{w}) - C(\boldsymbol{a}(\boldsymbol{D}), \boldsymbol{w}) = \nabla_z C_{\boldsymbol{w}}^T (\boldsymbol{a}(\boldsymbol{D} + \boldsymbol{E}) - \boldsymbol{a}(\boldsymbol{D})) + O(\|\boldsymbol{E}\|_F^2), \qquad (5.5)$$

where the term $O(\|\boldsymbol{E}\|_F^2)$ is based on the fact that $\boldsymbol{a}(\boldsymbol{D}, \boldsymbol{x})$ is uniformly Lipschitz and $\mathcal{X} \times \mathcal{D}$ is compact. It is then possible to show that

$$C(\boldsymbol{a}(\boldsymbol{D} + \boldsymbol{E}), \boldsymbol{w}) - C(\boldsymbol{a}(\boldsymbol{D}), \boldsymbol{w}) = \mathrm{Tr}(\boldsymbol{E}^T g(\boldsymbol{a}(\boldsymbol{D} + \boldsymbol{E}), \boldsymbol{w})) + O(\|\boldsymbol{E}\|_F^2), \qquad (5.6)$$

where g has the form given in Algorithm 5.1. This shows that C is differentiable on $\mathcal{D}$. $\qquad \square$

Built on the differentiability proof, we are able to solve (5.1) using a projected first order stochastic gradient descent (SGD) algorithm, whose detailed steps are outlined in Algorithm 5.1. At a high level overview, it consists of an outer stochastic gradient descent loop that incrementally samples the training data. It uses each sample to approximate gradients with respect to the classifier parameter $\boldsymbol{w}$ and the dictionary $\boldsymbol{D}$, which are then used to update them.

5.1.3.2 Maximum-Margin Loss

Xu et al. [3] proposed maximum margin clustering (MMC), which borrows the idea from the SVM theory. Their experimental results showed that the MMC technique could often obtain more accurate results than conventional clustering methods. Technically, MMC just finds a way to label the samples by running an SVM implicitly, and the SVM margin obtained is maximized over all possible labelings [5]. However, unlike supervised large margin methods, which are usually formulated as convex optimization problems, maximum margin clustering is a nonconvex integer optimization problem, which is much more difficult to solve. Li et al. [23] made several relaxations to the original MMC problem and reformulated it as a semidefinite programming (SDP) problem. The cutting plane maximum margin clustering (CPMMC) algorithm was presented in [5] to solve MMC with a much improved efficiency.

To develop the multiclass max-margin loss of clustering, we refer to the classical multiclass SVM formulation in [24]. Given the sparse code $\boldsymbol{a}_i$ comprises the features to

---

Algorithm 5.1: Stochastic gradient descent algorithm for solving (5.2), with $C(A, w)$ as defined in (5.4).

Require: X, σ; λ; D_0 and w_0 (initial dictionary and classifier parameter); ITER (number of iterations); t_0, ρ (learning rate)

1: Construct the matrix L from X and σ.

2: FOR $t = 1$ to ITER DO

3: Draw a subset (X_t, Y_t) from (X, Y)

4: Graph-regularized sparse coding: compute A^*:
$$A^* = \arg\min_A \tfrac{1}{2}\|X - DA\|_F^2 + \lambda \sum_i \|a_i\|_1 + \mathrm{Tr}(ALA^T).$$

5: Compute the active set S (the nonzero support of A^*)

6: Compute β^*: Set $\beta^*_{S^C} = 0$ and $\beta^*_S = (D_S^T D_S + \lambda_2 I)^{-1} \nabla_{A_S}[C(A, w)]$

7: Choose the learning rate $\rho_t = \min(\rho, \rho\tfrac{t_0}{t})$

8: Update D and W by a projected gradient step:
$$w = \prod_w [w - \rho_t \nabla_w C(A, w)]$$
$$D = \prod_D [D - \rho_t (\nabla_D (-D\beta^* A^T + (X_t - DA)\beta^{*T}))]$$
where $\prod_w$ and $\prod_D$ are respectively orthogonal projections on the embedding spaces of w and D.

9: END FOR

Ensure: D and w

---

be clustered, we define the multiclass model as

$$f(a_i) = \arg\max_{j=1,\ldots,K} f^j(a_i) = \arg\max_{j=1,\ldots,K} (w_j^T a_i), \qquad (5.7)$$

where f^j is the prototype for the jth cluster and w_j is its corresponding weight vector. The predicted cluster label of a_i is the cluster of the weight vector that achieves the maximum value $w_j^T a_i$. Let $w = [w_1, \ldots, w_K]$, the multiclass max-margin loss for a_i, be defined as

$$C(a_i, w) = \max(0, 1 + f^{r_i}(a_i) - f^{y_i}(a_i)),$$
$$\text{where} \quad y_i = \arg\max_{j=1,\ldots,K} f^j(a_i), \quad r_i = \arg\max_{j=1,\ldots,K, j \neq y_i} f^j(a_i). \qquad (5.8)$$

Note that different from training a multiclass SVM classier, where y_i is given as a training label, the clustering scenario requires us to jointly estimate y_i as a variable. The overall max-margin loss to be minimized is (λ as the coefficient)

$$C(A, w) = \tfrac{\lambda}{2}\|w\|^2 + \sum_{i=1}^n C(a_i, w). \qquad (5.9)$$

But to solve (5.8) or (5.9) with respect to the same framework as logistic loss will involve two additional concerns, which need to be handled specifically.

First, the hinge loss of the form (5.8) is not differentiable, with only a subgradient existing. This makes the objective function $C(A, w)$ nondifferentiable on $D \times w$, and further the analysis in the proof of Theorem 5.1 cannot be applied. We could have used the squared hinge loss or modified Huber loss for a quadratically smoothed loss function [25]. However, as we checked in the experiments, the quadratically smoothed loss is not as good as hinge loss in training time and sparsity. Also, though not theoretically guaranteed, using the subgradient of $C(A, w)$ works well in our case.

Second, given that w is fixed, it should be noted that y_i and r_i are both functions of a_i. Therefore, calculating the derivative of (5.8) with respect to a_i would involve expanding both r_i and y_i, making analysis quite complicated. Instead, we borrow ideas from the regularity of the elastic net solution [17], namely the set of nonzero coefficients of the elastic net solution should not change for small perturbations. Similarly, due to the continuity of the objective, it is assumed that a sufficiently small perturbation over the current a_i will not change y_i and r_i. Therefore in each iteration, we could directly precalculate y_i and r_i using the current w and a_i and fix them for a_i updates.[2]

Given the above two approaches, for a single sample a_i, if the hinge loss is larger than 0, the derivative of (5.8) with respect to w is

$$\Delta_i^j = \begin{cases} \lambda w_i^j - a_i & \text{if } j = y_i, \\ \lambda w_i^j + a_i & \text{if } j = r_i, \\ \lambda w_i^j & \text{otherwise,} \end{cases} \tag{5.10}$$

where Δ_i^j denotes the jth element of the derivative for the sample a_i. If the hinge loss is less than 0, then $\Delta_i^j = \lambda w_i^j$. The derivative of (5.8) with respect to a_i is $w^{r_i} - w^{y_i}$ if the hinge loss is larger than 0, and 0 otherwise. Note the above deduction can be conducted in a batch mode. It is then similarly solved using a projected SGD algorithm, whose steps are outlined in Algorithm 5.2.

5.1.4 Experiments

5.1.4.1 Datasets

We conduct our clustering experiments on four popular real datasets, which are summarized in Table 5.1. The ORL face database contains 400 facial images for 40 subjects, and each subject has 10 images of size 32×32. The images are taken at different times with varying lighting and facial expressions. The subjects are all in an upright, frontal position with a dark homogeneous background. The MNIST handwritten digit database consists of a total number of 70,000 images, with digits ranging from 0 to 9. The digits are normalized and centered in fixed-size images of 28×28. The COIL20 image library

---

[2] To avoid ambiguity, if y_i and r_i are the same, i.e., the max value is reached by two cluster prototypes simultaneously in current iteration, then we ignore the gradient update corresponding to a_i.

Algorithm 5.2: Stochastic gradient descent algorithm for solving (5.2), with $C(A, w)$ as defined in (5.9).

Require: $X, \sigma; \lambda; D_0$ and w_0 (initial dictionary and classifier parameter); ITER (number of iterations); t_0, ρ (learning rate)

1: Construct the matrix L from X and σ.
2: Estimate the initialization of y_i and r_i by preclustering, $i = 1, 2, \ldots, N$
3: FOR $t = 1$ to ITER DO
4: Conduct the same step 4–7 in Algorithm 5.1.
5: Update D and W by a projected gradient step, based on the derivatives of (5.9) with respect to a_i and w (5.10).
6: Update y_i and r_i using the current w and a_i, $i = 1, 2, \ldots, N$.
7: END FOR

Ensure: D and w

Table 5.1 Comparison of all datasets

Name	Number of Images	Class	Dimension
ORL	400	10	1024
MNIST	70,000	10	784
COIL20	1440	20	1024
CMU-PIE	41,368	68	1024

contains 1440 images of size 32×32, for 20 objects. Each object has 72 images, that were taken 5 degree apart as the object was rotated on a turntable. The CMU-PIE face database contains 68 subjects with 41,368 face images as a whole. For each subject, we have 21 images of size 32×32, under different lighting conditions.

5.1.4.2 Evaluation Metrics

We apply two widely-used measures to evaluate the performance of the clustering methods, the accuracy and the normalized mutual information (NMI) [8,12]. Suppose the predicted label of x_i is $\hat{y}_i$, which is produced by the clustering method, and y_i is the ground-truth label. The accuracy is defined as

$$\text{Acc} = \frac{I_{\Phi(\hat{y}_i) \neq y_i}}{n},$$ (5.11)

where I is the indicator function, and Φ is the best permutation mapping function [26]. On the other hand, suppose the clusters obtained from the predicted labels $\{\hat{y}_i\}_{i=1}^n$ and $\{y_i\}_{i=1}^n$ are $\hat{C}$ and C, respectively. The mutual information between $\hat{C}$ and C is defined as

$$\text{MI}(\hat{C}, C) = \sum_{\hat{c} \in \hat{C}, c \in C} p(\hat{c}, c) \log \frac{p(\hat{c}, c)}{p(\hat{c}) p(c)},$$ (5.12)

where $p(\hat{c})$ and $p(c)$ are the probabilities that a data point belongs to the clusters $\hat{C}$ and C, respectively, and $p(\hat{c}, c)$ is the probability that a data point jointly belongs to $\hat{C}$ and C. The normalized mutual information (NMI) is defined as

$$\mathrm{NMI}(\hat{C}, C) = \frac{\mathrm{MI}(\hat{C}, C)}{\max\{H(\hat{C}), H(C)\}}, \tag{5.13}$$

where $H(\hat{C})$ and $H(C)$ are the entropies of $\hat{C}$ and C, respectively. NMI takes values in [0,1].

5.1.4.3 Comparison Experiments

Comparison Methods

We compare the following eight methods on all four datasets:
- **KM**, K-means clustering on the input data.
- **KM + SC**, a dictionary D is first learned from the input data by K-SVD [20]. Then KM is performed on the graph-regularized sparse code features (5.1) over D.
- **EMC**, entropy-minimization clustering, by minimizing (5.4) on the input data.
- **EMC + SC**, EMC performed on the graph-regularized sparse codes over the pre-learned K-SVD dictionary D.
- **MMC**, maximum-margin clustering [5].
- **MMC + SC**, MMC performed on the graph-regularized sparse codes over the pre-learned K-SVD dictionary D.
- **Joint EMC**, the proposed joint optimization (5.2), with $C(A, w)$ as defined in (5.4).
- **Joint MMC**, the proposed joint optimization (5.2), with $C(A, w)$ as defined in (5.9).

All images are first reshaped into vectors, and PCA is then applied to reducing the data dimensionality by keeping 98% information, which is also used in [8] to improve efficiency. The multiclass MMC algorithm is implemented based on the publicly available CPMMC code for two-class clustering [5], following the multiclass case descriptions in the original paper. For all algorithms that involve graph-regularized sparse coding, the graph regularization parameter α is fixed to be 1, and the dictionary size p is 128 by default. For joint EMC and joint MMC, we set ITER as 30, ρ as 0.9, and t_0 as 5. Other parameters in competing methods are tuned in cross-validation experiments to our best efforts.

Comparison Analysis

All the comparison results (accuracy and NMI) are listed in Table 5.2, from which we could conclude the following:

Table 5.2 Accuracy and NMI performance comparisons on all datasets

		KM	KM + SC	EMC	EMC + SC	MMC	MMC + SC	joint EMC	joint MMC
ORL	Acc	0.5250	0.5887	0.6011	0.6404	0.6460	0.6968	0.7250	**0.7458**
	NMI	0.7182	0.7396	0.7502	0.7795	0.8050	0.8043	0.8125	**0.8728**
MNIST	Acc	0.6248	0.6407	0.6377	0.6493	0.6468	0.6581	0.6550	**0.6784**
	NMI	0.5142	0.5397	0.5274	0.5671	0.5934	0.6161	0.6150	**0.6451**
COIL20	Acc	0.6280	0.7880	0.7399	0.7633	0.8075	0.8493	0.8225	**0.8658**
	NMI	0.7621	0.9010	0.8621	0.8887	0.8922	0.8977	0.8850	**0.9127**
CMU-PIE	Acc	0.3176	0.8457	0.7627	0.7836	0.8482	0.8491	0.8250	**0.8783**
	NMI	0.6383	0.9557	0.8043	0.8410	0.9237	0.9489	0.9020	**0.9675**

1. The joint EMC and joint MMC methods each outperform their "non-joint" counterparts, e.g., EMC + SC and MMC + SC, respectively. For example, on the *ORL* dataset, joint MMC surpasses MMC + SC by around 5% in accuracy and 7% in NMI. This demonstrates that the key contribution of this section, i.e., joint optimization of the sparse coding and clustering steps, indeed leads to improved performances.

2. KM + SC, EMC + SC, and MMC + SC all outperform their counterparts using raw input data, which verifies that sparse codes are effective features that help improve the clustering discriminability.

3. The joint MMC obtains the best performance in all cases, outperforming the others, including joint EMC, with significant margins. The MMC + SC obtains the second best performance for the last three datasets (for ORL, it is joint EMC that ranks second). The above facts reveal the power of the max-margin loss (5.9).

Varying the Number of Clusters

On the COIL20 dataset, we reconduct the clustering experiments with the cluster number K ranging from 2 to 20, using EMC + SC, MMC + SC, joint EMC, and joint MMC. For each K, except for 20, 10 test runs are conducted on different randomly chosen clusters, and the final scores are obtained by averaging over the 10 tests. Fig. 5.1 shows the clustering accuracy and NMI measurements versus the number of clusters. It is revealed that the two joint methods consistently outperform their non-joint counterparts. When K increases, the performance of joint methods seems to degrade less slowly.

Initialization and Parameters

As a typical case in machine learning, we use SGD in a setting where it is not guaranteed to converge in theory, but behaves well in practice. As observed in our experiments, a good initialization of D and w can affect the final results notably. We initialize joint EMC by the D and w solved from EMC + SC, and joint MMC by the solutions from MMC + SC, respectively.

There are two parameters that we empirically set in ahead, the graph regularization parameter α and the dictionary size p. The regularization term imposes stronger smoothness constraints on the sparse codes when α grows larger. Also, while a compact dictionary is more desirable computationally, more redundant dictionaries may lead to less cluttered features that can be better discriminated. We investigate how the clustering performances EMC + SC, MMC + SC, joint EMC, and joint MMC change on the ORL dataset, with various α and p values. As depicted in Figs. 5.2 and 5.3, we observe that:

1. While α increases, the accuracy result will first increase then decrease (the peak is around $\alpha = 1$). This could be interpreted as when α is too small, the local manifold

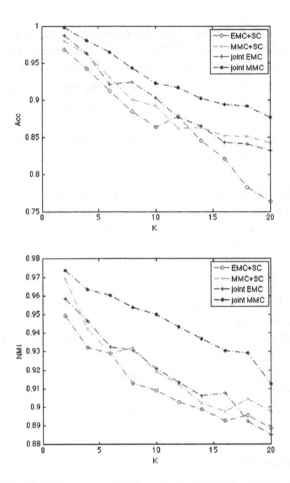

Figure 5.1 The clustering accuracy and NMI measurements versus the number of clusters K.

information is not sufficiently encoded. On the other hand, when α turns overly large, the sparse codes are "oversmoothed" with a reduced discriminability.

2. Increasing dictionary size p will first improve the accuracy sharply, which, however, soon reaches a plateau. Thus in practice, we keep a medium dictionary size $p = 128$ for all experiments.

5.1.5 Conclusion

We propose a joint framework to optimize sparse coding and discriminative clustering simultaneously. We adopt graph-regularized sparse codes as the feature to be learned, and design two clustering-oriented cost functions, by entropy-minimization and maximum-margin principles, respectively. The formulation of a task-driven bi-level optimization mutually reinforces both sparse coding and clustering steps. Experiments

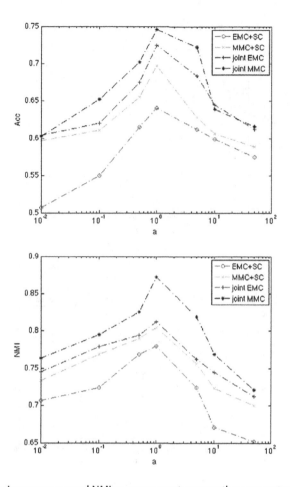

Figure 5.2 The clustering accuracy and NMI measurements versus the parameter choices of α.

on several benchmark datasets verify the remarkable performance improvement led by the proposed joint optimization.

5.1.6 Appendix

We recall the following lemma [5.2] in [17]:

Theorem 5.2 (Regularity of the elastic net solution). *Consider the formulation in* (5.1) *(we may drop the last term to obtain the exact elastic net form, without affecting the differentiability conclusions). Assume* $\lambda_2 > 0$, *and that* $\mathcal{X}$ *is compact. Then*

- *a is uniformly Lipschitz on* $\mathcal{X} \times \mathcal{D}$.

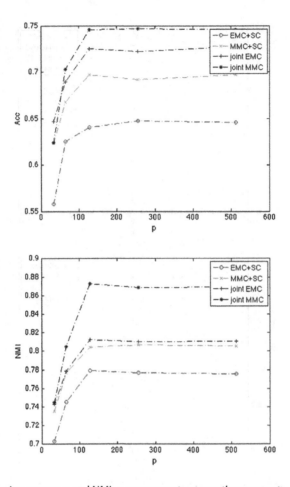

Figure 5.3 The clustering accuracy and NMI measurements versus the parameter choices of p.

- Let $\mathbf{D} \in \mathcal{D}$, σ be a positive scalar and $\mathbf{s}$ be a vector in $\{-1, 0, 1\}^p$. Define $K_s(\mathbf{D}, \sigma)$ as the set of vectors $\mathbf{x}$ satisfying for all j in $\{1, \dots, p\}$,

$$
\begin{aligned}
|\mathbf{d}_j^T(\mathbf{x} - \mathbf{D}\mathbf{a}) - \lambda_2 \mathbf{a}[j]| &\leq \lambda_1 - \sigma \quad \text{if} \quad \mathbf{s}[j] = 0, \\
\mathbf{s}[j]\mathbf{a}[j] &\geq \sigma \quad \text{if} \quad \mathbf{s}[j] \neq 0.
\end{aligned}
\tag{5.14}
$$

Then there exists $\kappa > 0$ independent of $\mathbf{s}, \mathbf{D}$ and σ so that for all $\mathbf{x} \in K_s(\mathbf{D}, \sigma)$, the function $\mathbf{a}$ is twice continuously differentiable on $B_{\kappa\sigma}(\mathbf{x}) \times B_{\kappa\sigma}(\mathbf{D})$, where $B_{\kappa\sigma}(\mathbf{x})$ and $B_{\kappa\sigma}(\mathbf{D})$ denote the open balls of radius $\kappa\sigma$ respectively centered at $\mathbf{x}$ and $\mathbf{D}$.

5.2. LEARNING A TASK-SPECIFIC DEEP ARCHITECTURE FOR CLUSTERING[3]

5.2.1 Introduction

While many classical clustering algorithms have been proposed, such as K-means, Gaussian mixture model (GMM) clustering [2], maximum-margin clustering [3] and information-theoretic clustering [6], most only work well when the data dimensionality is low. Since high-dimensional data exhibits dense grouping in low-dimensional embeddings [27], researchers have been motivated to first project the original data onto a low-dimensional subspace [10] and then cluster on the feature embeddings. Among many feature embedding learning methods, sparse codes [11] have proven to be robust and efficient features for clustering, as verified by many [12,8].

Effectiveness and scalability are two major concerns in designing a clustering algorithm under Big Data scenarios [28]. Conventional sparse coding models rely on iterative approximation algorithms, whose inherently sequential structure, as well as the data-dependent complexity and latency, often constitutes a major bottleneck in the computational efficiency [29]. This also results in the difficulty when one tries to jointly optimize the unsupervised feature learning and the supervised task-driven steps [17]. Such a joint optimization usually has to rely on solving complex bi-level optimization [30], such as in [31], which constitutes another efficiency bottleneck. What is more, to effectively model and represent datasets of growing sizes, sparse coding needs to refer to larger dictionaries [32]. Since the inference complexity of sparse coding increases more than linearly with respect to the dictionary size [31], the scalability of sparse coding-based clustering work turns out to be quite limited.

To conquer those limitations, we are motivated to introduce the tool of deep learning in clustering, to which there has been a lack of attention paid. The advantages of deep learning are achieved by its large learning capacity, the linear scalability with the aid of stochastic gradient descent (SGD), and the low inference complexity [33]. The feed-forward networks could be naturally tuned jointly with task-driven loss functions. On the other hand, generic deep architectures [34] largely ignore the problem-specific formulations and prior knowledge. As a result, one may encounter difficulties in choosing optimal architectures, interpreting their working mechanisms, and initializing the parameters.

In this section, we demonstrate how to **combine the sparse coding-based pipeline into deep learning models for clustering**. The proposed framework takes advantage of both sparse coding and deep learning. Specifically, the feature learning

---

layers are inspired by the graph-regularized sparse coding inference process, via refor-
mulating iterative algorithms [29] into a feed-forward network, named **TAGnet**. Those
layers are then jointly optimized with the task-specific loss functions from end to end.
Our technical novelty and merits are summarized as follows:

- As a deep feed-forward model, the proposed framework provides extremely efficient
  inference process and high scalability to large scale data. It allows learning more
  descriptive features than conventional sparse codes.
- We discover that incorporating the expertise of sparse code-based clustering
  pipelines [12,8] improves our performances significantly. Moreover, it greatly facilitates the model initialization and interpretation.
- We further enforce auxiliary clustering tasks on the hierarchy of features, we develop
  DTAGnet and observe further performance boosts on the CMU MultiPIE dataset
  [35].

5.2.2 Related Work

5.2.2.1 Sparse Coding for Clustering

Assuming data samples $X = [x_1, x_2, \ldots, x_n]$, where $x_i \in R^{m \times 1}$ and $i = 1, 2, \ldots, n$. They
are encoded into sparse codes $A = [a_1, a_2, \ldots, a_n]$, where $a_i \in R^{p \times 1}$ and $i = 1, 2, \ldots, n$,
using a learned dictionary $D = [d_1, d_2, \ldots, d_p]$, where $d_i \in R^{m \times 1}, i = 1, 2, \ldots, p$ are the
learned atoms. The sparse codes are obtained by solving the following convex optimiza-
tion (λ is a constant) problem:

$$A = \arg\min_A \tfrac{1}{2}||X - DA||_F^2 + \lambda \sum_i ||a_i||_1, \qquad (5.15)$$

In [12], the authors suggested that the sparse codes can be used to construct the similarity graph for spectral clustering [13]. Furthermore, to capture the geometric structure
of local data manifolds, the graph regularized sparse codes are further suggested in [8,19]
by solving

$$A = \arg\min_A \tfrac{1}{2}||X - DA||_F^2 + \lambda \sum_i ||a_i||_1 + \tfrac{\alpha}{2}\text{Tr}(ALA^T), \qquad (5.16)$$

where L is the graph Laplacian matrix and can be constructed from a prechosen pairwise
similarity (affinity) matrix P. More recently in [31], the authors suggested to simultaneously learn feature extraction and discriminative clustering, by formulating a task-driven
sparse coding model [17]. They proved that such joint methods consistently outper-
formed non-joint counterparts.

5.2.2.2 Deep Learning for Clustering

In [36], the authors explored the possibility of employing deep learning in graph clustering. They first learned a nonlinear embedding of the original graph by an auto encoder

(AE), followed by a K-means algorithm on the embedding to obtain the final clustering result. However, it neither exploits more adapted deep architectures nor performs any task-specific joint optimization. In [37], a deep belief network with nonparametric clustering was presented. As a generative graphical model, DBN provides a faster feature learning, but is less effective than AEs in terms of learning discriminative features for clustering. In [38], the authors extended the seminonnegative matrix factorization (Semi-NMF) model to a Deep Semi-NMF model, whose architecture resembles stacked AEs. Our proposed model is substantially different from all these previous approaches, due to its unique task-specific architecture derived from sparse coding domain expertise, as well as the joint optimization with clustering-oriented loss functions.

5.2.3 Model Formulation

The proposed pipeline consists of two blocks. As depicted in Fig. 5.4A, it is trained end-to-end in an unsupervised way. It includes a feed-forward architecture, termed *Task-specific And Graph-regularized Network* (**TAGnet**), to learn discriminative features, and the clustering-oriented loss function.

5.2.3.1 TAGnet: Task-specific And Graph-regularized Network

Different from generic deep architectures, TAGnet is designed in a way to take advantage of the successful sparse code-based clustering pipelines [8,31]. It aims to learn features that are optimized under clustering criteria, while encoding graph constraints (5.16) to regularize the target solution. TAGnet is derived from the following theorem:

Theorem 5.3. *The optimal sparse code A from (5.16) is the fixed point of*

$$A = h_{\frac{\lambda}{N}}[(I - \tfrac{1}{N}D^T D)A - A(\tfrac{\alpha}{N}L) + \tfrac{1}{N}D^T X], \tag{5.17}$$

where h_{θ} is an element-wise shrinkage function parameterized by θ:

$$[h_{\theta}(u)]_i = sign(u_i)(|u_i| - \theta_i)_+, \tag{5.18}$$

N is an upper bound on the largest eigenvalue of $D^T D$.

The complete proof of Theorem 5.3 can be found in the Appendix. Theorem 5.3 outlines an iterative algorithm to solve (5.16). Under quite mild conditions [39], after A is initialized, one may repeat the shrinkage and thresholding process in (5.17) until convergence. Moreover, the iterative algorithm could be alternatively expressed as the block diagram in Fig. 5.4B, where

$$W = \tfrac{1}{N}D^T, \; S = I - \tfrac{1}{N}D^T D, \; \theta = \tfrac{\lambda}{N}. \tag{5.19}$$

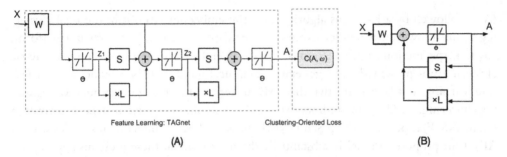

Figure 5.4 (A) The proposed pipeline, consisting of the TAGnet network for feature learning, followed by the clustering-oriented loss functions. The parameters W, S, θ and ω are all learnt end-to-end from training data. (B) The block diagram of solving (5.17).

In particular, we define the new operator "$\times L$": $A \to -\frac{\alpha}{N} AL$, where the input A is multiplied by the prefixed L from the right and scaled by the constant $-\frac{\alpha}{N}$.

By time–unfolding and truncating Fig. 5.4B to a fixed number of K iterations ($K = 2$ by default),[4] we obtain the TAGnet form in Fig. 5.4A; W, S and θ are all to be learnt jointly from data, while S and θ are tied weights for both stages.[5] It is important to note that the output A of TAGnet is not necessarily identical to the predicted sparse codes by solving (5.16). Instead, the goal of TAGnet is to learn discriminative embedding that is optimal for clustering.

To facilitate training, we further rewrite (5.18) as

$$[h_\theta(u)]_i = \theta_i \cdot \text{sign}(u_i)(|u_i|/\theta_i - 1)_+ = \theta_i h_1(u_i/\theta_i). \tag{5.20}$$

Equation (5.20) indicates that the original neuron with trainable thresholds can be decomposed into two linear scaling layers plus a unit-threshold neuron. The weights of the two scaling layers are diagonal matrices defined by θ and its element-wise reciprocal, respectively.

A notable component in TAGnet is the $\times L$ **branch** of each stage. The graph Laplacian L could be computed in advance. In the feed-forward process, a $\times L$ branch takes the intermediate Z_k ($k = 1, 2$) as the input, and applies the "$\times L$" operator defined above. The output is aggregated with the output from the learnable S layer. In the back propagation, L will not be altered. In such a way, the graph regularization is effectively encoded in the TAGnet structure as a prior.

---

[4] We tested larger K values (3 or 4), but they did not bring noticeable performance improvements in our clustering cases.

[5] Out of curiosity, we have also tried the architecture that treats W, S and θ in both stages as independent variables. We found that sharing parameters improves performance.

An appealing highlight of (D)TAGnet lies in its very effective and straightforward initialization strategy. With sufficient data, many latest deep networks train well with random initializations without pretraining. However, it has been discovered that poor initializations hamper the effectiveness of first-order methods (e.g., SGD) in certain cases [40]. For (D)TAGnet, it is, however, much easier to initialize the model in the right regime. This benefits from the analytical relationships between sparse coding and network hyperparameters defined in (5.19): we could initialize deep models from corresponding sparse coding components, the latter of which is easier to obtain. Such an advantage becomes much more important when the training data is limited.

5.2.3.2 *Clustering-Oriented Loss Functions*

Assuming K clusters, and $\boldsymbol{\omega} = [\boldsymbol{\omega}_1, \ldots, \boldsymbol{\omega}_K]$ as the set of parameters of the loss function, where $\boldsymbol{\omega}_i$ corresponds to the ith cluster, $i = 1, 2, \ldots, K$. In this section, we adopt the following two forms of clustering-oriented loss functions.

One natural choice of the loss function is extended from the popular softmax loss, and takes the entropy-like form as

$$C(\boldsymbol{A}, \boldsymbol{\omega}) = -\sum_{i=1}^{n} \sum_{j=1}^{K} p_{ij} \log p_{ij}, \tag{5.21}$$

where p_{ij} denotes the the probability that sample $\boldsymbol{x}_i$ belongs to cluster j, $i = 1, 2, \ldots, N$ and $j = 1, 2, \ldots, K$,

$$p_{ij} = p(j|\boldsymbol{\omega}, \boldsymbol{a}_i) = \frac{e^{-\boldsymbol{\omega}_j^T \boldsymbol{a}_i}}{\sum_{l=1}^{K} e^{-\boldsymbol{\omega}_l^T \boldsymbol{a}_i}}. \tag{5.22}$$

In testing, the predicted cluster label of input $\boldsymbol{a}_i$ is determined using the maximum likelihood criterion based on the predicted p_{ij}.

The maximum margin clustering (MMC) approach was proposed in [3]. MMC finds a way to label the samples by running an SVM implicitly, and the SVM margin obtained would be maximized over all possible labels [5]. By referring to the MMC definition, the authors of [31] designed the max-margin loss as

$$C(\boldsymbol{A}, \boldsymbol{\omega}) = \frac{\lambda}{2} ||\boldsymbol{\omega}||^2 + \sum_{i=1}^{n} C(\boldsymbol{a}_i, \boldsymbol{\omega}). \tag{5.23}$$

In the above equation, the loss for an individual sample $\boldsymbol{a}_i$ is defined as

$$C(\boldsymbol{a}_i, \boldsymbol{\omega}) = \max(0, 1 + f^{r_i}(\boldsymbol{a}_i) - f^{y_i}(\boldsymbol{a}_i)),$$
$$\text{where} \quad y_i = \underset{j=1,\ldots,K}{\arg\max} f^j(\boldsymbol{a}_i), \quad r_i = \underset{j=1,\ldots,K, j \neq y_i}{\arg\max} f^j(\boldsymbol{a}_i), \tag{5.24}$$

where f^j is the prototype for the jth cluster. In testing, the predicted cluster label of input $\boldsymbol{a}_i$ is determined by weight vector that achieves the maximum $\boldsymbol{\omega}_j^T \boldsymbol{a}_i$.

Model Complexity. The proposed framework can handle large-scale and high-dimensional data effectively via the stochastic gradient descent (SGD) algorithm. In each step, the back propagation procedure requires only operations of order $O(p)$ [29]. The training algorithm takes $O(Cnp)$ time (C is a constant in terms of the total numbers of epochs, stage numbers, etc.). In addition, SGD is easy to be parallelized and thus could be efficiently trained using GPUs.

5.2.3.3 Connections to Existing Models

There is a close connection between sparse coding and neural network. In [29], a feed-forward neural network, named LISTA, is proposed to efficiently approximate the sparse code a of input signal x, which is obtained by solving (5.15) in advance. The LISTA network learns the hyperparameters as a general regression model from training data to their pre-solved sparse codes using back-propagation.

LISTA overlooks the useful geometric information among data points [8], and therefore could be viewed as a **special case** of TAGnet in Fig. 5.4 when $\alpha = 0$ (i.e., removing the $\times L$ branches). Moreover, LISTA aims to approximate the "optimal" sparse codes preobtained from (5.15), and therefore requires the estimation of D and the tedious precomputation of A. The authors did not exploit its potential in supervised and task-specific feature learning.

5.2.4 A Deeper Look: Hierarchical Clustering by DTAGnet

Deep networks are well known for their capabilities to learn semantically rich representations by hidden layers [41]. In this section, we investigate how the intermediate features Z_k ($k = 1, 2$) in TAGnet (Fig. 5.4A) can be interpreted, and further utilized to improve the model, for specific clustering tasks. Compared to related non-deep models [31], such a hierarchical clustering property is another unique advantage of being deep.

Our strategy is mainly inspired by the algorithmic framework of deeply supervised nets [42]. As in Fig. 5.5, our proposed Deeply-Task-specific And Graph-regularized Network (**DTAGnet**) brings in additional deep feedbacks, by associating a clustering-oriented local auxiliary loss $C_k(Z_k, \omega_k)$ ($k = 1, 2$) with each stage. Such an auxiliary loss takes the same form as the overall $C(A, \omega)$, except that the expected cluster number may be different, depending on the auxiliary clustering task to be performed. The DTAGnet backpropagates errors not only from the overall loss layer, but also simultaneously from the auxiliary losses.

While seeking the optimal performance of the target clustering, DTAGnet is also driven by two auxiliary tasks that are explicitly targeted at clustering specific attributes. It enforces constraint at each hidden representation for directly making a good cluster prediction. In addition to the overall loss, the introduction of auxiliary losses gives another strong push to obtain discriminative and sensible features at each individual stage. As discovered in the classification experiments in [42], the auxiliary loss both acts

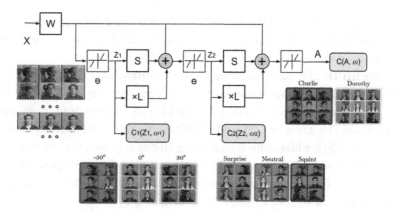

Figure 5.5 The DTAGnet architecture, taking the CMU MultiPIE dataset as an example. The model is able to simultaneously learn features for pose clustering (Z_1), for expression clustering (Z_2), and for identity clustering (A). The first two attributes are related to and helpful for the last (overall) task. Part of image sources are referred from [35] and [38].

as feature regularization to reduce generalization errors and results in faster convergence. We also find in Sect. 5.2.5 that every Z_k ($k = 1, 2$) is indeed most suited for its targeted task.

In [38], a Deep Semi-NMF model was proposed to learn hidden representations, which grant themselves an interpretation of clustering according to different attributes. The authors considered the problem of mapping facial images to their identities. A face image also contains attributes like pose and expression that help identify the person depicted. In their experiments, the authors found that by further factorizing this mapping in a way that each factor adds an extra layer of abstraction, the deep model could automatically learn latent intermediate representations that are implied for clustering identity-related attributes. Although there is a clustering interpretation, those hidden representations are not specifically optimized in clustering sense. Instead, the entire model is trained with only the overall reconstruction loss, after which clustering is performed using K-means on learnt features. Consequently, their clustering performance is not satisfactory. Our study shares the similar observation and motivation with [38], but in a more task-specific manner by performing the optimizations of auxiliary clustering tasks jointly with the overall task.

5.2.5 Experiment Results
5.2.5.1 Datasets and Measurements

We evaluate the proposed model on three publicly available datasets:

- **MNIST** [8] consists of a total number of 70,000 quasi–binary, handwritten digit images, with digits 0 to 9. The digits are normalized and centered in fixed-size images of 28×28.
- **CMU MultiPIE** [35] contains around 750,000 images of 337 subjects that are captured under varied laboratory conditions. A unique property of CMU MultiPIE lies in that each image comes with labels for the identity, illumination, pose and expression attributes. That is why CMU MultiPIE is chosen in [38] to learn multiattribute features (Fig. 5.5) for hierarchical clustering. In our experiments, we follow [38] and adopt a subset of 13,230 images of 147 subjects in 5 different poses and 6 different emotions. Notably, we do not preprocess the images by using piecewise affine warping as utilized by [38] to align these images.
- **COIL20** [43] contains 1440 32×32 gray scale images of 20 objects (72 images per object). The images of each object were taken 5 degrees apart.

Although the paper only evaluates the proposed method using image datasets, the methodology itself is not limited to only image subjects. We apply two widely-used measures to evaluate the clustering performances, the accuracy and the normalized mutual information (NMI) [8,12]. We follow the convention of many clustering works [8,19,31] and do not distinguish training from testing. We train our models on all available samples of each dataset, reporting the clustering performances as our testing results. Results are averaged from 5 independent runs.

5.2.5.2 Experiment Settings

The proposed networks are implemented using the cuda-convnet package [34]. The network takes $K = 2$ stages by default. We apply a constant learning rate of 0.01 with no momentum to all trainable layers. The batch size of 128. In particular, to encode graph regularization as a prior, we fix L during model training by setting its learning rate to be 0. Experiments run on a workstation with 12 Intel Xeon 2.67 GHz CPUs and 1 GTX680 GPU. The training takes approximately 1 hour on the MNIST dataset. It is also observed that the training efficiency of our model scales approximately linearly with data.

In our experiments, we set the default value of α to be 5, p to be 128, and λ to be chosen from [0.1, 1] by cross-validation.[6] A dictionary D is first learned from X by K-SVD [20]; W, S and θ are then initialized based on (5.19), while L is also pre-calculated from P, which is formulated by the Gaussian kernel, $P_{ij} = \exp(-\frac{\|x_i - x_j\|_2^2}{\delta^2})$ (δ is also selected by cross-validation). After obtaining the output A from the initial (D)TAGnet models, ω (or ω_k) could be initialized based on minimizing (5.21) or (5.23) over A (or Z_k).

---

[6] The default values of α and p are inferred from the related sparse coding literature [8], and validated in experiments.

5.2.5.3 Comparison Experiments and Analysis

Benefits of the Task-specific Deep Architecture

We denote the proposed model of TAGnet plus entropy-minimization loss (EML) (5.21) as TAGnet-EML, and the one plus maximum-margin loss (MML) (5.23) as TAGnet-MML, respectively. We include the following comparison methods:

- We refer to the initializations of the proposed joint models as their **"Non-Joint"** counterparts, denoted as NJ-TAGnet-EML and NJ-TAGnet-MML (NJ is short for non-joint), respectively.

- We design a **baseline encoder** (BE), which is a fully-connected feed-forward network, consisting of three hidden layers of dimension p with ReLU neuron. It is obvious that the BE has the *same parameter complexity* as TAGnet.[7] The BEs are also tuned by EML or MML in the same way, denoted as BE-EML or BE-MML, respectively. We intend to verify our important claim that *the proposed model benefits from the task-specific TAGnet architecture, rather than just the large learning capacity of generic deep models*.

- We compare the proposed models with their closest **"shallow"** competitors, i.e., the joint optimization methods of graph-regularized sparse coding and discriminative clustering in [31]. We reimplement their work using both (5.21) or (5.23) losses, denoted as SC-EML and SC-MML (SC is short for sparse coding). Since in [31] the authors already revealed that SC-MML outperforms the classical methods such as MMC and ℓ_1-graph methods, we do not compare with them again.

- We also include **Deep Semi-NMF** [38] as a state-of-the-art deep learning-based clustering method. We mainly compare our results with their reported performances on CMU MultiPIE.[8]

As revealed by the full comparison results in Table 5.3, the proposed task-specific deep architectures outperform others with a noticeable margin. The underlying domain expertise guides the data-driven training in a more principled way. In contrast, the "general-architecture" baseline encoders (BE-EML and BE-MML) appear to produce much worse (even worst) results. Furthermore, it is evident that the proposed end-to-end optimized models outperform their "non-joint" counterparts. For example, on the MNIST dataset, TAGnet-MML surpasses NJ-TAGnet-MML by around 4% in accuracy and 5% in NMI.

By comparing the TAGnet-EML/TAGnet-MML with SC-EML/SC-MML, we draw a promising conclusion: adopting a more parameterized deep architecture allows a larger feature learning capacity compared to conventional sparse coding. Although similar points are well made in many other fields [34], we are interested in a closer

---

[7] Except for the "θ" layers, each of which contains only p free parameters and is thus ignored.
[8] With various component numbers tested in [38], we choose their best cases (60 components).

Table 5.3 Accuracy and NMI performance comparisons on all three datasets

		TAGnet-EML	TAGnet-MML	NJ-TAGnet-EML	NJ-TAGnet-MML	BE-EML	BE-MML	SC-EML	SC-MML	Deep Semi-NMF
MNIST	Acc	0.6704	0.6922	0.6472	0.5052	0.5401	0.6521	0.6550	0.6784	/
	NMI	0.6261	0.6511	0.5624	0.6067	0.5002	0.5011	0.6150	0.6451	/
CMU	Acc	0.2176	0.2347	0.1727	0.1861	0.1204	0.1451	0.2002	0.2090	0.17
MultiPIE	NMI	0.4338	0.4555	0.3167	0.3284	0.2672	0.2821	0.3337	0.3521	0.36
COIL20	Acc	0.8553	0.8991	0.7432	0.7882	0.7441	0.7645	0.8225	0.8658	/
	NMI	0.9090	0.9277	0.8707	0.8814	0.8028	0.8321	0.8850	0.9127	/

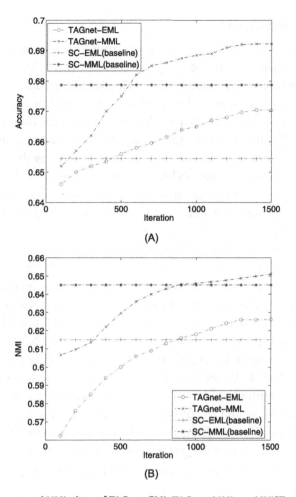

Figure 5.6 The accuracy and NMI plots of TAGnet-EML/TAGnet-MML on MNIST, starting from the initialization, and tested every 100 iterations. The accuracy and NMI of SC-EML/SC-MML are also plotted as baselines.

look between the two. Fig. 5.6 plots the clustering accuracy and NMI curves of TAGnet-EML/TAGnet-MML on the MNIST dataset, along with iteration numbers. Each model is well initialized at the very beginning, and the clustering accuracy and NMI are computed every 100 iterations. At first, the clustering performances of deep models are even slightly worse than sparse-coding methods, mainly since the initialization of TAGnet hinges on a truncated approximation of graph-regularized sparse coding. After a small number of iterations, the performance of the deep models surpass sparse coding ones, and continue rising monotonically until reaching a higher plateau.

Effects of Graph Regularization

In (5.16), the graph regularization term imposes stronger smoothness constraints on the sparse codes with a larger α. It also happens to the TAGnet. We investigate how the clustering performances of TAGnet–EML/TAGnet–MML are influenced by various α values. From Fig. 5.7, we observe the identical general tendency on all three datasets. While α increases, the accuracy/NMI result will first rise then decrease, with the peak appearing for $\alpha \in [5, 10]$. As an interpretation, the local manifold information is not sufficiently encoded when α is too small ($\alpha = 0$ will completely disable the $\times L$ branch of TAGnet, and reduce it to the LISTA network [29] fine-tuned by the losses). On the other hand, when α is large, the sparse codes are "oversmoothed" with a reduced discriminative ability. Note that similar phenomena are also reported in other relevant literature, e.g., [8,31].

Furthermore, comparing Fig. 5.7A–F, it is noteworthy to observe how graph regularization behaves differently on three of them. We notice that the COIL20 dataset is the most sensitive to the choice of α. Increasing α from 0.01 to 50 leads to an improvement of more than 10%, in terms of both accuracy and NMI. It verifies the significance of graph regularization when training samples are limited [19]. On the MNIST dataset, both models obtain a gain of up to 6% in accuracy and 5% in NMI, by tuning α from 0.01 to 10. However, unlike COIL20 that almost always favors larger α, the model performance on the MNIST dataset tends to be not only saturated, but even significantly hampered when α continues rising to 50. The CMU MultiPIE dataset witnesses moderate improvements of around 2% in both measurements. It is not as sensitive to α as the other two. Potentially, it might be due to the complex variability in original images that makes the graph W unreliable for estimating the underlying manifold geometry. We suspect that more sophisticated graphs may help alleviate the problem, and we will explore it in the future.

Scalability and Robustness

On the MNIST dataset, we reconduct the clustering experiments with the cluster number N_c ranging from 2 to 10, using TAGnet–EML/TAGnet–MML. Fig. 5.8 shows that the clustering accuracy and NMI change by varying the number of clusters. The clustering performance transits smoothly and robustly when the task scale changes.

To examine the proposed models' robustness to noise, we add various Gaussian noise, whose standard deviation s ranges from 0 (noiseless) to 0.3, to retrain our MNIST model. Fig. 5.9 indicates that both TAGnet–EML and TAGnet–MML own certain robustness to noise. When s is less than 0.1, there is even little visible performance degradation. While TAGnet–MML constantly outperforms TAGnet–EML in all experiments (as MMC is well-known to be highly discriminative [3]), it is interesting to observe in Fig. 5.9 that the latter is slightly more robust to noise than the former. It is

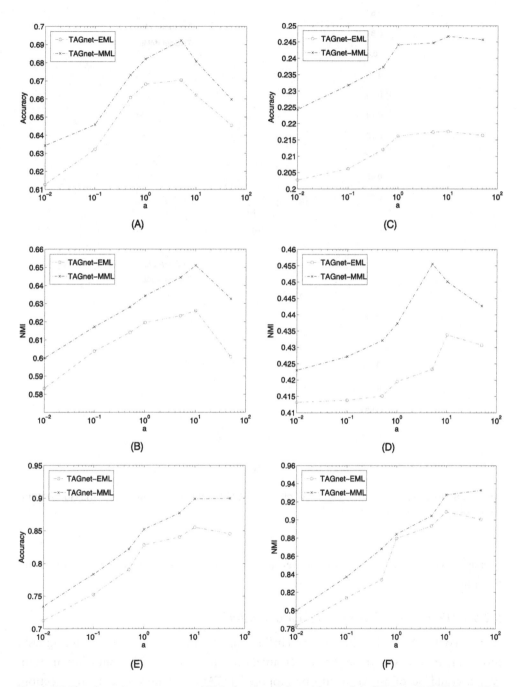

Figure 5.7 The clustering accuracy and NMI plots (x-axis logarithm scale) of TAGnet-EML/TAGnet-MML versus the parameter choices of α, on: (A)–(B) MNIST; (C)–(D) CMU MultiPIE; (E)–(F) COIL20.

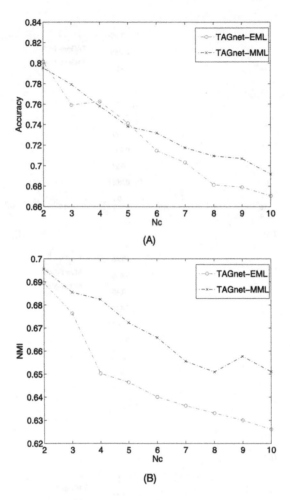

Figure 5.8 The clustering accuracy and NMI plots of TAGnet-EML/TAGnet-EML versus the cluster number N_c ranging from 2 to 10 on MNIST.

perhaps owing to the probability–driven loss form (5.21) of EML that allows for more flexibility.

5.2.5.4 Hierarchical Clustering on CMU MultiPIE

As observed, CMU MultiPIE is very challenging for the basic identity clustering task. However, it comes with several other attributes: pose, expression, and illumination, which could be of assistance in our proposed DTAGnet framework. In this section, we apply a similar setting of [38] on the same CMU MultiPIE subset, by setting pose clustering as the Stage I auxiliary task, and expression clustering as the Stage II auxiliary

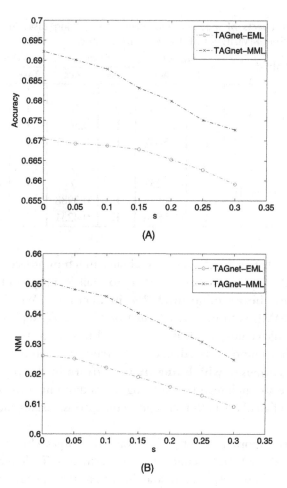

Figure 5.9 The clustering accuracy and NMI plots of TAGnet-EML/TAGnet-MML versus the noise level s on MNIST.

task.[9] In that way, we target $C_1(\mathbf{Z}_1, \boldsymbol{\omega}_1)$ at 5 clusters, $C_2(\mathbf{Z}_2, \boldsymbol{\omega}_2)$ at 6 clusters, and finally, $C(\mathbf{A}, \boldsymbol{\omega})$ at 147 clusters.

The training of DTAGnet-EML/DTAGnet-MML follows the same aforementioned process except for considered extra back-propagated gradients from task $C_k(\mathbf{Z}_k, \boldsymbol{\omega}_k)$ in Stage k ($k = 1, 2$). After that we test each $C_k(\mathbf{Z}_k, \boldsymbol{\omega}_k)$ separately on their targeted task. In DTAGnet, each auxiliary task is also jointly optimized with its intermediate feature $\mathbf{Z}_k$, which differentiates our methodology substantially from [38]. It is thus no surprise

[9] In fact, although claimed to be applicable to multiple attributes, [38] only examined the first level features for pose clustering without considering expressions, since it relied on a warping technique to preprocess images, which gets rid of most expression variability.

Table 5.4 Effects of incorporating auxiliary clustering tasks in DTAGnet-EML/DTAGnet-MML (P, Pose; E, Expression; I, Identity)

Method	Stage I		Stage II		Overall	
	Task	Acc	Task	Acc	Task	Acc
DTAGnet–EML	/	/	/	/	I	0.2176
	P	0.5067	/	/	I	0.2303
	/	/	E	0.3676	I	0.2507
	P	0.5407	E	0.7027	I	0.2833
DTAGnet–MML	/	/	/	/	I	0.2347
	P	0.5251	/	/	I	0.2635
	/	/	E	0.3988	I	0.2858
	P	0.5538	E	0.4231	I	0.3021

to see in Table 5.4 that each auxiliary task obtains much improved performances than [38].[10] Most notably, the performances of the overall identity clustering task witness **a very impressive boost of around 7% in accuracy**. We also test DTAGnet-EML/DTAGnet-MML with only $C_1(\boldsymbol{Z}_1, \boldsymbol{\omega}_1)$ or $C_2(\boldsymbol{Z}_2, \boldsymbol{\omega}_2)$ kept. Experiments verify that by adding auxiliary tasks gradually, the overall task keeps being benefited. Those auxiliary tasks, when enforced together, can also reinforce each other mutually.

One might be curious of **which one matters more in the performance boost**: the deeply task-specific architecture that brings extra discriminative feature learning, or the proper design of auxiliary tasks that capture the intrinsic data structure characterized by attributes.

To answer this important question, we vary the target cluster number in either $C_1(\boldsymbol{Z}_1, \boldsymbol{\omega}_1)$ or $C_2(\boldsymbol{Z}_2, \boldsymbol{\omega}_2)$, and reconduct the experiments. Table 5.5 reveals that more auxiliary tasks, even those without any straightforward task-specific interpretation (e.g., partitioning the Multi-PIE subset into 4, 8, 12 or 20 clusters hardly makes semantic sense), may still help gain better performances. It is comprehensible that they simply promote more discriminative feature learning in a low-to-high, coarse-to-fine scheme. In fact, it is a complementary observation to the conclusion found in classification [42]. On the other hand, at least in this specific case, while the target cluster numbers of auxiliary tasks get closer to the ground truth (5 and 6 here), the models seem to achieve the best performances. We conjecture that when properly "matched", every hidden representation in each layer is in fact most suited for clustering the attributes corresponding to the layer of interest. The whole model can be resembled to the problem of sharing low-level feature filters among several relevant high-level tasks in convolutional networks [44], but in a distinct context.

---

[10] In [38], Table 2 it reports that the best accuracy of pose clustering task falls around 28%, using the most suited layer features.

Table 5.5 Effects of varying target cluster numbers of auxiliary tasks in DTAGnet-EML/DTAGnet-MML

Method	#clusters in Stage I	#clusters in Stage II	Overall Accuracy
DTAGnet–EML	4	4	0.2827
	8	8	0.2813
	12	12	0.2802
	20	20	0.2757
DTAGnet–MML	4	4	0.3030
	8	8	0.3006
	12	12	0.2927
	20	20	0.2805

We hence conclude that the deeply-supervised fashion shows to be helpful for the deep clustering models, even when there are no explicit attributes for constructing a practically meaningful hierarchical clustering problem. However, it is preferable to exploit those attributes when available, as they lead to not only superior performances but more clearly interpretable models. The learned intermediate features can be potentially utilized for multitask learning [45].

5.2.6 Conclusion

In this section, we present a deep learning-based clustering framework. Trained from end to end, it features a task-specific deep architecture inspired by the sparse coding domain expertise, which is then optimized under clustering-oriented losses. Such a well-designed architecture leads to more effective initialization and training, and significantly outperforms generic architectures of the same parameter complexity. The model could be further interpreted and enhanced, by introducing auxiliary clustering losses to the intermediate features. Extensive experiments verify the effectiveness and robustness of the proposed models.

REFERENCES

[1] Duda RO, Hart PE, Stork DG. Pattern classification. John Wiley & Sons; 1999.
[2] Biernacki C, Celeux G, Govaert G. Assessing a mixture model for clustering with the integrated completed likelihood. IEEE TPAMI 2000;22(7):719–25.
[3] Xu L, Neufeld J, Larson B, Schuurmans D. Maximum margin clustering. In: NIPS; 2004. p. 1537–44.
[4] Zhao B, Wang F, Zhang C. Efficient multiclass maximum margin clustering. In: Proceedings of the 25th international conference on Machine learning. ACM; 2008. p. 1248–55.
[5] Zhao B, Wang F, Zhang C. Efficient maximum margin clustering via cutting plane algorithm. In: SDM; 2008.
[6] Li X, Zhang K, Jiang T. Minimum entropy clustering and applications to gene expression analysis. In: CSB. IEEE; 2004. p. 142–51.

[7] Barber D, Agakov FV. Kernelized infomax clustering. In: Advances in neural information processing systems; 2005. p. 17–24.

[8] Zheng M, Bu J, Chen C, Wang C, Zhang L, Qiu G, et al. Graph regularized sparse coding for image representation. IEEE TIP 2011;20(5):1327–36.

[9] Zhang Li, Zhou Wei-da, Jiao Li-cheng. Kernel clustering algorithm. Chinese Journal of Computers 2002;6:004.

[10] Roth V, Lange T. Feature selection in clustering problems. In: NIPS; 2003.

[11] Wright J, Yang AY, Ganesh A, Sastry SS, Ma Y. Robust face recognition via sparse representation. IEEE TPAMI 2009;31(2):210–27.

[12] Cheng B, Yang J, Yan S, Fu Y, Huang TS. Learning with l1 graph for image analysis. IEEE TIP 2010;19(4).

[13] Ng AY, Jordan MI, Weiss Y, et al. On spectral clustering: analysis and an algorithm. NIPS 2002;2:849–56.

[14] Sprechmann P, Sapiro G. Dictionary learning and sparse coding for unsupervised clustering. In: Acoustics speech and signal processing (ICASSP), 2010 IEEE international conference on. IEEE; 2010. p. 2042–5.

[15] Chen YC, Sastry CS, Patel VM, Phillips PJ, Chellappa R. In-plane rotation and scale invariant clustering using dictionaries. IEEE Transactions on Image Processing 2013;22(6):2166–80.

[16] Yang Y, Wang Z, Yang J, Han J, Huang TS. Regularized ℓ_1-graph for data clustering. In: Proceedings of the British machine vision conference; 2014.

[17] Mairal J, Bach F, Ponce J. Task-driven dictionary learning. IEEE TPAMI 2012;34(4):791–804.

[18] Wang Z, Nasrabadi NM, Huang TS. Semisupervised hyperspectral classification using task-driven dictionary learning with Laplacian regularization. IEEE Transactions on Geoscience and Remote Sensing 2015;53(3):1161–73.

[19] Yang Y, Wang Z, Yang J, Wang J, Chang S, Huang TS. Data clustering by Laplacian regularized ℓ_1-graph. In: AAAI; 2014.

[20] Aharon M, Elad M, Bruckstein A. K-SVD: an algorithm for designing overcomplete dictionaries for sparse representation. IEEE TSP 2006.

[21] Yang J, Wang Z, Lin Z, Shu X, Huang T. Bilevel sparse coding for coupled feature spaces. In: CVPR. IEEE; 2012. p. 2360–7.

[22] Dai B, Hu B. Minimum conditional entropy clustering: a discriminative framework for clustering. In: ACML; 2010. p. 47–62.

[23] Li YF, Tsang IW, Kwok JT, Zhou ZH. Tighter and convex maximum margin clustering. In: International conference on artificial intelligence and statistics; 2009. p. 344–51.

[24] Crammer K, Singer Y. On the algorithmic implementation of multiclass kernel-based vector machines. The Journal of Machine Learning Research 2002;2:265–92.

[25] Lee CP, Lin CJ. A study on l2-loss (squared hinge-loss) multiclass SVM. Neural Computation 2013;25(5):1302–23.

[26] Lovász L, Plummer M. Matching theory, vol. 367. American Mathematical Soc.; 2009.

[27] Nie F, Xu D, Tsang IW, Zhang C. Spectral embedded clustering. In: IJCAI; 2009. p. 1181–6.

[28] Chang S, Han W, Tang J, Qi G, Aggarwal C, Huang TS. Heterogeneous network embedding via deep architectures. In: ACM SIGKDD; 2015.

[29] Gregor K, LeCun Y. Learning fast approximations of sparse coding. In: ICML; 2010. p. 399–406.

[30] Bertsekas DP. Nonlinear Programming 1999.

[31] Wang Z, Yang Y, Chang S, Li J, Fong S, Huang TS. A joint optimization framework of sparse coding and discriminative clustering. In: IJCAI; 2015.

[32] Lee H, Battle A, Raina R, Ng AY. Efficient sparse coding algorithms. In: NIPS; 2006. p. 801–8.

[33] Bengio Y. Learning deep architectures for ai. Foundations and Trends in Machine Learning 2009;2(1):1–127.

[34] Krizhevsky A, Sutskever I, Hinton GE. Imagenet classification with deep convolutional neural networks. In: NIPS; 2012. p. 1097–105.

[35] Gross R, Matthews I, Cohn J, Kanade T, Baker S. Multi-PIE. Image and Vision Computing 2010;28(5).

[36] Tian F, Gao B, Cui Q, Chen E, Liu TY. Learning deep representations for graph clustering. In: AAAI; 2014.

[37] Chen G. Deep learning with nonparametric clustering. arXiv preprint arXiv:1501.03084, 2015.

[38] Trigeorgis G, Bousmalis K, Zafeiriou S, Schuller B. A deep semi-NMF model for learning hidden representations. In: ICML; 2014. p. 1692–700.

[39] Beck A, Teboulle M. A fast iterative shrinkage-thresholding algorithm for linear inverse problems. SIAM Journal on Imaging Sciences 2009;2(1):183–202.

[40] Sutskever I, Martens J, Dahl G, Hinton G. On the importance of initialization and momentum in deep learning. In: ICML; 2013. p. 1139–47.

[41] Donahue J, Jia Y, Vinyals O, Hoffman J, Zhang N, Tzeng E, et al. DeCAF: a deep convolutional activation feature for generic visual recognition. arXiv preprint arXiv:1310.1531, 2013.

[42] Lee CY, Xie S, Gallagher P, Zhang Z, Tu Z. Deeply-supervised nets. arXiv preprint arXiv:1409.5185, 2014.

[43] Nene SA, Nayar SK, Murase H, et al. Columbia object image library (COIL-20). Tech. rep.

[44] Glorot X, Bordes A, Bengio Y. Domain adaptation for large-scale sentiment classification: a deep learning approach. In: ICML; 2011. p. 513–20.

[45] Wang Y, Wipf D, Ling Q, Chen W, Wassail I. Multi-task learning for subspace segmentation. In: ICML; 2015.

CHAPTER 6

Signal Processing

Zhangyang Wang*, Ding Liu†, Thomas S. Huang‡

*Department of Computer Science and Engineering, Texas A&M University, College Station, TX, United States
†Beckman Institute for Advanced Science and Technology, Urbana, IL, United States
‡Department of Electrical and Computer Engineering, University of Illinois at Urbana-Champaign, Champaign, IL, United States

Contents

6.1.	Deeply Optimized Compressive Sensing	121
	6.1.1 Background	121
	6.1.2 An End-to-End Optimization Model of CS	122
	6.1.3 DOCS: Feed-Forward and Jointly Optimized CS	124
	6.1.4 Experiments	126
	6.1.5 Conclusion	129
6.2.	Deep Learning for Speech Denoising	130
	6.2.1 Introduction	130
	6.2.2 Neural Networks for Spectral Denoising	131
	6.2.3 Experimental Results	134
	6.2.4 Conclusion and Future Work	139
References		140

6.1. DEEPLY OPTIMIZED COMPRESSIVE SENSING

6.1.1 Background

Consider a signal $x \in R^d$, which admits a sparse representation $\alpha \in R^n$ over a dictionary $D \in R^{d \times n}$, with $||\alpha||_0 \ll n$. Compressive sensing (CS) [1,2] revealed the amazing fact that a signal, which is sparsely represented over an appropriate D, can be recovered from much fewer measurements, than what the Nyquist–Shannon sampling theorem requires. It implied the potential for joint signal sampling and compression, and triggered the development of many novel devices [3]. Given the sampling (or sensing) matrix $P \in R^{m \times d}$ with $m \ll d$, CS samples $y \in R^m$ from x as

$$y = Px = PD\alpha. \tag{6.1}$$

Considering real CS cases where the acquisition and transmission of y could be noisy, the original x could be recovered from its compressed form y, by finding its sparse representation α through solving the convex problem (λ is a constant)

$$\min_{\alpha} \lambda||\alpha||_1 + \frac{1}{2}||y - PD\alpha||^2. \tag{6.2}$$

Deep Learning Through Sparse and Low-Rank Modeling
DOI: 10.1016/B978-0-12-813659-1.00006-8

CS is feasible only if P and D are chosen wisely, so that (1) x could be sufficiently sparsely represented over D; and (2) P and D have a low mutual coherence [4]. The choice of D has been studied. While earlier literature adopted standard orthogonal transformations for D, such as Discrete Cosine Transform (DCT), where $d = n$, an overcomplete dictionary where $d < n$ becomes dominant for a more sparse x. An overcomplete D can either be a prespecified set of functions (overcomplete DCT, wavelets, etc.), or designed by adapting its content to fit a given set of signal examples. Lately, the development of *dictionary learning* led to powerful data-driven methods to learn D from training data, which achieved the improved performance.

Another fundamental problem in CS is how to choose P so that x could be recovered with a high probability. Previously, the random projection matrix was shown to be a good choice since its columns are almost incoherent with any basis D. Rules of designing deterministic projection matrices are studied in, for example, [5–8]. Elad [5] minimized the t-averaged mutual coherence, which was suboptimal. Xu et al. [6] proposed a solver based on the Welch bound [9]. Lately, [8] developed an ℓ_∞-based minimization metric.

While obtaining CS measurements is trivial (multiplication with P), the computational load is transferred to the recovery process. Equation (6.2) can be solved via either convex optimization [10] or greedy algorithms [11]. Their iterative nature leads to an inherently sequential structure and thus remarkable latency, causing a major bottleneck of the computational efficiency. Very recently, it has been advocated to replace iterative algorithms with feed-forward deep neural networks, in order for improved and much faster signal recovery [12,13]. However, existing limited works appear to largely overlook the classical CS pipeline: they instead simply applied data-driven deep models like "black boxes", to map y back to x (or equivalently, its sparse representation α) through regression.

6.1.2 An End-to-End Optimization Model of CS

Assume we are given a targeted class of signals $X \in R^{d \times t}$, where each column $x_i \in R^d$ stands for an input signal. Correspondingly, $Y \in R^{m \times t}$ and $A \in R^{n \times t}$ denote the collections of CS samples $y_i = PDx_i$, and their sparse representations α_i, respectively. CS samples Y from the input signal X on the **encoder** side, followed by reconstructing $\hat{X}$ from Y on the **decoder** side, so that $\hat{X}$ is as close to X as possible. Conceptually, such a pipeline is divided into the following four stages:

- **Stage I: Representation** seeks the sparsest representation $A \in R^{n \times t}$ of X, with regard to D.
- **Stage II: Measurement** obtains Y from A by $Y = PDA$.
- **Stage III: Recovery** recovers $\hat{A}$ from Y by solving (6.2).
- **Stage IV: Reconstruction** reconstructs $\hat{X} = D\hat{A}$.

Stages I and IV are usually implemented by designing a transformation pair. We introduce an *analysis dictionary* $G \in R^{n \times d}$, jointly with the *synthesis dictionary* D, so that $X = DA$ and $A = GX$. Stages I and II belong to the encoder, and are usually combined to one step in practice, $Y = PX$. While A can be treated as an "auxiliary" variable in real CS pipelines, we separately write Stages I and II in order to emphasize the inherent sparsity prior of X. Stages III and IV constitute the decoder. It is assumed that P is shared by the encoder and the decoder.[1]

To the best of our knowledge, a joint end-to-end optimization of the CS encoder and decoder is absent. During the *training* stage, given the training data X, we express the entire CS pipeline as

$$\min_{D,G,P} \quad ||X - D\hat{A}||_F^2 + \gamma ||X - DGX||_F^2,$$

$$\text{s.t.} \quad \hat{A} \in \arg\min \frac{1}{2}||Y - PD\hat{A}||_F^2 + \lambda ||\hat{A}||_1, \ Y = PX. \tag{6.3}$$

Equation (6.3) is a bi-level optimization problem [14]. In the lower level, we incorporate Stage II as the equality constraint, and Stage III as the ℓ_1-minimization. In the upper level, the reconstruction error between the original X and the reconstruction $D\hat{A}$ is measured. Furthermore, the synthesis and analysis dictionary pair has to satisfy the reconstruction constraint. Their weighted sum is the overall objective of (6.3) to be minimized. The obtained D, G, and P are applied to *testing*, where an input x goes through Stages I–IV and is reconstructed at the end.

Many classical CS pipelines use fixed sparsifying transformation pairs, such as DCT and Inverse DCT (IDCT), as D and G, while P is some random matrix. More recently, there have been some model variants that try to learn a fraction of parameters:

- **Case I: given D and G, learning P.** This method is used in [6,8]. The authors did not pay much attention to sparsifying the representation α.
- **Case II: given $Y = PX$, learning D and P.** This corresponds to optimizing the decoder only. Duarte-Carvajalino and Sapiro [7] proposed the coupled-SVD algorithm to jointly optimize P and D, but their formulation (Eq. (19) in [7]) did not minimize the CS reconstruction error directly. The recently-proposed deep learning-based CS models, such as in [13,12], are also categorized into this case (see Sect. 3.5 for their comparisons with DOCS).

In contrast, our primary goal is to jointly optimize the pipeline from end to end, which is supposed to outperform the above "partial" solutions. Solving the bi-level optimization (6.3) is possible [15], yet requires heavy computation, and without much theoretical guarantee.[2] Further, to apply the learned CS model to testing, it is still inevitable to iteratively solve (6.2) for recovering α.

---

[1] Practical CS systems transmit the "seed" of the random sampling matrix, together with the CS samples.

[2] A nonconvex convergence proof of SGD assumes three times differentiable cost functions [14].

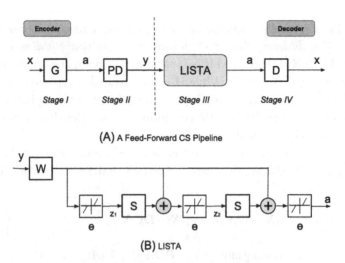

(A) A Feed-Forward CS Pipeline

(B) LISTA

Figure 6.1 (A) An illustration of the feed-forward CS pipeline, where Stages I, II and IV are all simplified to matrix multiplications. Panel (B) depicts the LISTA structure for implementing the iterative sparse recovery algorithm in a feed-forward network way (with a truncation of $K = 3$ stages).

6.1.3 DOCS: Feed-Forward and Jointly Optimized CS

We aim to transform the insights of (6.3) into a fully *feed-forward* CS pipeline, which is expected to ensure a reduced computational complexity. The general feed-forward pipeline is called *Deeply Optimized Compressive Sensing* (**DOCS**), as illustrated in Fig. 6.1A.

While Stages I, II and IV are naturally feed-forward, Stage III usually refers to iterative algorithms for solving (6.2) and makes the major computation bottleneck. Letting $M = PD \in R^{m \times n}$ for notation simplicity, we exploit a feed-forward approximation, called *Learned Iterative Shrinkage and Thresholding Algorithm* (LISTA) [16]. LISTA was proposed to efficiently approximate the ℓ_1-based sparse code α of the input x. It could be implemented as a neural network, as illustrated in Fig. 6.1B. The network architecture is obtained by unfolding and truncating the iterative shrinkage and thresholding algorithm (ISTA). The network has K stages, each of which updates the intermediate sparse code z^k ($k = 0, \ldots, K - 1$) according to

$$z^{k+1} = s_\theta(Wx + Sz^k), \text{ where } W = M^T, S = I - M^T M, \theta = \lambda, \quad (6.4)$$

and s_θ is an element-wise soft shrinkage operator.[3] The network parameters, W, S and θ, could be initialized from M and λ of the original model (6.2), and tuned further by back-propagation [16]. For convenience, we disentangle M and D and treat them as two separate variables during training.

---

[3] $[s_\theta(u)]_i = \text{sign}(u_i) \max(u_i| - \theta_i, 0)$ (u is a vector and u_i is its ith element).

After the transforming (6.2) into the LISTA form, and combining Stages I and II into one linear layer P, the four-stage CS pipeline now becomes a fully-connected, feed-forward network. Note that Stages I, II and IV are all linear, while only Stage III contains nonlinear "neurons" s_θ. We are certainly aware of possibilities to add non-linear neurons in Stages I, II and IV, but choose to stick to their "original forms" to be faithful to (6.2). We then tune all parameters jointly from training data, G, W, S, and D, using back-propagation.[4] The overall loss function of DOCS is $||X - D\hat{A}||_F^2 + \gamma ||X - DGX||_F^2$. The entire DOCS model is jointed updated by the stochastic gradient descent algorithm.

Complexity

Since the training of DOCS can be performed offline, we are mainly concerned about its testing complexity. On the encoder side, the time complexity is simply $O(md)$. On the decoder side, there are $K + 1$ trainable layers, making the time complexity $O(mn + (K - 1)n^2 + nd)$.

Related Work

There have been abundant works investigating CS models [1,2], and developing iterative algorithms [10,11]. However, few of them benefit from optimizing all the stages from end to end. It was only recently that feed-forward networks were considered for the efficient recovery. Mousavi et al. [12] applied a stacked denoising auto-encoder (SDA) to map CS measurements back to the original signals. Kulkarni et al. [13] proposed a convolutional neural network (CNN) to reconstruct images via CS, followed by an extra global enhancement step as post-processing. However, both prior works relied on off-the-shelf deep architectures that are not customized much for CS, and focused on optimizing the decoder. Lately, [17,18] started looking at encoder–decoder networks for the specific problem of video CS.

Beyond LISTA, there are also more blooming interests in bridging iterative optimization algorithms and deep learning models. In [19], the authors leveraged a similar idea on fast trainable regressors and constructed feed-forward network approximations. It was later extended in [20] to develop an algorithm of learned deterministic fixed-complexity pursuits, for sparse and low rank models. Lately, [21] modeled ℓ_0 sparse approximation as feed-forward neural networks. The authors extended the strategy to the graph-regularized ℓ_1-approximation in [22], and to ℓ_∞-based minimization in [23]. However, there has been no systematic study along this direction on the CS pipeline.

---

[4] During tuning, W and S are **not** tied, but S is shared by the k LISTA stages.

6.1.4 Experiments

Settings

We implement all deep models with CUDA ConvNet [24], on a workstation with 12 Intel Xeon 2.67 GHz CPUs and 1 Titan X GPU. For all models, we use a batch size of 256 and a momentum of 0.9. The training is regularized by weight decay (ℓ_2-penalty multiplier set to 5×10^{-4}) and dropout (ratio set to 0.5); γ is set as 5, while D and G are initialized using overcomplete DCT and IDCT, respectively. We also initialize P as a Gaussian random matrix. In addition to DOCS, we further design the following two baselines:

- **Baseline I:** D and G are fixed to be overcomplete DCT and IDCT, respectively. While Stage II utilizes a random P, Stage III relies on running the state-of-the-art iterative sparse recovery algorithm, TVAL3 [25].[5] This constitutes the most basic CS baseline, with no parameters learned from data.
- **Baseline II:** D and G are replaced with the learned matrices by DOCS. Stages II and III remain the same as for Baseline I.

Simulation

We generate $x \in R^d$, which ensures an inherent sparse structure, for the simulation purpose: we produce T-sparse vectors $\alpha_0 \in R^n$ (e.g., only T out of n entries are nonzero). The locations of nonzero entries are chosen randomly, and their values are drawn from a uniform distribution in $[-1, 1]$. We then generate $x = G\alpha_0 + n_0$, where $G \in R^{d \times n}$ is an overcomplete DCT dictionary [26], and $n_0 \in R^d$ is a random Gaussian noise with zero mean and standard deviation of 0.02. For training each model, we generate 10,000 samples for training, and 1000 for testing, on which the overall results are averaged. The default number of LISTA layers $K = 3$.

Reconstruction Error

The DOCS model and two baselines are tested on three schemes: (1) $m = [6 : 2 : 16]$; $d = 30$; $n = 60$; (2) $m = [10 : 5 : 35]$; $d = 60$; $n = 120$; (3) $m = [10 : 10 : 60]$; $d = 90$; $n = 180$; T is fixed to 5 in all cases. As shown in Fig. 6.2, DOCS is able to reduce reconstruction errors dramatically, compared to the Baselines I and II. It is evident that the learned D and G lead to a more sparse representation than prefixed transformations and thus allow for more accurate recovery, by comparing Baselines I and II. Then thanks to the end-to-end tuning, DOCS is capable of producing smaller errors than Baseline II, reflecting that learned sparse recovery could potentially fit the data better.

---

[5] We use the code made available by the authors on their website. Parameters are set to the default values.

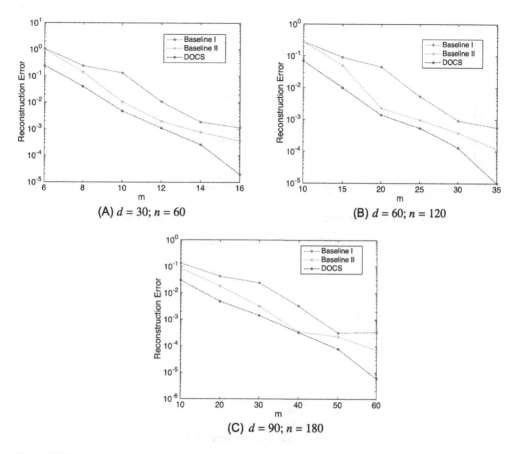

Figure 6.2 Reconstruction error w.r.t. the number of measurements *m*.

Efficiency

We compare the *running time* of different models during testing, all of which are collected in the CPU mode. The decoders of DOCS recover the original signals in no more than 1 ms, in all above test cases. It is almost **1000 times faster** than the iterative recovery algorithms [25] in Baselines I and II. The speedup achieved by DOCS is not solely because of the different implementations. DOCS relies on only a *fixed* number of the simplest matrix multiplications, whose computational complexity grows strictly linearly with the number of measurements *m*, given that *n*, *d* and *K* are fixed. Besides, the feed-forward nature makes DOCS amenable to GPU parallelization.

We further evaluate how the reconstruction errors, testing and training times vary with different *K* values, for DOCS in the specific setting: $m = 60$; $d = 90$; $n = 180$; $T = 30$, as demonstrated in Fig. 6.3. With end-to-end tuning, Fig. 6.3A shows that smaller *K* values, such as 2 and 3, perform sufficiently well with reduced complexi-

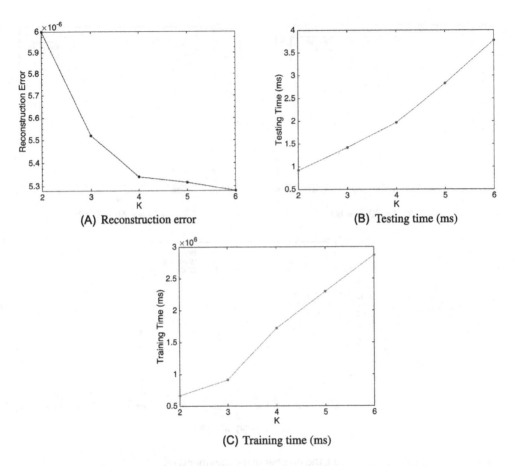

(A) Reconstruction error

(B) Testing time (ms)

(C) Training time (ms)

Figure 6.3 Reconstruction error, testing time, and training time (ms) w.r.t. K.

ties. Fig. 6.3B depicts the actual testing time that grows nearly linearly with K, which conforms to the model complexity.

Experiments on Image Reconstruction

We use the disjoint training set (200 images) and test set (200 images) of BSD500 database [27] as our training set; its validation set (100 images) is used for validation. During training, we first divide each original image into overlapped 8×8 patches as the input x, e.g., $d = 64$. Parameter n is fixed as 128, while $m = [4 : 4 : 16]$. For a testing image, we sample 8×8 blocks with a stride of 4, and apply models in a patch-wise manner. The final result is obtained via aggregating all patches, with the overlapping regions averaged. We use the 29 images in the LIVE1 dataset [28] (converted to the gray scale) to evaluate both the quantitative and qualitative performances. Parameter K is chosen to be 3.

Table 6.1 Averaged PSNR comparison (dB) on the LIVE1 dataset

m	CNN	SDA	DOCS	m	CNN	SDA	DOCS
4	20.76	20.62	**22.66**	12	24.62	24.40	**27.58**
8	23.34	23.84	**26.43**	16	24.86	25.42	**29.05**

(A) CNN(PSNR = 24.06 dB) (B) SDA(PSNR = 26.13 dB) (C) DOCS(PSNR = 27.18 dB)

Figure 6.4 Visual comparison of various methods on *Parrots* at $m = 8$. PSNR values are also shown.

We implement a CNN baseline based on [13],[6] and a SDA baseline following [12].[7] Both of them only optimize Stages III and IV. They are fed with $Y = PX$ and try to reconstruct X, where the random P is the same for training and testing. On the other hand, DOCS jointly optimizes Stages I to IV from end to end. All comparison methods are **not** postprocessed. The training details of CNN and SDA remain the same as their original sections, except for those specified above. Table 6.1 compares the averaged PSNR results on the LIVE1 dataset. We observe that ReconNet performs not well, possibly because its CNN-based architecture favors larger d, and also due to the absence of BM3D postprocessing. SDA obtains a better reconstruction performance than ReconNet when m grows larger. DOCS gains a 4–5 dB margin over ReconNet or SDA. Figs. 6.4 and 6.5 further compare two groups of reconstructed images visually. DOCS is able to retain more subtle features, such as the fine textures on the Monarch wings, with suppressed artifacts. In contrast, both CNN and SDA introduce many undesirable blurs and distortions.

6.1.5 Conclusion

We discuss the DOCS model in the section, which enhances CS in the context of deep learning. Our methodology could be extended to derive the deep feed-forward network solutions to many less tractable optimization problems, with both efficiency and

---

[6] We adopt the feature map size of 8×8, and hereby replace the 7×7 and 11×11 filters with 3×3 filters.
[7] We construct a 3-layer SDA, which are of the same dimensions as the three trainable layers of LISTA.

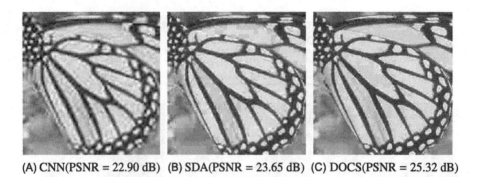

(A) CNN(PSNR = 22.90 dB) (B) SDA(PSNR = 23.65 dB) (C) DOCS(PSNR = 25.32 dB)

Figure 6.5 Visual comparison of various methods on *Monarch* at $m = 8$. PSNR values are also shown.

performance gains. We advocate such task-specific design of deep networks to be considered for a wider variety of applications. Our immediate future works concern how to incorporate more domain-specific priors of CS, such as the incoherence between P and D.

6.2. DEEP LEARNING FOR SPEECH DENOISING[8]

6.2.1 Introduction

The goal of speech denoising is to produce noise-free speech signals from noisy recordings, while improving the perceived quality of the speech component and increasing its intelligibility. Here we investigate the problem of monaural source speech denoising. It is a challenging and ill-posed problem since given only one single channel of information available, an infinite number of solutions are possible. Speech denoising can be utilized in various applications where we experience the presence of background noise in communications. The accuracy of automatic speech recognition (ASR) can be enhanced by speech denoising. A number of techniques have been proposed based on different assumptions on the signal and noise characteristics, including spectral subtraction [29] statistical model-based estimation [30], Wiener filtering [31], subspace method [32] and non-negative matrix factorization (NMF) [33]. In this section we introduce a lightweight learning-based approach to remove noise from single-channel recordings using a deep neural network structure.

Neural networks as a nonlinear filter have been applied to this problem in the past, for example, the early work by [34] utilizing shallow neural networks (SNNs) for speech denoising. However, at that time constraints in computational power and size of training

---

[8] ©Reprinted, with permission, from Liu, Ding Liu, Smaragdis, Paris, and Kim, Minje. "Experiments on deep learning for speech denoising." Fifteenth Annual Conference of the International Speech Communication Association (2014).

data resulted in relatively small neural network implementations that limited denoising performance.

Over the last few years, the development of computer hardware and advanced machine learning algorithms enabled people to increase the depth and width of neural networks. The deep neural networks (DNNs) have achieved many state-of-the-art results in the field of speech recognition [35] and speech separation [36]. DNNs containing multiple hidden layers of nonlinearity have shown great potential to better capture the complex relationships between noisy and clean utterances across various speakers, noise types and noise levels. More recently, Xu et al. [37] proposed a regression-based speech enhancement framework of DNNs using restricted Boltzmann machines (RBMs) for pretraining.

In this section we explore the use of DNNs for speech denoising, and propose a simpler training and denoising procedure that does not necessitate RBM pretraining or complex recurrent structures. We use a DNN that operates on the spectral domain of speech signals, and predicts clean speech spectra when presented with noisy input spectra. A series of experiments is conducted to compare the denoising performance under different parameter settings. Our results show that our simplified approach can perform better than other popular supervised single-channel denoising approaches and that it results in a very efficient processing model which forgoes computationally costly estimation steps.

6.2.2 Neural Networks for Spectral Denoising

In the following sections we introduce our model's structure, some domain-specific choices that we make, and a training procedure optimized for this task.

6.2.2.1 Network Architecture

The core concept in this section is to compute a regression between a noisy signal frame and a clean signal frame in the frequency domain. To do so we start with the obvious choice of using frames from a magnitude short-time Fourier transform (STFT). Using these features allows us to abstract many of the phase uncertainties and to focus on "turning off" parts of the input spectral frames that are purely noise [34].

More precisely, for a speech signal $s(t)$ and a noise signal $n(t)$ we construct a corresponding mixture signal $m(t) = s(t) + n(t)$. We compute the STFTs of the above time series to obtain the vectors s_t, n_t and m_t, which are the spectral frames corresponding to time t (each element of these vectors corresponds to a frequency bin). These vectors will constitute our training data set, with m_t being the input and its corresponding s_t being the target output.

We then proceed to design a neural network with L layers which would output a spectral frame prediction y_t when it is presented with $\|m_t\|$. This is akin to a Denoising Autoencoder (DAE) [38], although in this case we do not care to find an efficient hidden representation, instead we strive to predict the spectra of a clean signal when provided with the spectra of a noisy signal. The runtime denoising process is defined by

$$h_t^{(l)} = f_l\left(W^{(l)} \cdot h_t^{(l-1)} + b^{(l)}\right) \tag{6.5}$$

with l signifying the layer index (from 1 to L), and with $h_t^{(0)} = \|m_t\|$ and $y_t = h_t^{(L)}$. The function $f_l(\cdot)$ is known as the activation function and can take various forms depending on our goals, but it is traditionally a sigmoid or some piecewise linear function. We will explore this selection in a later section. Likewise the number of layers L can range from 1 (which forms a shallow network), or as many as we deem necessary (which comes with a higher computational burden and the need for more training data).

For $L = 1$ and $f_l(\cdot)$ being the identity function, this model collapses to a linear regression, whereas when using nonlinear $f_l(\cdot)$'s and multiple layers we perform a deep nonlinear regression (or a regression deep neural network).

6.2.2.2 Implementation Details

The parameters that need to be estimated in order to obtain a functioning system are the set of $W^{(l)}$ matrices and $b^{(l)}$ vectors, known as the layer weights and biases, respectively. Fixed parameters that we will not learn include the number of layers L and the choice of activation functions $f_l(\cdot)$. In order to perform training, we need to specify a cost function between the network predictions and the target outputs which we will need to optimize, and that will provide a means to see how well our model has adapted to the training data.

Activation Function

For the activation function the most common choices are the hyperbolic tangent and the logistic sigmoid function. However, we note that the outputs that we wish to predict are spectral magnitude values which would lie in the interval $[0, \infty)$. This means that we should prefer an activation function that produces outputs in that interval. A popular choice that satisfies this preference is the rectified linear activation, which is defined as $y = \sup\{x, 0\}$, i.e., the maximum between the input and 0. In our experience, however, this is a particularly difficult function to work with since it exhibits a zero derivative for negative values and is very likely to result in nodes that get "stuck" with a zero output once they reach that state. Instead we use a modified version which is defined as

$$f(x) = \begin{cases} x & \text{if } x \geq \epsilon, \\ \frac{-\epsilon}{x-1-\epsilon} & \text{if } x < \epsilon, \end{cases}$$

where ϵ is a sufficiently small number (in our simulations set to 10^{-5}). This modification introduces a slight ramp starting from $-\infty$ to ϵ, which guarantees that the derivative will point (albeit weakly) towards positive values and will provide a way to escape a zero state once a node is in it.

Cost Function

For the cost function we select the mean squared error (MSE) between the target and predicted vectors, $E \propto \left\| \boldsymbol{y}_t - \| \boldsymbol{s}_t \| \right\|^2$. Although a choice such as the KL divergence or the Itakura–Saito divergence would have been more appropriate for measuring differences between spectra, in our experiments we find them to ultimately perform worse than the MSE.

Training Strategy

Once the above network characteristics have been specified, we can use a variety of methods to estimate the model parameters. Traditional choices include the backpropagation algorithm, as well as more sophisticated procedures such as conjugate gradient methods and optimization approaches such as Levenberg–Marquardt [39]. Additionally, there is a trend towards including a pretraining step using an RBM analogy for each layer [40]. In our experiments for this specific task, we find many of the sophisticated approaches to be either numerically unstable, computationally too expensive, or plainly redundant. We obtain the most rapid and reliable convergence behavior using the resilient back-propagation algorithm [41]. Combined with the use of the modified activation function that we present above, it requires no pretraining and converges in roughly the same number of iterations as conjugate gradient algorithms with far fewer computational requirements. The initial parameter values are set using the Nguyen– Widrow procedure [42]. For most of the experiments we train our models for 1000 iterations which are usually sufficient to achieve convergence. The details regarding the training data are discussed in the experimental results section.

6.2.2.3 Extracting Denoised Signals

After training a model, the denoising is performed as follows: the magnitude spectral frames from noisy speech signals are extracted and presented as inputs. If the model is properly trained, we obtain a prediction of the clean signal's magnitude spectrum for each noisy spectrum that we analyze. In order to invert that magnitude spectrum back to the time domain, we apply the phase of the mixture spectrum on it and we use the inverse STFT with overlap-add to synthesize the denoised signal in the time domain. For all our experiments we use a square-root Hann window for both the analysis and synthesis transforms, and a hop size of 25% of the Fourier window length.

6.2.2.4 Dealing with Gain

One potential problem with this scheme is that this network might not be able to extrapolate when presented with data at significantly large scales (e.g., 10x louder).

When using large data sets, there is a high probability that we will see enough spectra at various low gains to adequately perform regression at lower scales, but we will not observe spectra louder than some threshold which means that we will not be able to denoise very loud signals. One approach is to standardize the gain of the involved spectra to lie inside a specific range, but we can instead employ some simple modifications to help us extrapolate better.

In order to do so we perform the following steps. We first normalize all the input and output spectra to have the same ℓ_1-norm (we arbitrarily choose unit norm). In the training process we add one more output node that is trained to predict the output gain of the speech signal. The target output gain values are also normalized to have unit variance over an utterance in order to impose invariance on the scale of the desired output signal. With this modification, in order to obtain the spectrum of the denoised signal we would have to multiply the output of that gain node with the speech spectrum predicted from all the other nodes. Because of the normalization on the predicted gain we will not recover the clean input signal with the exact gain, but rather a denoised signal that has roughly the same amplitude modulation with a constant scaling factor. In the next section we show how this method compares to simply training on unnormalized spectra.

6.2.3 Experimental Results

We now present the results of experiments that explore the effects of relevant signal and network parameters, as well as the degradation in performance when the training data set does not adequately represent the testing data.

6.2.3.1 Experimental Setup

The experiments are set up using the following recipe. We use 100 utterances from the TIMIT database, spanning ten different speakers. We also maintain a set of five noises specified as: Airport, Train, Subway, Babble, and Drill. We then generate a number of noisy speech recordings by selecting random subsets of noise and overlaying them with speech signals. While constructing the noisy mixtures we also specify the signal to noise ratio for each recording. Once we complete the generation of the noisy signals we split them into a training set and a test set.

During the denoising process we can specify multiple parameters that have a direct effect on separation quality and are linked to the network's structure. In this section we present the subset that we find to be most important. These include the number of input nodes, the number of hidden layers and the number of their nodes, the activation functions, and the number of prior input frames to take into account.

Of course, the number of parameters is quite large and considering all the possible combinations is an intractable task. In the following experiments we perform single parameter searches while keeping the rest of the parameters fixed in a set of sensible choices according to our observations. The fixed parameters are: input frame size 1024 pts, a single hidden layer with 2000 units, the rectified linear activation with the modification described above, 0 dB SNR inputs, no input normalization, and no temporal memory.

For all parameter sweeps we show the resulting signal-to-distortion ratio (SDR), signal-to-interference ratio (SIR) and signal-to-artifacts ratio (SAR) as computed from the BSS-EVAL toolbox [43]. We additionally compute the short-time objective intelligibility measure (STOI) which is a quantitative estimate of the intelligibility of the denoised speech [44]. For all these measures higher values are better.

6.2.3.2 Network Structure Analysis

In this section we present the effects of the network's structure on performance. We focus on four parameters that we find to be the most crucial, namely input window size, number of layers, activation function and temporal memory.

The number of input nodes is directly related to the size of the analysis window that we use, which is the same as the size of the FFT that we use to transform the time domain data to the frequency domain. In Fig. 6.6 we show the effects of different window sizes. We see that a window of about 64 ms (1024 pts) produces the best result.

Another important parameter is that of the depth and width of the network, i.e., the number of hidden layers and their corresponding nodes. In Fig. 6.7 we show the results over various settings ranging from a simple shallow network to a two-hidden layer network with 2000 nodes per layer. We note that with more units we tend to see an increase in the SIR, but that this trend stops after a while. It is not clear if this is an effect that relates to the number of training data points that we use or not.

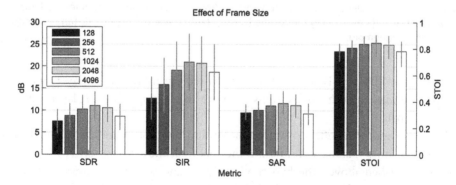

Figure 6.6 Comparing different input FFT sizes we see that for speech signals sampled at 16 kHz we obtain the optimal results with 1024 pts. As with all figures in this section, the bars show average values and the vertical lines on the bars denote minimum and maximum observed values from our experiments.

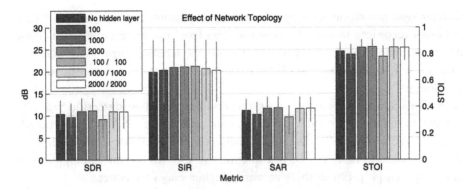

Figure 6.7 Comparing different network structures we see that a single hidden layer with 2000 units seems to perform best. Entries corresponding to a single legend number denote a single hidden layer with that many hidden units. Entries corresponding to two legend numbers denote a two hidden layer network with the two numbers being the units in the first and second hidden layer, respectively.

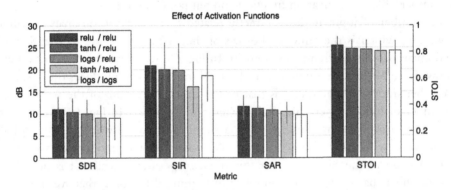

Figure 6.8 Comparing different activation functions we see that the rectified linear activation outperforms other common functions. The legend entries show the activation function for the hidden and the output layer, with "relu" being the rectified linear, "tanh" being the hyperbolic tangent and "logs" being the logistic sigmoid.

Regardless the SDR, SAR and STOI seem to require more hidden layers with more units. Consolidating both observations we note that a single hidden layer with 2000 units is optimal.

We also examine the effect of various activation functions with the results shown in Fig. 6.8. The ones that we consider are the rectified linear activation (with the modifications described above), the hyperbolic tangent and the logistic sigmoid functions. For all cases it seems that the modified rectified linear activation is consistently the best performer.

Finally, we examine the effects of a convolutive structure on the input as shown in Fig. 6.9. We do so using a model that receives as input the current analysis window

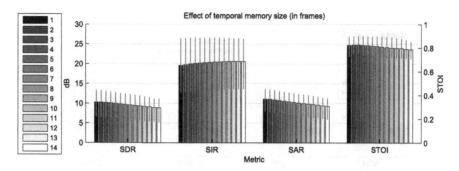

Figure 6.9 Using a convolutive form that takes into account prior input frames, we note that although SIR performance increases as we include more past frames there is an overall degradation in quality after more than two frames.

as well as an arbitrary number of past windows. The number of past windows ranges from 0 to 14 in our experiments. We observe a familiar pattern in the measured results, where the SIR improves at the expense of a diminishing SDR/SAR/STOI. Overall we conclude that the input of two consecutive frames is a good choice, although even a simple memoryless model would perform reasonably well enough.

6.2.3.3 Analysis of Robustness to Variations

In order to evaluate the robustness of this model, we test it under a variety of situations in which it is presented with unseen data, such as unseen SNRs, speakers and noise types.

In Fig. 6.10 we show the robustness of this model under various SNRs. The model is trained on 0 dB SNR mixtures and it is evaluated on mixtures ranging from 20 dB SNR to −18 dB SNR. We additionally test both the method to train on the raw input data and the method using the gain prediction model described above. In Fig. 6.10 these two methods are compared with the use of the front and back bars. Note that the shown values are absolute, not the improvement from the input mixture. As we see, for positive SNRs we get a much improved SIR and a relatively constant SDR/SIR/STOI, and training on the raw inputs seems to work better. For negative SNRs we still get an improvement although it is not as drastic as before. We also note that in these cases training with gain prediction tends to perform better.

Next we evaluate this method's robustness to data that is unseen in the training process. These tests provide a glimpse of how well we can expect this approach to work when applied on noise and speakers on which it is not trained. We perform three experiments for this, one where the testing noise is not seen in training, one where the testing speaker is not seen in training, and one where both the testing noise and the testing speaker are not seen in training. For the unseen noise case we train the model on mixtures with Babble, Airport, Train and Subway noises, and evaluate it on mixtures that include a Drill noise (which is significantly different from the training noises in both

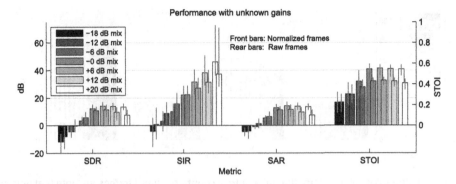

Figure 6.10 Using multiple SNR inputs and testing on a network that is trained on 0 dB SNR. Note that the results are absolute, i.e., we do not show the improvement. All results are shown using pairs of bars. The left/back bars in each pair show the results when we train on raw data, and the right/front bars show the results when we do the gain prediction.

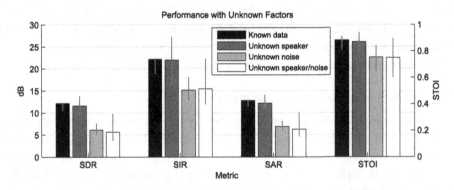

Figure 6.11 In this figure we compare the performance of our network when used on data that is not represented in training. We show the results of separation with known speakers and noise, with unseen speakers, with unseen noise, and with unseen speakers and noise.

spectral and temporal structure). For the unknown speaker case we simply hold out from the training data some of the speakers, and for the case where both the noise and the speaker are unseen we use a combination of the above. The results of these experiments are shown in Fig. 6.11. For the case where the speaker is unknown we see only a mild degradation in performance, which means that this approach can be easily used in speaker variant situations. With the unseen noise we observe a larger degradation in results, which is expected due to the drastically different nature of the noise type. Even then, the result is still good enough as compared to other single-channel denoising approaches. The result of the case where both the noise and the speaker are unknown seems to be at the same level as that of the case of the unseen noise, which once again reaffirms our conclusion that this approach is very good at generalizing across speakers.

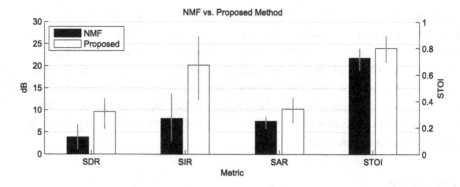

Figure 6.12 Comparison of the proposed approach with NMF-based denoising.

6.2.3.4 Comparison with NMF

We present one more plot that shows how this approach compares to another popular supervised single-channel denoising approach. In Fig. 6.12 we compare our performance to a nonnegative matrix factorization (NMF) model trained on the speakers and noise at hand [33]. For the NMF model we use what we find to be the optimal number of basis functions for this task. It is clear that our proposed method significantly outperforms this approach.

Based on the above experiments we can draw a series of conclusions. Primarily we see that this approach is a viable one, being adequately robust to unseen mixing situations (both with gains and types of sources). We also see that a deep or convolutive structure is not crucial, although it does offer a minor performance advantage. In terms of activation functions we note that the rectified linear activation seems to perform the best. Our proposed approach provides a very efficient runtime denoising process which is comprised of only a linear transform on the size of the input frame followed by a max operation. This brings our approach in the same level of computational complexity as spectral subtraction, while offering a significant advantage in denoising performance. Unlike methods such as NMF-based denoising there is no estimation performed at runtime which makes for a significantly more lightweight process.

Of course, our experiments are not exhaustive, but they do provide some guidelines on what structure to use to achieve good denoising results. We expect that with further experiments measuring many more of the available options, in both training and postprocessing, we can achieve even better performance.

6.2.4 Conclusion and Future Work

We build a deep neural network to learn the mapping between the noisy speech signal to its clean counterpart, and conduct a series of experiments to investigate its usefulness. Our proposed model demonstrates clear advantage over the NMF competitor.

Speech denoising is one problem of monaural source separation, i.e., source separation from monaural recordings, which includes other related tasks, such as speech separation and singing voice separation. Monaural source separation is important for a number of real world applications. For example, in the singing voice separation, extracting the singing voice from the music accompaniment can enhance the accuracy of chord recognition and pitch estimation. In the future, we may make use of the connection between speech denoising and other related tasks, and adopt similar deep learning-based approaches to solve these tasks.

REFERENCES

[1] Candès EJ, Romberg J, Tao T. Robust uncertainty principles: exact signal reconstruction from highly incomplete frequency information. Information Theory, IEEE Transactions on 2006;52(2):489–509.

[2] Candès EJ, Wakin MB. An introduction to compressive sampling. Signal Processing Magazine, IEEE 2008;25(2):21–30.

[3] Wakin MB, Laska JN, Duarte MF, Baron D, Sarvotham S, Takhar D, et al. An architecture for compressive imaging. In: ICIP. IEEE; 2006.

[4] Gribonval R, Nielsen M. Sparse representations in unions of bases. Information Theory, IEEE Transactions on 2003;49(12):3320–5.

[5] Elad M. Optimized projections for compressed sensing. IEEE TIP 2007.

[6] Xu J, Pi Y, Cao Z. Optimized projection matrix for compressive sensing. EURASIP Journal on Advances in Signal Processing 2010;2010:43.

[7] Duarte-Carvajalino JM, Sapiro G. Learning to sense sparse signals: simultaneous sensing matrix and sparsifying dictionary optimization. IEEE TIP 2009;18(7):1395–408.

[8] Lin Z, Lu C, Li H. Optimized projections for compressed sensing via direct mutual coherence minimization. arXiv preprint arXiv:1508.03117, 2015.

[9] Welch LR. Lower bounds on the maximum cross correlation of signals (corresp.). Information Theory, IEEE Transactions on 1974;20(3):397–9.

[10] Candes EJ, Tao T. Near-optimal signal recovery from random projections: universal encoding strategies? Information Theory, IEEE Transactions on 2006;52(12):5406–25.

[11] Beck A, Teboulle M. A fast iterative shrinkage-thresholding algorithm for linear inverse problems. SIAM Journal on Imaging Sciences 2009;2(1):183–202.

[12] Mousavi A, Patel AB, Baraniuk RG. A deep learning approach to structured signal recovery. arXiv preprint arXiv:1508.04065, 2015.

[13] Kulkarni K, Lohit S, Turaga P, Kerviche R, Ashok A. ReconNet: non-iterative reconstruction of images from compressively sensed random measurements. arXiv preprint arXiv:1601.06892, 2016.

[14] Mairal J, Bach F, Ponce J. Task-driven dictionary learning. IEEE TPAMI 2012.

[15] Wang Z, Yang Y, Chang S, Li J, Fong S, Huang TS. A joint optimization framework of sparse coding and discriminative clustering. In: IJCAI; 2015.

[16] Gregor K, LeCun Y. Learning fast approximations of sparse coding. In: ICML; 2010. p. 399–406.

[17] Iliadis M, Spinoulas L, Katsaggelos AK. Deep fully-connected networks for video compressive sensing. arXiv preprint arXiv:1603.04930, 2016.

[18] Iliadis M, Spinoulas L, Katsaggelos AK. DeepBinaryMask: learning a binary mask for video compressive sensing. arXiv preprint arXiv:1607.03343, 2016.

[19] Sprechmann P, Litman R, Yakar TB, Bronstein AM, Sapiro G. Supervised sparse analysis and synthesis operators. In: NIPS; 2013.

[20] Sprechmann P, Bronstein A, Sapiro G. Learning efficient sparse and low rank models. IEEE TPAMI 2015.

[21] Wang Z, Ling Q, Huang T. Learning deep ℓ_0 encoders. AAAI 2016.

[22] Wang Z, Chang S, Zhou J, Wang M, Huang TS. Learning a task-specific deep architecture for clustering. SDM 2016.

[23] Wang Z, Yang Y, Chang S, Ling Q, Huang T. Learning a deep ℓ_∞ encoder for hashing. IJCAI 2016.

[24] Krizhevsky A, Sutskever I, Hinton GE. ImageNet classification with deep convolutional neural networks. In: NIPS; 2012.

[25] Li C, Yin W, Jiang H, Zhang Y. An efficient augmented lagrangian method with applications to total variation minimization. Computational Optimization and Applications 2013;56(3):507–30.

[26] Aharon M, Elad M, Bruckstein A. K-SVD: an algorithm for designing overcomplete dictionaries for sparse representation. IEEE TSP 2006;54(11):4311–22.

[27] Arbelaez P, Maire M, Fowlkes C, Malik J. Contour detection and hierarchical image segmentation. IEEE TPAMI 2011;33(5):898–916.

[28] Sheikh HR, Wang Z, Cormack L, Bovik AC. Live image quality assessment database release 2. 2005.

[29] Boll S. Suppression of acoustic noise in speech using spectral subtraction. Acoustics, Speech and Signal Processing, IEEE Transactions on 1979;27(2):113–20.

[30] Ephraim Y, Malah D. Speech enhancement using a minimum-mean square error short-time spectral amplitude estimator. Acoustics, Speech and Signal Processing, IEEE Transactions on 1984;32(6):1109–21.

[31] Scalart P, et al. Speech enhancement based on a priori signal to noise estimation. In: Acoustics, speech, and signal processing, 1996. ICASSP-96. Conference proceedings., 1996 IEEE international conference on. IEEE; 1996. p. 629–32.

[32] Ephraim Y, Van Trees HL. A signal subspace approach for speech enhancement. Speech and Audio Processing, IEEE Transactions on 1995;3(4):251–66.

[33] Wilson KW, Raj B, Smaragdis P, Divakaran A. Speech denoising using nonnegative matrix factorization with priors. In: ICASSP; 2008. p. 4029–32.

[34] Wan EA, Nelson AT. Networks for speech enhancement. In: Handbook of neural networks for speech processing. Boston, USA: Artech House; 1999.

[35] Hinton G, Deng L, Yu D, Dahl GE, Mohamed Ar, Jaitly N, et al. Deep neural networks for acoustic modeling in speech recognition: the shared views of four research groups. Signal Processing Magazine, IEEE 2012;29(6):82–97.

[36] Huang PS, Kim M, Hasegawa-Johnson M, Smaragdis P. Deep learning for monaural speech separation. In: ICASSP. p. 1562–6; 2014.

[37] Xu Y, Du J, Dai L, Lee C. An experimental study on speech enhancement based on deep neural networks. IEEE Signal Processing Letters 2014;21(1).

[38] Vincent P, Larochelle H, Bengio Y, Manzagol PA. Extracting and composing robust features with denoising autoencoders. In: Proceedings of the 25th international conference on machine learning. ACM; 2008. p. 1096–103.

[39] Haykin SS. Neural networks and learning machines, vol. 3. Pearson; 2009.

[40] Erhan D, Bengio Y, Courville A, Manzagol PA, Vincent P, Bengio S. Why does unsupervised pre-training help deep learning? The Journal of Machine Learning Research 2010;11:625–60.

[41] Riedmiller M, Braun H. A direct adaptive method for faster backpropagation learning: the RPROP algorithm. In: IEEE international conference on neural networks; 1993. p. 586–91.

[42] Nguyen D, Widrow B. Improving the learning speed of 2-layer neural networks by choosing initial values of the adaptive weights. In: Neural networks, 1990., 1990 IJCNN international joint conference on, vol. 3; 1990. p. 21–6.

[43] Févotte C, Gribonval R, Vincent E. BSS Eval, a toolbox for performance measurement in (blind) source separation. URL: http://bass-db.gforge.inria.fr/bss_eval, 2010.

[44] Taal CH, Hendriks RC, Heusdens R, Jensen J. An algorithm for intelligibility prediction of time–frequency weighted noisy speech. Audio, Speech, and Language Processing, IEEE Transactions on 2011;19(7):2125–36.

CHAPTER 7

Dimensionality Reduction

Shuyang Wang*, Zhangyang Wang†, Yun Fu‡

*Department of Electrical and Computer Engineering, Northeastern University, Boston, MA, United States
†Department of Computer Science and Engineering, Texas A&M University, College Station, TX, United States
‡Department of Electrical and Computer Engineering and College of Computer and Information Science (Affiliated), Northeastern University, Boston, MA, United States

Contents

7.1. Marginalized Denoising Dictionary Learning with Locality Constraint	143
7.1.1 Introduction	143
7.1.2 Related Works	145
7.1.3 Marginalized Denoising Dictionary Learning with Locality Constraint	147
7.1.4 Experiments	157
7.1.5 Conclusion	164
7.1.6 Future Works	165
7.2. Learning a Deep ℓ_∞ Encoder for Hashing	165
7.2.1 Introduction	166
7.2.2 ADMM Algorithm	168
7.2.3 Deep ℓ_∞ Encoder	168
7.2.4 Deep ℓ_∞ Siamese Network for Hashing	170
7.2.5 Experiments in Image Hashing	172
7.2.6 Conclusion	178
References	178

7.1. MARGINALIZED DENOISING DICTIONARY LEARNING WITH LOCALITY CONSTRAINT[1]

7.1.1 Introduction

Learning good representation for images is always a hot topic in machine learning and pattern recognition fields. Among the numerous algorithms, dictionary learning is a well-known strategy for effective feature extraction. Recently, more discriminative sub-dictionaries have been built by Fisher discriminative dictionary learning with specific class labels. Different types of constraints, such as sparsity, low-rankness and locality, are also exploited to make use of global and local information. On the other hand, as the

---

[1] ©2017 IEEE. Reprinted, with permission, from Wang, Shuyang, Zhengming Ding, and Yun Fu. "Marginalized denoising dictionary learning with locality constraint." IEEE Transactions on Image Processing 27.1 (2018): 500–510.

basic building block of a deep structure, the auto-encoder has demonstrated its promising performance in extracting new feature representation. In this section, a unified feature learning framework by incorporating the marginalized denoising auto-encoder into a locality-constrained dictionary learning scheme, named Marginalized Denoising Dictionary Learning (MDDL), will be introduced [1]. Overall, the introduced method deploys a low-rank constraint on each sub-dictionary and locality constraint instead of sparsity on coefficients, in order to learn more concise and pure feature spaces while inheriting the discrimination from sub-dictionary learning. Experimental results listed in this section demonstrate the effectiveness and efficiency of MDDL by comparing with several state-of-the-art methods.

Sparse representation has experienced rapid growth in both theory and application from recent researches and led to interesting results in image classification [2–4], speech denoising [5], and bioinformatics [6], etc. For each input signal, the key idea is to find a linear combination using atoms from a given over-complete dictionary D as a new representation. Therefore, sparse representation is capable of revealing the underlying structure of high dimensional data. Among the large range of problems sparse representations has been applied to, in this section we focus on image classification which demands a discriminative representation.

The generative and discriminative ability of dictionary, apparently, is a major factor for sparse representation. Directly using the original training samples as dictionary in [7] will raise a problem that the noise and ambiguity contained in the training set could impede the test sample being faithfully represented. In addition, the discerning information hidden behind the training samples will be ignored in this strategy. Actually, the aforementioned problem can be solved through dictionary learning which intends to learn a proper dictionary from the original training samples. After the training process is finished, a new coming signal can be well represented by a set of basis learned from the original training set. Solutions to problems like face recognition have been dramatically improved with a well-learned dictionary. In order to distinctively represent the test samples, a lot of research efforts have been made to seek a well-adapted dictionary. Recently, a discriminative constraint was added to the dictionary learning model based on K-SVD [8], in which the classification error was considered in order to gain discriminability [9]. In Jiang et al.'s paper, the discerning ability of dictionary is enforced by associating label information with each dictionary atom [10]. In order to learn a structured dictionary, the Fisher criterion was introduced to learn a set of sub-dictionaries with specific class labels [11]. The above algorithms are designed to deal with clear signals or those corrupted only by small noise. For the situation in which training samples are corrupted by large noise, the learned dictionary will include corruptions resulting in a failure to represent the test samples.

7.1.2 Related Works

In this section, we mainly discuss two lines of related works, namely dictionary learning and auto-encoder.

7.1.2.1 Dictionary Learning

Recent researches on dictionary learning have demonstrated that a well-learned dictionary will greatly boost the performance by yielding a better representation in human action recognition [21], scene categorization [22], and image coloration [23].

A sparse representation based classification (SRC) method for robust face recognition was proposed by Wright et al. Suppose we have c classes of a training set $X = [X_1, X_2, \ldots, X_c]$ where $X_i \in \mathbb{R}^{d \times n_i}$ is the training sample from the ith class with dimension d and n_i samples. The SRC procedure is specified by two phases to classify a given test sample x_{test}. First, in the coding phase, we solve the following l_1-norm minimization problem:

$$\bar{a} = \arg\min_a \|x_{\text{test}} - Xa\|_2^2 + \lambda\|a\|_1, \tag{7.1}$$

in which the l_1-norm is used as the convex envelope to replace the l_0-norm in order to avoid an NP-hard problem and while keeping the sparsity. Then the classification is done via

$$\text{identity}(x_{\text{test}}) = \arg\min_i \varepsilon_i, \tag{7.2}$$

where $\varepsilon_i = \|x_{\text{test}} - X_i \bar{a}_i\|_2$, and $\bar{a}_i$ is the coefficient vector associated with class i. SRC classifies a test image by picking the smallest reconstruction error ε_i.

Instead of directly using the training set itself as a dictionary, several algorithms and regularizations have been introduced into the dictionary learning framework to learn a compact dictionary with more representation power. In FDDL, a set of class-specified sub-dictionaries whose atoms correspond to the class labels are updated iteratively based on the Fisher discrimination criterion to include discriminative information. Jiang et al. presented a Label Consistent K-SVD (LC-KSVD) algorithm to make a learned dictionary more discriminative for sparse coding [10]. These methods have shown that a structured dictionary could dramatically improve classification performance. However, the performance of these methods will drop a lot if the training data is largely corrupted.

Recently introduced low-rank dictionary learning [12] aims to uncover the global structure by grouping similar samples into one cluster, which has been successfully applied to many applications, e.g., object detection [13], multi-view learning [14], unsupervised subspace segmentation [15], and 3D visual recovery [16]. The goal is to generate a low-rank matrix from corrupted original input data. That is, if a given data matrix X is corrupted by a sparse matrix E while the samples share a similar pattern, then the sparse noisy matrix can be separated to practically recover X via rank minimization. According to the previous research, many works have integrated low-rank regularization into sparse representation for separating the sparse noises from inputs signals while simultaneously optimizing the dictionary atoms in order to faithfully reconstruct the de-noised data. Moreover, a low-rank dictionary addresses the noisy data well by adding an error term with different norms, e.g., l_1-norm, $l_{2,1}$-norm.

Applying augmented Lagrange multipliers (ALM), Lin [24] proposed Robust PCA, with which the corrupted data in a single subspace can be recovered by solving a matrix rank minimization problem. In [12], Liu et al. proposed converting the task of face image clustering into a subspace segmentation problem with the assumption that face images from different individuals lie in different, nearly independent subspaces.

7.1.2.2 Auto-encoder

Most recently, deep learning has attracted a lot of interest when looking for better feature extraction. In this direction, auto-encoder [18] is one of the most popular building blocks to form a lite-version deep learning framework. The auto-encoder has drawn increasing attention in the feature learning area and has been considered as a simulation of the way the human visual system processes imagery. The auto-encoder architecture explicitly involves two modules, an encoder and a decoder. The encoder outputs a group of hidden representation units, which is realized by a linear deterministic mapping with a weight matrix and a nonlinear transformation employing a logistic sigmoid. The decoder reconstructs the input data based on the received sparse hidden representation. The aforementioned dictionary learning model can be formalized as a decoder module.

As a typical single hidden layer neural network with identical input and target, the auto-encoder [29] aims to discover data's intrinsic structure by encouraging the output to be as similar to the target as possible. Essentially, the neurons in the hidden layer can be seen as a good representation since they are able to reconstruct the input data. To encourage structured feature learning, further constraints have been imposed on the parameters during training.

Moreover, there is a well-known trick of the trade to deal with noisy data, that is, manually injecting noise into the training samples thereby learning with artificially corrupted data. Denoising auto-encoders (DAEs) [30,18], learned with artificial corrupted data as input, have been successfully applied to a wide range of machine learning tasks by learning a new denoising representation. During the training process, DAEs reconstruct the input data from partial corruption with a pre-specified corrupting distribution to its original clean version. This process learns a robust representation which ensures tolerance to certain distortions in input data.

On this basis, stacked denoising auto-encoders (SDAs) [19] have been successfully used to learn new representations and attained record accuracy on standard benchmark for domain adaptation. However, there are two crucial limitations of SDAs: (i) high computational cost, and (ii) lack of scalability to high-dimensional features. To address these two problems, the authors of [20] proposed marginalized SDAs (mSDA). Different from SDAs, mSDA marginalizes noise and thus the parameters can be computed in closed-form rather than using stochastic gradient descent or other optimization algorithms. Consequently, mSDA significantly speeds up SDAs by two orders of magnitude.

Wang et al. [28] explored the effectiveness of a locality constraint on two different types of feature learning technique, i.e., dictionary learning and auto-encoder, respectively. A locality-constrained collaborative auto-encoder (LCAE) was proposed to extract features with local information for enhancing the classification ability. In order to introduce locality into the coding procedure, the input data is first reconstructed by LLC coding criteria and then used as the target of the auto-encoder. That is, target $\hat{x}$ is replaced by a locality reconstruction as follows:

$$\min_{C} \sum_{i=1}^{N} \|x_i - Dc_i\|^2 + \lambda \|l_i \odot c_i\|^2$$
$$\text{s.t. } \mathbf{1}^T c_i = 1, \forall i, \tag{7.3}$$

where dictionary D will be initialized by PCA on the input training matrix X. The proposed LCAE can be trained using the backpropagation algorithm, which updates W and b by backpropagating the reconstruction error gradient from the output layer to the locality coded target layer.

In this chapter, the introduced model jointly learns the auto-encoder and dictionary to benefit from both techniques. To make the model fast, a lite version of the auto-encoder, i.e., marginalized denoising auto-encoder [20] is adapted, which has shown appealing performance and efficiency. Furthermore, there are several benchmarks used to evaluate the proposed algorithm, and the experimental results show its better performance compared to the state-of-the-art methods.

7.1.3 Marginalized Denoising Dictionary Learning with Locality Constraint

In this section, we first revisit locality-constrained dictionary learning and marginalized denoising auto-encoder. Then we introduce marginalized denoising dictionary learning with a locality constraint along with an efficient solution to optimize the proposed algorithm. The proposed overall framework could be viewed in Fig. 7.1.

7.1.3.1 Preliminaries and Motivations

In this section, we introduce a feature learning model by unifying the marginalized denoising auto-encoder and locality-constrained dictionary learning (MDDL) together to simultaneously benefit from their merits [1]. Specifically, dictionary learning manages handling corrupted data from the sample space, while marginalized auto-encoder at-

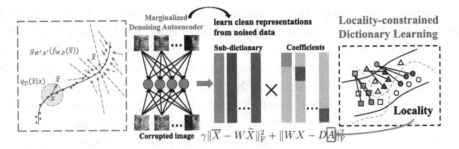

Figure 7.1 Illustration of MDDL. A marginalized denoising auto-encoder is adopted in dictionary learning (DL) schemes. The weights in the auto-encoder and sub-dictionaries in DL are trained jointly. Each sub-dictionary learned is of low-rank, which can narrow the negative effect of noise contained in training samples. For the marginalized denoising auto-encoder, the input is manually added with noise.

tempts to deal with the noisy data from the feature space. Thus, the algorithm aims to work well with the corrupted data from both the sample and feature spaces by integrating dictionary learning and auto-encoder into a unified framework. The key points of this introduced framework are listed as follows:

- The MDDL model seeks a transformation matrix to filter out the noise inside the data with a marginalized denoising auto-encoder, which avoids forward and backward propagation, thereby working both efficiently and effectively.
- Secondly, with the transformed data, the model aims to build a set of supervised sub-dictionaries with a locality constraint. In this way, the sub-dictionaries are discriminative for each class, which makes the new representation preserve the manifold structure while uncovering the global structure.
- The marginalized denoising transformation and locality-constrained dictionary are jointly learned in a unified framework. In this way, the model can integrate auto-encoders and dictionary learning to produce features with denoising ability and discriminative information.

Consider a matrix $X = [x_1, x_2, \ldots, x_n]$ having n samples. A low-rank representation tries to segment these samples by minimizing the following objective function:

$$\arg\min_{Z} \|Z\|_* \quad \text{s.t. } X = DZ, \tag{7.4}$$

where $\| \cdot \|_*$ denotes the nuclear norm and Z is the coefficient matrix. In the subspace segmentation problem, a low-rank approximation is enforced by minimizing the error between X and its low rank representation ($D = X$). By applying low-rank regularization in dictionary updating, the DLRD [25] algorithm achieved impressive results, especially when corruption exists. Jiang et al. proposed a sparse and dense hybrid representation based on a supervised low-rank dictionary decomposition to learn a class-specific dictionary and erase non-class-specific information [26]. Furthermore, supervised information has been well utilized to seek a more discriminative dictionary [11,17]. In the introduced model, supervised information is also adopted to learn multiple sub-dictionaries so that samples from the same class are drawn from one low-dimensional subspace.

Above-mentioned sparse representation based methods consider each sample as an independent sparse linear combination, this assumption fails to exploit the spatial consistency between neighboring samples. Recent research efforts have yielded more promising results on the task of classification by using the idea of locality [27]. The method named Local Coordinate Coding (LCC), which specifically encourages the coding to rely on local structure, has been presented as a modification to sparse coding. In [27] the author also theoretically proved that locality is more essential than sparsity under certain assumptions. Inspired by the above learning techniques, Wang et al. proposed a Locality-Constrained Low-Rank Dictionary Learning (LC-LRD) to enhance the identification capability by using the geometric structure information [28].

7.1.3.2 LC-LRD Revisited

Given a set of training data $X = [X_1, X_2, \ldots, X_c] \in \mathbb{R}^{d \times n}$, where d is the feature dimensionality, n is the number of total training samples, c is the number of classes, and $X_i \in \mathbb{R}^{d \times n_i}$ is the sample from class i which has n_i samples. The goal of dictionary learning is to learn an m atoms dictionary $D \in \mathbb{R}^{d \times m}$ which yields a sparse representation matrix $A \in \mathbb{R}^{m \times n}$ from X for future classification tasks. Then we can write $X = DA + E$, where E is the sparse noise matrix. Rather than learning the dictionary as a whole from all the training samples, we learn sub-dictionary D_i for the ith class separately. Then A and D could be written as $A = [A_1, A_2, \ldots, A_c]$ and $D = [D_1, D_2, \ldots, D_c]$, where A_i is the sub-matrix that denotes the coefficients for X_i over D.

In [28], Wang et al. have proposed the following LC-LRD model for each sub-dictionary:

$$
\min_{D_i, A_i, E_i} R(D_i, A_i) + \alpha \|D_i\|_* + \beta \|E_i\|_1
$$

$$
+ \lambda \sum_{k=1}^{n_i} \|l_{i,k} \odot a_{i,k}\|^2, \quad \text{s.t. } X_i = DA_i + E_i, \tag{7.5}
$$

where $R(D_i, A_i)$ is the Fisher discriminant regularization on each sub-dictionary, $\|D_i\|_*$ is the nuclear norm to enforce low-rank properties, and $\|l_{i,k} \odot a_{i,k}\|^2$ is a locality constraint to replace sparsity on the coding coefficient matrix; $a_{i,k}$ denotes the kth column in A_i, which means the coefficient for the kth sample in class i. This model was broken down into the following modules: discriminative sub-dictionaries, low-rank regularization term, and the locality constraint on the coding coefficients.

Sub-dictionary D_i should be endowed with the discrimination power to well represent samples from the ith class. Mathematically, the coding coefficients of X_i over D can be written as $A_i = [A_i^1; A_i^2; \ldots; A_i^c]$, where A_i^j is the coefficient matrix of X_i over D_j.

The discerning power of D_i is produced by following two aspects: first, it is expected that X_i should be well represented by D_i but not by D_j, $j \neq i$. Therefore, we will have to minimize $\|X_i - D_i A_i^i - \mathcal{E}_i\|_F^2$, where $\mathcal{E}_i$ is the residual. Meanwhile, D_i should not be good at representing samples from other classes, that is, each A_j^i, where $j \neq i$, should have nearly zero coefficients so that $\|D_i A_j^i\|_F^2$ is as small as possible. Thus we denote the discriminative fidelity term for sub-dictionary D_i as follows:

$$R(D_i, A_i) = \|X_i - D_i A_i^i - \mathcal{E}_i\|_F^2 + \sum_{j=1, j \neq i}^{c} \|D_i A_j^i\|_F^2. \tag{7.6}$$

In the task of image classification, the within–class samples are linearly correlated and lie in a low dimensional manifold. Therefore, we want to find the dictionary with the most concise atoms by minimizing the rank of D_i, which suggests replacing it by $\|D_i\|_*$ [31], where $\|\cdot\|_*$ denotes the nuclear norm of a matrix (i.e., the sum of its singular values).

In addition, a locality constraint is deployed on the coefficient matrix instead of the sparsity constraint. As suggested by LCC [32], locality is more essential than sparsity under certain assumptions, as locality must lead to sparsity but not necessary vice versa. Specifically, the locality constraint uses the following criterion:

$$\min_A \sum_{i=1}^{n} \|l_i \odot a_i\|^2, \text{ s.t. } \mathbf{1}^T a_i = 1, \forall i, \tag{7.7}$$

where $\odot$ denotes the element-wise multiplication, and $l_i \in \mathbb{R}^m$ is the locality adaptor which gives different freedom for each basis vector proportional to its similarity to the input sample x_i. Specifically,

$$l_i = \exp(\frac{\text{dist}(x_i, D)}{\delta}), \tag{7.8}$$

where $\text{dist}(x_i, D) = [\text{dist}(x_i, d_1), \text{dist}(x_i, d_2), \ldots, \text{dist}(x_i, d_m)]^T$, and $\text{dist}(x_i, d_j)$ is the Euclidean distance between sample x_i and the jth dictionary atom d_j; δ controls the bandwidth of the distribution.

Generally speaking, LC-LRD is based on the following three observations: (i) locality is more essential than sparsity to ensure obtain the similar representations for similar samples; (ii) each sub-dictionary should have discerning ability by introducing the discriminative term; (iii) low-rank is introduced to each sub-dictionary to separate noise from samples and discover the latent structure.

7.1.3.3 Marginalized Denoising Auto-encoder (mDA)

Consider a vector input $x \in \mathbb{R}^d$, with d as the dimensionality of the visual descriptor. There are two important transformations, which can be considered as encoder and decoder processes involved in the auto-encoder, namely "input→hidden units" and "hidden units→output", which are given by

$$h = \sigma(Wx + b_h), \quad \hat{x} = \sigma(W^T h + b_o), \tag{7.9}$$

where $h \in \mathbb{R}^z$ is the hidden representation unit, and $\hat{x} \in \mathbb{R}^d$ is interpreted as a reconstruction of input x. The parameter set includes a weight matrix $W \in \mathbb{R}^{z \times d}$, and two offset vectors $b_h \in \mathbb{R}^z$ and $b_o \in \mathbb{R}^d$ for hidden units and output, respectively; σ is a nonlinear mapping such as the sigmoid function of the form $\sigma(x) = (1 + e^{-x})^{-1}$. In general, an auto-encoder is a single layer hidden neural network, with identical input and target, meaning the auto-encoder encourages the output to be as similar to the target as possible, namely,

$$\min_{W, b_h, b_o} L(x) = \min_{W, b_h, b_o} \frac{1}{2n} \sum_{i=1}^{n} \|x_i - \hat{x}_i\|_2^2, \tag{7.10}$$

where n is the number of images, x_i is the target, and $\hat{x}_i$ is the reconstructed input. In this way, the neurons in the hidden layer can be seen as a good representation for the input, since they are able to reconstruct the data.

Since an auto-encoder deploys nonlinear functions, it takes more time to train the model, especially when the dimension of the data is very high. Recently, marginalized denoising auto-encoder (mDA) [20] was developed to address the data reconstruction in a linear fashion and achieved comparable performance with the original auto-encoder. The general idea of mDA is to learn a linear transformation matrix W to reconstruct the data with the transformation matrix by minimizing the squared reconstruction loss

$$\frac{1}{2n} \sum_{i=1}^{n} \|x_i - W\tilde{x}_i\|^2, \tag{7.11}$$

where $\tilde{x}_i$ is a corrupted version of x_i. The above objective solution is correlated to the randomly corrupted features of each input. To make the variance lower, a marginal denoising auto-encoder was proposed to minimize the overall squared loss from t different corrupted versions

$$\frac{1}{2tn} \sum_{j=1}^{t} \sum_{i=1}^{n} \|x_i - W\tilde{x}_{i,(j)}\|^2, \tag{7.12}$$

where $\tilde{x}_{i,(j)}$ denotes the jth corrupted version of the original input x_i. Define $X = [x_1, \ldots, x_n]$ and its t-times repeated version as $\overline{X} = [X, \ldots, X]$ with its t-times differently corrupted version $\tilde{X} = [\tilde{X}_{(1)}, \ldots, \tilde{X}_{(t)}]$, where $\tilde{X}_{(i)}$ denotes the ith corrupted version of X. In this way, Eq. (7.12) can be reformulated as

$$\frac{1}{2tn} \|\overline{X} - W\tilde{X}\|_F^2, \tag{7.13}$$

which has the well-known closed-form solution for ordinary least squares. When $t \rightarrow \infty$, it can be solved by the expectation, using the weak law of large numbers [20].

7.1.3.4 Proposed MDDL Model

Previous discussion on mDA gives a brief idea that, with a linear transformation matrix, mDA can be implemented in several lines of Matlab code and works very efficiently. The learned transformation matrix can well reconstruct the data and extract the noisy data.

Inspired by this, the proposed model aims at jointly learning a dictionary and a marginalized denoising transformation matrix in a unified framework. We formulate the objective function as

$$
\min_{D_i, A_i, E_i, W} \mathcal{F}(D_i, A_i, E_i) + \|\overline{X} - W\tilde{X}\|_F^2,
$$
$$
\text{s.t. } WX_i = DA_i + E_i
$$

(7.14)

where $\mathcal{F}(D_i, A_i, E_i) = R(D_i, A_i) + \alpha\|D_i\|_* + \beta_1\|E_i\|_1 + \lambda \sum_{k=1}^{n_i} \|l_{i,k} \odot a_{i,k}\|^2$ is the locality-constrained dictionary learning part in Eq. (7.5) and $R(D_i, A_i) = \|WX_i - D_iA_i^i - \mathcal{E}_i\|_F^2 + \sum_{j=1, j\neq i}^{c} \|D_iA_j^i\|_F^2$ is the discriminative term in Eq. (7.6); α, β_1, and λ are trade-off parameters.

Discussion. The proposed marginalized denoising regularized dictionary learning (MDDL) aims to learn a more discriminative dictionary on transformed data. Since the marginalized denoising regularizer could generate a better transformation matrix to address corrupted data, the dictionary could be learned on denoised clean data. The framework unifies the marginal denoising auto-encoder and locality-constrained dictionary learning together. Generally, dictionary learning seeks a well-represented basis in order to achieve more discriminative coefficients for original data. Therefore, dictionary learning can handle noisy data to some extent, while the denoising auto-encoder has demonstrated its power in many applications. To this end, the joint learning scheme can benefit from both the marginal denoising auto-encoder and locality-constrained dictionary learning.

7.1.3.5 Optimization

The proposed objective function in Eq. (7.14) could be optimized by dividing into two sub-problems: first, updating one by one each coefficient A_i $(i = 1, 2, \ldots, c)$ and W, by fixing dictionary D and all other A_j $(j \neq i)$, and then putting together to get the coding coefficient matrix A; second, updating D_i by fixing other variables. These two steps are iteratively repeated to get the discriminative low-rank sub-dictionaries D, the marginal denoising transformation W, and the locality-constrained coefficients A. One problem arises in the second sub-problem, though. Recall that in Eq. (7.6) the coefficients A_i^i corresponding to X_i over D_i should be updated to minimize the term $\|X_i - D_iA_i^i - \mathcal{E}_i\|_F^2$.

Therefore, when we update D_i in the second sub-problem, the related variance A_i^i is also updated.

Sub-problem I. Assume that the structured dictionary D is given, the coefficient matrices A_i $(i = 1, 2, \ldots, c)$ are updated one by one, then the original objective function in Eq. (7.14) reduces to the following locality-constrained coding problem for the coefficient of each class and W:

$$\min_{A_i, E_i, W} \lambda \sum_{k=1}^{n_i} \|l_{i,k} \odot a_{i,k}\|^2 + \beta_1 \|E_i\|_1 + \|\overline{X} - W\widetilde{X}\|_F^2 \tag{7.15}$$
$$\text{s.t. } WX_i = DA_i + E_i,$$

which can be solved by the following Augmented Lagrange Multiplier method [33]. We transform Eq. (7.15) into its Lagrange function as follows:

$$\min_{A_i, E_i, W, T_1} \sum_{i=1}^{c} \left(\lambda \sum_{k=1}^{n_i} \|l_{i,k} \odot a_{i,k}\|^2 + \beta_1 \|E_i\|_1 + \right.$$
$$\left. \langle T_1, WX_i - DA_i - E_i \rangle + \frac{\mu}{2} \|WX_i - DA_i - E_i\|_F^2 \right) \tag{7.16}$$
$$+ \|\overline{X} - W\widetilde{X}\|_F^2,$$

where T_1 is the Lagrange multiplier, and μ is a positive penalty parameter. Different from traditional locality-constrained linear coding (LLC) [27], MDDL adds an error term which could handle large noise in samples. In the following, we perform iterative optimization on A_i, E_i, and W.

Updating A_i:

$$A_i = \arg\min_{A_i} \frac{\mu}{2} \|Z_i - DA_i\|_F^2 + \lambda \sum_{k=1}^{n_i} \|l_{i,k} \odot a_{i,k}\|^2,$$

$$\Rightarrow A_i = \text{LLC}(Z_i, D, \lambda, \delta), \tag{7.17}$$

where $Z_i = WX_i - E_i + \frac{T_1}{\mu}$, and $l_{i,k} = \exp(\text{dist}(z_{i,k}, D)/\delta)$. Function $\text{LLC}(\cdot)$ is a locality-constrained linear coding function[2] [27].

Updating E_i:

$$E_i = \arg\min_{E_i} \frac{\beta_1}{\mu} \|E_i\|_1 + \frac{1}{2} \|E_i - (WX_i - DA_i + \frac{T_1}{\mu})\|_F^2, \tag{7.18}$$

which can be solved by the shrinkage operator [34].

---

[2] Z_i, D, λ and σ are set as the input of function LLC [27], and the code can be downloaded from http://www.ifp.illinois.edu/~jyang29/LLC.htm.

Updating W:

$$W = \arg\min_{W} \sum_{i=1}^{c} \left(\frac{\mu}{2} \| WX_i - DA_i - E_i + \frac{T_1}{\mu} \|_F^2 \right)$$
$$+ \| \overline{X} - W\widetilde{X} \|_F^2 \tag{7.19}$$
$$= \arg\min_{W} \frac{\mu}{2} \| WX - D_A \|_F^2 + \| \overline{X} - W\widetilde{X} \|_F^2,$$

where $X = [X_1, \ldots, X_c]$ and $D_A = [DA_1 + E_1 - \frac{T_{1,1}}{\mu}, \ldots, DA_c + E_c - \frac{T_{1,c}}{\mu}]$. Equation (7.19) has a well-known closed-form solution

$$W = (\mu D_A X^T + 2\overline{X}\widetilde{X}^T)(\mu XX^T + 2\widetilde{X}\widetilde{X}^T)^{-1} \tag{7.20}$$

where $\overline{X}$ is the t-times repeated version of X and $\widetilde{X}$ consists of its t-times corrupted version. We define $P = \mu D_A X^T + 2\overline{X}\widetilde{X}^T$ and $Q = \mu XX^T + 2\widetilde{X}\widetilde{X}^T$. And we would like the repetition number t to be ∞. Therefore, the denoising transformation W could be effectively learned from infinitely many copies of noisy data. Practically, we cannot generate $\widetilde{X}$ with infinitely many corrupted versions; however, matrices P and Q converge to their expectations when t becomes very large. In this way, we can derive the expected values of P and Q, and calculate the corresponding mapping W as:

$$W = \mathbb{E}[P]\mathbb{E}[Q]^{-1}$$
$$= \mathbb{E}[\mu D_A X^T + 2\overline{X}\widetilde{X}^T]\mathbb{E}[\mu XX^T + 2\widetilde{X}\widetilde{X}^T]^{-1} \tag{7.21}$$
$$= \left(\mu D_A X^T + 2\mathbb{E}[\overline{X}\widetilde{X}^T] \right)\left(\mu XX^T + 2\mathbb{E}[\widetilde{X}\widetilde{X}^T] \right)^{-1}$$

where D_A and μ are treated as constant values when optimizing W. The expectations $\mathbb{E}[\overline{X}\widetilde{X}^T]$ and $\mathbb{E}[\widetilde{X}\widetilde{X}^T]$ are easy to be computed through mDA [20].

Sub-problem II. For the procedure of sub-dictionary updating, MDDL uses the same method as in [25]. Considering the second sub-problem, when A_i is fixed, sub-dictionaries $D_i (i = 1, 2, \ldots, c)$ are updated one by one. The objective function in Eq. (7.14) is converted into the following problem:

$$\min_{D_i, \mathcal{E}_i, A_i^i} \sum_{j=1, j \neq i}^{c} \| D_i A_j^i \|_F^2 + \alpha \| D_i \|_* + \beta_2 \| \mathcal{E}_i \|_1$$
$$+ \lambda \sum_{k=1}^{n_i} \| l_{i,k}^i \odot a_{i,k}^i \|^2, \quad \text{s.t. } WX_i = D_i A_i^i + \mathcal{E}_i \tag{7.22}$$

where $a_{i,k}^i$ is the kth column in A_i^i, which means the coefficient for the kth sample in class i over D_i. Problem Eq. (7.22) can be solved using the Augmented Lagrange

Multiplier method [33] by introducing a relaxing variable J:

$$
\begin{aligned}
\min_{D_i,\mathcal{E}_i,A_i^i} & \lambda \sum_{k=1}^{n_i} \|l_{i,k}^i \odot a_{i,k}^i\|^2 + \sum_{j=1,j\neq i}^{c} \|D_i A_j^i\|_F^2 + \alpha\|J\|_* \\
& + \beta_2\|\mathcal{E}_i\|_1 + \langle T_2, WX_i - D_i A_i^i - \mathcal{E}_i\rangle + \langle T_3, D_i - J\rangle \\
& + \tfrac{\mu}{2}(\|WX_i - D_i A_i^i - \mathcal{E}_i\|_F^2 + \|D_i - J\|_F^2),
\end{aligned}
\tag{7.23}
$$

where T_2 and T_3 are Lagrange multipliers, and μ is a positive penalty parameter. In the following, we describe the iterative optimization of D_i and A_i^i.

Updating A_i^i:

Similar as for Eq. (7.17), we have the solution for A_i^i as follows:

$$
A_i^i = \mathrm{LLC}((WX_i - \mathcal{E}_i + \frac{T_2}{\mu}), D_i, \lambda, \delta),
\tag{7.24}
$$

where function $\mathrm{LLC}(\cdot)$ is a locality-constrained linear coding function [27].

Updating J and D_i:

Here we convert Eq. (7.23) to a problem for J and D_i as:

$$
\begin{aligned}
\min_{J,D_i} & \sum_{j=1,j\neq i}^{c} \|D_i A_j^i\|_F^2 + \alpha\|J\|_* \\
& + \langle T_2, WX_i - D_i A_i^i - \mathcal{E}_i\rangle + \langle T_3, D_i - J\rangle \\
& + \tfrac{\mu}{2}(\|D_i - J\|_F^2 + \|WX_i - D_i A_i^i - \mathcal{E}_i\|_F^2),
\end{aligned}
\tag{7.25}
$$

where $J = \arg\min_J \alpha\|J\|_* + \langle T_3, D_i - J\rangle + \tfrac{\mu}{2}(\|D_i - J\|_F^2)$, and the solution for D_i is:

$$
\begin{aligned}
D_i &= (J + WX_i A_i^{iT} - \mathcal{E}_i A_i^{iT} + (T_2 A_i^{iT} - T_3)/\mu) \\
& (I + A_i^i A_i^{iT} + V)^{-1}, \text{ where } V = \tfrac{2\lambda}{\mu}\sum_{j=1,j\neq i}^{c} A_j^i A_j^{iT};
\end{aligned}
\tag{7.26}
$$

Updating $\mathcal{E}_i$:

$$
\begin{aligned}
\mathcal{E}_i = \arg\min_{\mathcal{E}_i} & \beta_2\|E_i\|_1 + \langle T_2, WX_i - D_i A_i^i - \mathcal{E}_i\rangle \\
& + \tfrac{\mu}{2}\|WX_i - D_i A_i^i - \mathcal{E}_i\|_F^2,
\end{aligned}
\tag{7.27}
$$

which can be solved by the shrinkage operator [34]. The details of the optimization problem solution for the proposed model can be referred to as Algorithm 7.1.

7.1.3.6 Classification Based on MDDL

MDDL uses a linear classifier for classification. After the dictionary is learned, the locality-constrained coefficients A of training data X and A_{test} of test data X_{test} are

---

Algorithm 7.1: Optimization for MDDL.

Input: Training data $X = [X_1, \ldots, X_c]$, Parameters α, λ, δ, β_1, β_2
Output: W, $A = [A_1, A_2, \ldots, A_c]$, $D = [D_1, D_2 \ldots, D_c]$

1 Initialize: $W = I$, PCA initialized D, $E_i = \mathcal{E}_i = 0$, $T_1 = 0$, $T_2 = 0$, $T_3 = 0$, $\mu_{max} = 10^{30}$, $\rho = 1.1$,
   $\epsilon = 10^{-8}$, $maxiter = 10^4$

2 **repeat**

3       $iter = 0$, $\mu = 10^{-6}$

4   `%Solving Eq. (7.16) via ALM`

5   **while** *not converge and iter $\leq$ maxiter* **do**

6   Fix others and update A_i with Eq. (7.17)

7   Fix others and update E_i with Eq. (7.18)

8   Fix others and update W with Eq. (7.21)

9   Update multipliers T_1 by

10          $T_1 = T_1 + \mu(WX_i - DA_i - E_i)$

11   Update parameter μ by

12          $\mu = \min(\rho\mu, \mu_{max})$

13   Check the convergence conditions

14          $\|WX_i - DA_i - E_i\|_\infty < \epsilon$

15   **end**

16      $iter = 0$, $\mu = 10^{-6}$

17   `%Solving Eq. (7.22) via ALM`

18   **while** *not converge and iter $\leq$ maxiter* **do**

19   Fix others and update A_i^i with Eq. (7.24)

20   Fix others and update J and D_i with Eq. (7.25) and Eq. (7.26)

21   Fix others and update $\mathcal{E}_i$ with Eq. (7.27)

22   Update multipliers T_2 and T_3 by

23          $T_2 = T_2 + \mu(WX_i - D_iA_i^i - \mathcal{E}_i)$

24          $T_3 = T_3 + \mu(D_i - J)$

25   Update parameter μ by

26          $\mu = \min(\rho\mu, \mu_{max})$

27   Check the convergence conditions

28          $\|WX_i - DA_i^i - E_i\|_\infty < \epsilon$

29   **end**

30 **until** The subdictionary converges or the maximal iteration is reached;

---

calculated. The representation a_i for test sample i is the ith column vector in A_{test}. The multivariate ridge regression model [35] was used to obtain a linear classifier $\hat{P}$:

$$\hat{P} = \arg\min_P \|L - PA\|_F^2 + \gamma\|P\|_F^2 \qquad (7.28)$$

where L is the class label matrix. This yields $\hat{P} = LA^T(AA^T + \gamma I)^{-1}$. When testing points A_{test} come in, we first compute $\hat{P}A_{test}$. Then the label for sample i is assigned by the position corresponding to the largest value in the label vector, that is, $label = \arg\max_{label}(\hat{P}a_i)$.

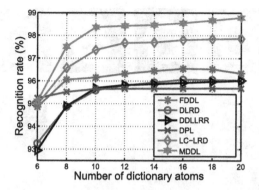

Figure 7.2 The recognition rates of six DL based methods versus the number of dictionary atoms with 20 training samples per class on Extend YaleB dataset.

7.1.4 Experiments

7.1.4.1 Experimental Settings

To verify the effectiveness and generality of the introduced MDDL, we show experiments conducted on various visual classification applications in this section. The method is tested on five datasets, including three face datasets: ORL [36], Extend YaleB [37], and CMU PIE [38], one object categorization dataset COIL-100 [39], and digits recognition dataset MNIST [40].

We show the experiments in comparison with LDA [41], linear regression classification (LRC) [42] and several latest dictionary learning based classification methods, i.e., FDDL [11], DLRD [25], $D^2L^2R^2$ [43], DPL [44], and LC-LRD [28]. Moreover, for verifying the advantage of joint learning, a simple combination framework was proposed as a baseline, named as AE+DL, which first uses a traditional SAE to learn a new representation, then feeds in LC-LRD framework [28].

Parameter selection. The number of atoms in a sub-dictionary, which is denoted as m_i, is one of the most important parameters in most of dictionary learning algorithms. We conduct the experiment on Extended YaleB with different number of dictionary atoms m_i and analyze its effect on the performance of MDDL model and other competitors. Fig. 7.2 shows that all comparisons obtain better performance with larger dictionary size. In the experiments, the number of the dictionary columns of each class is fixed as the training size for ORL, Extend YaleB, AR and COIL-100 datasets, while it is fixed as 30 for CMU PIE and MNIST datasets. All the dictionaries are initialized with PCA on input data.

There are five parameters in Algorithm 7.1: α, λ, δ along with β_1, β_2 as two error term parameters, respectively, for updating dictionary and coefficients. These five are associated with the dictionary learning part in MDDL and are chosen by 5-fold cross

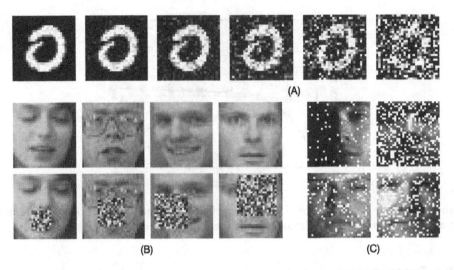

Figure 7.3 Example images from three datasets: (A) images with 30, 20, 10, 5, and 1 db SNR addition of white Gaussian noise from MNIST digit dataset; (B) ORL with 10%, 20%, 30% block occlusion; (C) Extended YaleB with 10%, 15%, 20%, 25% random pixel corruption.

validation. Experiments show that β_1 and β_2 play more important roles than the other parameters; therefore, the other parameters are set as $\alpha = 1$, $\lambda = 1$ and $\delta = 1$. For Extended YaleB, $\beta_1 = 15$, $\beta_2 = 100$; for ORL, $\beta_1 = 5$, $\beta_2 = 50$; for CMU PIE, $\beta_1 = 5$, $\beta_2 = 1.5$; for COIL-100, $\beta_1 = 3$, $\beta_2 = 150$; for MNIST, $\beta_1 = 2.5$, $\beta_2 = 2.5$.

7.1.4.2 Face Recognition

ORL Face Database. The ORL dataset contains 400 images of 40 individuals, so that there are 10 images for each subject with varying pose and illumination. The subjects of the images are in frontal and upright posture while the background is dark and uniform. The images are resized to 32×32, converted to gray scale, normalized and the pixels are concatenated to form a vector. Each image is manually corrupted by a randomly located and unrelated block image. Fig. 7.3B shows four examples of images with increasing block corruptions. For each subject, 5 samples are selected for training and the rest for testing, and the experiment is repeated on 10 random splits for evaluation. Furthermore, SIFT and Gabor filter features are extracted to evaluate MDDL generality.

We illustrate the recognition rates under different percentages of occlusion in Table 7.1. From the table, we can observe two phenomena: first, locality-constrained dictionary learning based methods achieve the best results for all the settings; second, the MDDL model performs best when the data is clean; however, along with the percentage of occlusion increase, MDDL drops behind with LC-LRD. This makes sense because

Table 7.1 Average recognition rate (%) of different algorithms on ORL dataset for various occlusion percentage (%). Red denotes the best results, while blue means the second best results. (For interpretation of the colors in the tables, the reader is referred to the web version of this chapter)

Occlusion	LDA [41]	LRC [42]	FDDL [11]	DLRD [25]	$D^2L^2R^2$ [43]
0	92.5±1.8	91.8±1.6	96.0±1.2	93.5±1.5	93.9±1.8
0 (SIFT)	95.8±1.3	92.9±2.1	95.2±1.3	93.7±1.3	93.9±1.5
0 (Gabor)	89.0±3.3	93.4±1.7	96.0±1.3	96.3±1.3	96.6±1.3
10	71.7±3.2	82.2±2.2	86.6±1.9	91.3±1.9	91.0±1.9
20	54.3±2.0	71.3±2.8	75.3±3.4	82.8±3.0	82.8±3.3
30	40.5±3.7	63.7±3.1	63.8±2.7	78.9±3.1	78.8±3.3
40	25.7±2.5	48.0±3.0	48.1±2.4	67.3±3.2	67.4±3.4
50	20.7±3.0	40.9±3.7	36.7±1.2	58.6±3.1	58.7±3.3

Occlusion	DPL [44]	LC-LRD [28]	AE+DL [1]	MDDL [1]
0	94.1±1.8	96.7±1.4	96.2±1.2	96.8±1.3
0 (SIFT)	95.3±1.4	93.5±2.6	96.0±1.3	96.3±1.4
0 (Gabor)	97.0±1.4	94.6±1.8	96.7±1.1	97.4±1.2
10	84.5±2.7	92.3±1.3	91.5±1.2	92.0±1.4
20	71.2±1.7	83.9±2.3	83.6±1.8	84.3±2.2
30	59.8±3.9	80.2±2.9	78.9±2.9	78.0±2.4
40	43.0±2.9	68.0±3.0	67.3±2.6	67.5±3.2
50	32.2±3.5	58.9±3.5	58.7±3.2	58.5±3.0

in this experiment occlusion noise is added on to the images, while the denoising auto-encoder module in MDDL is introduced to tackle Gaussian noise. In conclusion, first, MDDL can achieve top results in a no occlusion situation because of the locality term; second, in a larger occlusion situation, the low-rank term outweighs DAEs.

The effects from two parameters of the error term, β_1 and β_2, are demonstrated in Fig. 7.4. From the six sub-figures under increasing percentage of corrupted pixels, parameter β_1 in the coefficient updating procedure makes a larger impact. As more occlusion is applied, the best result appears when the parameter β_1 is smaller, which means the error term plays a more important role when noise exists.

The results show that MDDL introduces significant improvement on some datasets, and for some other datasets, its significance increases along with the noise level in the input data.

Extended YaleB Dataset. The Extended Yale Face Database B contains 2414 frontal-face images from 38 human subjects captured under various laboratory-controlled illumination conditions. The size of image is cropped to 32 × 32. Two experiments are deployed on this dataset. First, random subsets with $p (= 5, 10, \ldots, 40)$ images per individual taken with labels are chosen to form the training set, and the rest

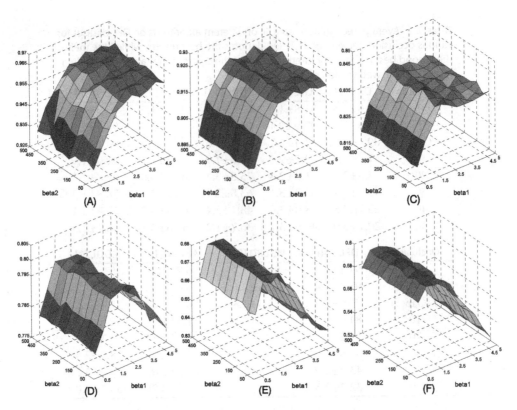

Figure 7.4 MDDL's performance (A)–(F) on ORL dataset under increasing percentage of corrupted pixels versus different parameters. As more occlusion is applied, the best result appears when the parameter β_1 is smaller, which means the error term plays a more important role when noise exists.

of the dataset is considered to be the testing set. For each given p, there are 10 random splits. Second, a certain percentage of randomly selected pixels from the images are replaced with pixel value of 255 (shown in Fig. 7.3C). Then we randomly take 30 images as training samples, with the rest reserved for testing, and the experiment is repeated 10 times. The experimental results are given in Tables 7.2 and 7.3, respectively.

We can observe from Table 7.2 that in different training size settings three locality-constrained based methods (LC-LRD, AE+DL, MDDL) archive the best accuracy, and the proposed MDDL performs the best. MDDL's robustness to noise is demonstrated in Table 7.3. As the percentage of corruption increases, MDDL still produces the best recognition results. The performance of LDA, as well as LRC, FDDL and DPL, drops rapidly with larger corruption, while LC-LRD, MDDL, $D^2L^2R^2$ and DLRD can still get much better recognition accuracy. This demonstrates the effectiveness of low-rank regularization and the error term when noise exists. LC-LRD and simple AE+DL equally match in different situations, while MDDL constantly performs the best.

Table 7.2 Average recognition rate (%) of different algorithms on Extended YaleB dataset with different number of training samples per class. Red denotes the best results, while blue means the second best results

Training	5	10	20	30	40
LDA [41]	74.1±1.5	86.7±0.9	90.6±1.1	86.8±0.9	95.3±0.8
LRC [42]	60.2±2.0	83.00±0.8	91.8±1.0	94.6±0.6	96.1±0.6
FDDL [11]	77.8±1.3	91.2±0.9	96.2±0.7	97.9±0.3	98.8±0.5
DLRD [25]	76.2±1.2	89.9±0.9	96.0±0.8	97.9±0.5	98.8±0.4
$D^2L^2R^2$ [43]	76.0±1.2	89.6±0.9	96.0±0.9	97.9±0.4	98.1±0.4
DPL [44]	75.2±1.9	89.3±0.6	95.7±0.9	97.8±0.4	98.7±0.4
LC-LRD [28]	78.6±1.2	92.1±0.9	97.9±0.9	99.2±0.5	99.5±0.4
AE+DL [1]	78.6±1.1	92.1±0.9	96.6±0.9	98.6±0.5	99.2±0.4
MDDL [1]	79.1±1.2	92.2±0.8	98.8±0.7	99.3±0.2	99.8±0.2

Table 7.3 Average recognition rate (%) of different algorithms on Extended YaleB dataset with various corruption percentage (%). Red denotes the best results, while blue means the second best results

Corruption	0	5	10	15	20
LDA [41]	86.8±0.9	29.0±0.8	18.5±1.2	13.6±0.5	11.3±0.5
LRC [42]	94.6±0.6	80.5±1.1	67.6±1.3	56.8±1.2	47.2±1.6
FDDL [11]	97.9±0.4	63.6±0.9	44.7±1.2	32.7±1.0	25.3±0.4
DLRD [25]	97.9±0.5	91.8±1.1	85.8±1.5	80.9±1.4	73.6±1.6
$D^2L^2R^2$ [43]	97.8±0.4	91.9±1.1	85.7±1.5	80.5±1.6	73.6±1.5
DPL [44]	97.8±0.4	78.3±1.2	64.6±1.1	53.8±0.9	44.9±1.4
LC-LRD [28]	99.2±0.5	93.3±0.7	87.0±0.9	81.7±0.8	74.1±1.0
AE+DL [1]	98.6±0.5	93.2±0.7	87.0±0.9	81.6±0.9	74.2±1.6
MDDL [1]	99.4±0.2	93.6±0.4	87.5±0.8	82.1±0.6	76.3±1.5

CMU PIE Dataset. The CMU PIE dataset contains a total of 41,368 face images from 68 people, each with 13 different poses, 4 different expressions, and 43 different illumination conditions. Two experiments are deployed on two subsets of CMU PIE. First of all, five near frontal poses (C05, C07, C09, C27, C29) are selected as a first subset of PIE, and all the images under different illuminations and expressions (11,554 samples in total) are used. Thus, there are about 170 images for each person, and each image is normalized to have size of 32 × 32 pixels; 60 images per person are selected for training. Secondly, a relatively large-scale dataset is built by choosing more poses, which contains 24,245 samples in total. Overall, there are around 360 images for each person, and each image is normalized to have size of 32 × 32 pixels. The training set is constructed by randomly selecting 200 images per person, while the rest is used for evaluation. Table 7.4 shows that MDDL achieves good results and outperforms the compared methods.

Table 7.4 Classification error rates (%) on CMU PIE dataset. Five poses (C05, C07, C09, C27, C29) are selected as near frontal poses

Methods	CMU (near frontal poses)	CMU (all poses)
LRC [42]	4.12	9.65
FDDL [11]	3.30	11.20
DLRD [25]	3.33	10.64
$D^2L^2R^2$ [43]	3.29	10.14
DPL [44]	3.47	9.30
LC-LRD [28]	3.01	8.98
MDDL [1]	**2.74**	**7.64**

Table 7.5 Average recognition rate (%) with standard deviations of different algorithms on COIL-100 dataset with different number of classes. Red denotes the best results, while blue means the second best results

Class No.	20	40	60	80	100
LDA [41]	81.9±1.2	76.7±0.3	66.2±1.0	59.2±0.7	52.5±0.5
LRC [42]	90.7±0.7	89.0±0.5	86.6±0.4	85.1±0.3	83.2±0.6
FDDL [11]	85.7±0.8	82.1±0.4	77.2±0.7	74.8±0.6	73.6±0.6
DLRD [25]	88.6±1.0	86.4±0.5	83.5±0.1	81.5±0.5	79.9±0.6
$D^2L^2R^2$ [43]	91.0±0.4	88.3±0.4	86.4±0.5	84.7±0.5	83.1±0.4
DPL [44]	87.6±1.3	85.1±0.2	81.2±0.2	78.8±0.9	76.3±0.9
LC-LRD [28]	92.2±0.3	89.9±0.5	87.1±0.7	85.4±0.6	84.2±0.4
AE+DL [1]	91.3±0.5	89.1±0.7	87.2±0.3	85.1±0.5	84.1±0.4
MDDL [1]	91.6±0.4	92.2±0.3	88.1±0.3	86.2±0.3	85.3±0.3

7.1.4.3 Object Recognition

COIL–100 Dataset. In this section, MDDL is evaluated on object categorization by using the COIL-100 dataset. The training set is constructed by randomly selecting 10 images per object, and the testing set contains the rest of the images. This random selection is repeated 10 times, and the average results of all the compared methods are reported. To evaluate the scalability of different methods, the experiment separately utilizes images of 20, 40, 60, 80 and 100 objects from the dataset. Table 7.5 shows the average recognition rates with standard deviations of all compared methods. The results show that MDDL could not only work on face recognition but also on object categorization.

7.1.4.4 Digits Recognition

MNIST Dataset. This section tests MDDL on the subset of MNIST handwritten digit dataset downloaded from CAD website, which includes first 2000 training images and first 2000 test images, with the size of each digit image being 16×16. This experimental

Table 7.6 Average recognition rate (%) & running time (s) on MNIST dataset

Methods	Accuracy	Training time	Testing time
LDA [41]	77.45	0.164	0.545
LRC [42]	82.70	227.192	–
FDDL [11]	85.35	240.116	97.841
DLRD [25]	86.05	156.575	48.373
$D^2L^2R^2$ [43]	84.65	203.375	48.154
DPL [44]	84.65	1.773	0.847
LC-LRD [28]	88.25	80.581	48.970
AE+DL [1]	87.95	176.525	49.230
MDDL [1]	**89.75**	81.042	49.781

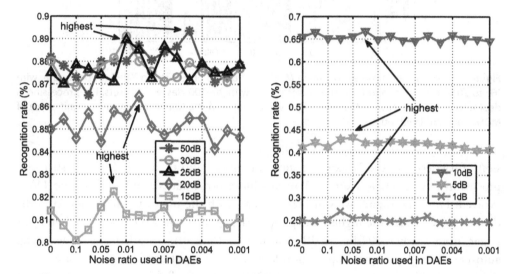

Figure 7.5 The performance on MNIST datasets with different snr noise. As snr gets lower, the best result appears when the noise on reconstruction process is larger, which means the DAEs play a more important role when noise becomes more persistent on MNIST dataset.

setting follows [43], and the experiments get consistent results. The recognition rates and training/testing time by different algorithms on MNIST dataset are summarized in Table 7.6. MDDL achieves the highest accuracy relative to its competitors. Compared within LC-LRD, MDDL costs only slightly more computational time thanks to the easy updating of marginalized auto-encoder.

Another experiment setting is conducted on this dataset to evaluate the effect from denoising auto-encoders. All the training and testing images in MNIST are corrupted with additive white Gaussian noise having signal-to-noise ratio (snr) from 50 to 1 dB (shown in Fig. 7.3A). Fig. 7.5 illustrates the recognition rate curves on 8 noised ver-

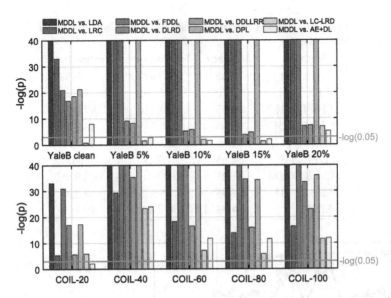

Figure 7.6 p-values of the t-test between MDDL and others on Extended YaleB (upper figure, with 0% to 20% corruption) and COIL-100 (lower figure, with 20–100 classes) datasets. Pre-processing is deployed using $-\log(p)$, so that the larger value shown in the figure means more significance of MDDL compared with others.

sion of datasets. On the X-axis, we show the noise ratio used in input reconstruction process in DAEs, where close to 1 means more noise added, and 0 means no DAEs evolved. From the figure, we can observe that, with the increasing noise added in the datasets (50 to 1 dB), the highest recognition rate appears when the noise parameter gets larger (from nearly 0.004 for 50 dB to nearly 0.1 for 1 dB). In other words, denoising auto-encoders play a more important role when the dataset contains more noise.

To verify if the improvement from MDDL is statistically significant, a significance test (t-test) is further conducted for the results shown in Fig. 7.6. A significance level of 0.05 was used, that is, when the p-value is less than 0.05, the performance difference of two methods is statistically significant. The p-values of MDDL and other competitors are listed in Fig. 7.6. Since we do $-\log(p)$ processing, the comparison shows that MDDL outperforms the others significantly if the values are greater than $-\log(0.05)$. The results show that MDDL introduces significant improvement on COIL-100 dataset, and for Extended YaleB dataset, the significance increases along with the noise level in the input data.

7.1.5 Conclusion

In this section, we introduced an efficient marginalized denoising dictionary learning (MDDL) framework with a locality constraint. The proposed algorithm was designed to take advantage of two feature learning schemes, dictionary learning and auto-encoder. Specifically, MDDL adopted a lite version of the auto-encoder to seek a denoising transformation matrix. Then, dictionary learning with a locality constraint was built on the transformed data. These two strategies were iteratively optimized so that a marginalized denoising transformation and a locality-constrained dictionary were jointly learned. Experiments on several image datasets, e.g., face, object, digits, demonstrated the superiority of our proposed algorithm by comparing with other existing dictionary algorithms.

7.1.6 Future Works

In this chapter, we described MDDL model with a desire to seek a transformation matrix to filter out the noise inside the data with a marginalized denoising auto-encoder. However, one problem arises with the auto-encoder representation architecture, which forces the network to learn an approximation to the identity by encouraging the output to be as similar to the input as possible. This scheme leads to the problem that the majority of the learned high-level features may be blindly used to compresses not only the discriminative information but also lots of redundancies or even noise in data. In the following training procedure, it is unreasonable to endow the discriminablity to this kind of task-irrelevant units.

In the future, we plan to explore the auto-encoder high-level features further to extract the discriminative from the task-irrelevant ones. By compressing more task-relevant information on the hidden units, we hope the dictionary learning module could exploit more discriminative information and learn a more compact and pure dictionary. Furthermore, we plan to explore multi-view data classification where the auto-encoder could serve as a domain adaptation to learn a latent feature space for cross-domain datasets. We believe that low-rank and locality term could also play an important role in cross-domain applications.

7.2. LEARNING A DEEP ℓ_∞ ENCODER FOR HASHING[3]

We investigate the ℓ_∞-constrained representation, which demonstrates robustness to quantization errors, utilizing the tool of deep learning. Based on the Alternating Direction Method of Multipliers (ADMM), we formulate the original convex minimization problem as a feed-forward neural network, named *Deep ℓ_∞ Encoder*, by introducing a novel Bounded Linear Unit (BLU) neuron and modeling the Lagrange multipliers as network biases. Such a structural prior acts as an effective network regularization, and facilitates model initialization. We then investigate the effective use of the proposed model

---

[3] Reprinted, with permission, from Wang, Zhangyang, Yang, Yingzhen, Chang, Shiyu, Ling, Qing, and Huang, Thomas S. "Learning a deep ℓ_∞ encoder for hashing." IJCAI (2016).

in the application of hashing, by coupling the proposed encoders under a supervised pairwise loss, to develop a *Deep Siamese ℓ_∞ Network*, which can be optimized from end to end. Extensive experiments demonstrate the impressive performance of the proposed model. We also provide an in-depth analysis of its behaviors against the competitors.

7.2.1 Introduction
7.2.1.1 Problem Definition and Background

While ℓ_0 and ℓ_1 regularizations have been well-known and successfully applied in sparse signal approximations, utilizing the ℓ_∞ norm to regularize signal representations has been less explored. In this section, we are particularly interested in the following ℓ_∞-constrained least squares problem:

$$\min_x ||\boldsymbol{D}x - y||_2^2 \quad \text{s.t.} \quad ||x||_\infty \leq \lambda, \tag{7.29}$$

where $y \in R^{n\times 1}$ denotes the input signal, $\boldsymbol{D} \in R^{n\times N}$ the (overcomplete) the basis (often called frame or dictionary) with $N < n$, and $x \in R^{N\times 1}$ the learned representation. Further, the maximum absolute magnitude of x is bounded by a positive constant λ, so that each entry of x has the smallest dynamic range [45]. As a result, model (7.29) tends to spread the information of y approximately evenly among the coefficients of x. Thus, x is called "democratic" [46] or "anti-sparse" [47], as all of its entries are of approximately the same importance.

In practice, x usually has most entries reaching the same absolute maximum magnitude [46], therefore resembling an antipodal signal in an N-dimensional Hamming space. Furthermore, the solution x to (7.29) withstands errors in a very powerful way: the representation error gets bounded by the average, rather than the sum, of the errors in the coefficients. These errors may be of arbitrary nature, including distortion (e.g., quantization) and losses (e.g., transmission failure). This property was quantitatively established in Section II.C of [45]:

Theorem 7.1. *Assume $||x||_2 < 1$ without loss of generality, and that each coefficient of x is quantized separately by performing a uniform scalar quantization of the dynamic range $[-\lambda, \lambda]$ with L levels. The overall quantization error of x from (7.29) is bounded by $\frac{\lambda\sqrt{N}}{L}$. In comparison, a least squares solution x_{LS}, by minimizing $||\boldsymbol{D}x_{LS} - y||_2^2$ without any constraint, would only give the bound $\frac{\sqrt{n}}{L}$.*

In the case of $N \ll n$, the above will yield great robustness for the solution to (7.29) with respect to noise, in particular, quantization errors. Also note that its error bound will not grow with the input dimensionality n, a highly desirable stability property for high-dimensional data. Therefore, (7.29) appears to be favorable for the applications such as vector quantization, hashing and approximate nearest neighbor search.

In this section, we investigate (7.29) in the context of deep learning. Based on the Alternating Direction Methods of Multipliers (**ADMM**) algorithm, we formulate

(7.29) as a feed-forward neural network [48], called **Deep ℓ_∞ Encoder**, by introducing a novel Bounded Linear Unit (**BLU**) neuron and modeling the Lagrange multipliers as network biases. The major technical merit to be presented is how a specific optimization model (7.29) could be translated to designing a task-specific deep model, which displays the desired quantization-robust property. We then study its application in hashing, by developing a **Deep Siamese ℓ_∞ Network** that couples the proposed encoders under a supervised pairwise loss, which could be optimized from end to end. Impressive performances are observed in our experiments.

7.2.1.2 Related Work

Similar to the case of ℓ_0/ℓ_1 sparse approximation problems, solving (7.29) and its variants (e.g., [46]) relies on iterative solutions. The authors of [49] proposed an active set strategy similar to that of [50]. In [51], the authors investigated a primal–dual path-following interior-point method. Albeit effective, the iterative approximation algorithms suffer from their inherently sequential structures, as well as the data-dependent complexity and latency, which often constitute a major bottleneck in the computational efficiency. In addition, the joint optimization of the (unsupervised) feature learning and the supervised steps has to rely on solving complex bi-level optimization problems [52]. Further, to effectively represent datasets of growing sizes, larger dictionaries D are usually needed. Since the inference complexity of those iterative algorithms increases more than linearly with respect to the dictionary size [53], their scalability turns out to be limited. Last but not least, while the hyperparameter λ sometimes has physical interpretations, e.g., for signal quantization and compression, it remains unclear how to set or adjust it for many application cases.

Deep learning has recently attracted great attention [54]. The advantages of deep learning lie in its composition using multiple nonlinear transformations to yield more abstract and descriptive embedding representations. The feed-forward networks could be naturally tuned jointly with task-driven loss functions [55]. With the aid of gradient descent, it also scales linearly in time and space with the number of training samples.

There has been a blooming interest in bridging "shallow" optimization and deep learning models. In [48], a feed-forward neural network, named LISTA, was proposed to efficiently approximate the sparse codes, whose hyperparameters were learned from general regression. In [56], the authors leveraged a similar idea on fast trainable regressors and constructed feed-forward network approximations of the learned sparse models. It was later extended in [57] to develop a principled process of learned deterministic fixed-complexity pursuits, for sparse and low rank models. Lately, [55] proposed Deep ℓ_0 Encoders, to model ℓ_0 sparse approximation as feed-forward neural networks. The authors extended the strategy to the graph-regularized ℓ_1 approximation in [58], and the dual sparsity model in [59]. Despite the above progress, to the best of our knowledge, few efforts have been made beyond sparse approximation (e.g., ℓ_0/ℓ_1) models.

7.2.2 ADMM Algorithm

ADMM has been popular for its remarkable effectiveness in minimizing objectives with linearly separable structures [53]. We first introduce an auxiliary variable $z \in R^{N \times 1}$, and rewrite (7.29) as

$$\min_{x,z} \tfrac{1}{2}||Dx - y||_2^2 \quad \text{s.t.} \quad ||z||_\infty \le \lambda, \quad z - x = 0. \tag{7.30}$$

The augmented Lagrangian function of (7.30) is

$$\tfrac{1}{2}||Dx - y||_2^2 + p^T(z - x) + \tfrac{\beta}{2}||z - x||_2^2 + \Phi_\lambda(z). \tag{7.31}$$

Here $p \in R^{N \times 1}$ is the Lagrange multiplier attached to the equality constraint, β is a positive constant (with a default value of 0.6), and $\Phi_\lambda(z)$ is the indicator function which goes to infinity when $||z||_\infty > \lambda$, and is 0 otherwise. ADMM minimizes (7.31) with respect to x and z in an alternating direction manner, and updates p accordingly. It guarantees global convergence to the optimal solution to (7.29). Starting from any initialization points of x, z, and p, ADMM iteratively solves ($t = 0, 1, 2, \ldots$ denotes the iteration number):

$$(x \text{ update}) \quad \min_{x_{t+1}} \tfrac{1}{2}||Dx - y||_2^2 - p_t^T x + \tfrac{\beta}{2}||z_t - x||_2^2, \tag{7.32}$$

$$(z \text{ update}) \quad \min_{z_{t+1}} \tfrac{\beta}{2}||z - (x_{t+1} - \tfrac{p_t}{\beta})||_2^2 + \Phi_\lambda(z), \tag{7.33}$$

$$(p \text{ update}) \quad p_{t+1} = p_t + \beta(z_{t+1} - x_{t+1}). \tag{7.34}$$

Furthermore, both (7.32) and (7.33) enjoy closed-form solutions:

$$x_{t+1} = (D^T D + \beta I)^{-1}(D^T y + \beta z_t + p_t), \tag{7.35}$$

$$z_{t+1} = \min(\max(x_{t+1} - \tfrac{p_t}{\beta}, -\lambda), \lambda). \tag{7.36}$$

The above algorithm could be categorized to the primal–dual scheme. However, discussing the ADMM algorithm in more details is beyond the focus of this section. Instead, the purpose of deriving (7.30)–(7.36) is to prepare us for the design of the task-specific deep architecture, as presented below.

7.2.3 Deep ℓ_∞ Encoder

We first substitute (7.35) into (7.36), in order to derive an update form explicitly dependent on only z and p:

$$z_{t+1} = B_\lambda((D^T D + \beta I)^{-1}(D^T y + \beta z_t + p_t) - \tfrac{p_t}{\beta}), \tag{7.37}$$

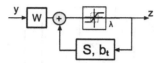

Figure 7.7 The block diagram of solving (7.29).

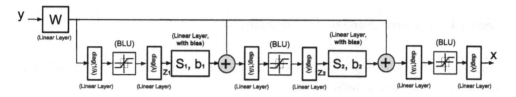

Figure 7.8 Deep ℓ_∞ Encoder with two time-unfolded stages.

where B_λ is defined as a box-constrained element-wise operator (u denotes a vector and u_i is its ith element):

$$[B_\lambda(u)]_i = \min(\max(u_i, -\lambda), \lambda). \tag{7.38}$$

Equation (7.37) could be alternatively rewritten as

$$z_{t+1} = B_\lambda(\boldsymbol{W}y + \boldsymbol{S}z_t + b_t), \text{ where}$$
$$\boldsymbol{W} = (\boldsymbol{D}^T\boldsymbol{D} + \beta\boldsymbol{I})^{-1}\boldsymbol{D}^T, \boldsymbol{S} = \beta(\boldsymbol{D}^T\boldsymbol{D} + \beta\boldsymbol{I})^{-1}, b_t = [(\boldsymbol{D}^T\boldsymbol{D} + \beta\boldsymbol{I})^{-1} - \tfrac{1}{\beta}\boldsymbol{I}]p_t, \tag{7.39}$$

and expressed as the block diagram in Fig. 7.7, which outlines a recurrent structure when solving (7.29). Note that in (7.39), while $\boldsymbol{W}$ and $\boldsymbol{S}$ are pre-computed hyperparameters shared across iterations, b_t remains a variable dependent on p_t, and has to be updated throughout iterations, too (b_t's update block is omitted in Fig. 7.7).

By time-unfolding and truncating Fig. 7.7 to a fixed number of K iterations ($K = 2$ by default),[4] we obtain a feed-forward network structure in Fig. 7.8, named **Deep ℓ_∞ Encoder**. Since the threshold λ is less straightforward to update, we repeat the same trick as in [55] to rewrite (7.38) as $[B_\lambda(u)]_i = \lambda_i B_1(u_i/\lambda_i)$. The original operator is thus decomposed into two linear diagonal scaling layers, plus a unit-threshold neuron; the latter is called a Bounded Linear Unit (**BLU**) by us. All the hyperparameters $\boldsymbol{W}$, $\boldsymbol{S}_k$ and b_k ($k = 1, 2$), as well as λ, are all to be learnt from data by backpropagation. Although the equations in (7.39) do not directly apply to solving the deep ℓ_∞ encoder, they can serve as high-quality initializations.

It is crucial to notice the modeling of the Lagrange multipliers p_t as the biases, and to incorporate their updates into network learning. This provides important clues on

---

[4] We tested larger K values (3 or 4). In several cases they did bring performance improvements, but added complexity, too.

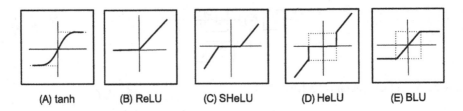

Figure 7.9 A comparison among existing neurons and BLU.

how to relate deep networks to a larger class of optimization models, whose solutions rely on dual domain methods.

Comparing BLU with existing neurons. As shown in Fig. 7.9E, BLU tends to suppress large entries while not penalizing small ones, resulting in dense, nearly antipodal representations. A first look at the plot of BLU easily reminds of the tanh neuron (Fig. 7.9A). In fact, with its output range being $[-1, 1]$ and a slope of 1 at the origin, tanh could be viewed as a smoothed differentiable approximation of BLU.

We further compare BLU with other popular and recently proposed neurons: Rectifier Linear Unit (ReLU) [54], Soft-tHresholding Linear Unit (SHeLU) [58], and Hard thrEsholding Linear Unit (HELU) [55], as depicted in Fig. 7.9B–D, respectively. Contrary to BLU and tanh, they all introduce sparsity in the outputs, and thus prove successful and outperform tanh in classification and recognition tasks. Interestingly, HELU seems to rival BLU, as it does not penalize large entries but suppresses small ones down to zero.

7.2.4 Deep ℓ_∞ Siamese Network for Hashing

Rather than solving (7.29) first and then training the encoder as general regression, as [48] did, we instead concatenate encoder(s) with a task-driven loss, and optimize the pipeline **from end to end**. In this section, we focus on discussing its application in hashing, although the proposed model is not limited to one specific application.

Background. With the ever-growing large-scale image data on the Web, much attention has been devoted to nearest neighbor search via hashing methods [60]. For big data applications, compact bitwise representations improve the efficiency in both storage and search speed. The state-of-the-art approach, learning-based hashing, learns similarity-preserving hash functions to encode input data into binary codes. Furthermore, while earlier methods, such as linear search hashing (LSH) [60], iterative quantization (ITQ) [61] and spectral hashing (SH) [62], do not refer to any supervised information, it has been lately discovered that involving the data similarities/dissimilarities in training benefits the performance [63,64].

Prior Work. Traditional hashing pipelines first represent each input image as a (hand-crafted) visual descriptor, followed by separate projection and quantization steps to

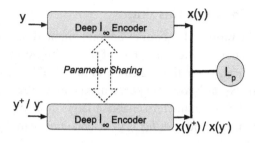

Figure 7.10 Deep ℓ_∞ Siamese Network, by coupling two parameter-sharing encoders, followed by a pairwise loss (7.40).

encode it into a binary code. The authors of [65] first applied the *Siamese network* [66] architecture to hashing, which fed two input patterns into two parameter-sharing "encoder" columns and minimized a pairwise-similarity/dissimilarity loss function between their outputs, using pairwise labels. The authors further enforced the sparsity prior on the hash codes in [67], by substituting a pair of LISTA-type encoders [48] for the pair of generic feed-forward encoders in [65] while [68,69] utilized tailored convolution networks with the aid of pairwise labels. The authors of [70] further introduced a triplet loss with a divide-and-encode strategy applied to reduce the hash code redundancy. Note that for the final training step of quantization, [67] relied on an extra hidden layer of tanh neurons to approximate binary codes, while [70] exploited a piecewise linear and discontinuous threshold function.

Our Approach. In view of its robustness to quantization noise, as well as BLU's property as a natural binarization approximation, we construct a Siamese network as in [65], and adopt a pair of parameter-sharing deep ℓ_∞ encoders as the two columns. The resulting architecture, named the **Deep ℓ_∞ Siamese Network**, is illustrated in Fig. 7.10. Assume y and y^+ make a similar pair while y and y^- make a dissimilar pair, and further denote $x(y)$ the output representation by inputting y. The two coupled encoders are then optimized under the following pairwise loss (the constant m represents the margin between dissimilar pairs):

$$L_p := \tfrac{1}{2}||x(y) - x(y^+)||^2 - \tfrac{1}{2}(\max(0, m - ||x(y) - x(y^-)||))^2. \qquad (7.40)$$

The representation is learned to make similar pairs as close as possible and dissimilar pairs to be at least a distance m away. In this section, we follow [65] to use a default $m = 5$ for all experiments.

Once a deep ℓ_∞ Siamese network is learned, we apply its encoder part (i.e., a deep ℓ_∞ encoder) to a new input. The computation is extremely efficient, involving only a few matrix multiplications and element-wise thresholding operations, with a total complexity of $O(nN + 2N^2)$. One can obtain an N-bit binary code by simply quantizing the output.

7.2.5 Experiments in Image Hashing

Implementation. The proposed networks are implemented with the CUDA ConvNet package [54]. We use a constant learning rate of 0.01 with no momentum, and a batch size of 128. Different from prior findings such as in [55,58], we discover that untying the values of S_1, b_1 and S_2, b_2 boosts the performance more than sharing them. It is not only because that more free parameters enable a larger learning capacity, but also due to the important fact that p_t (and thus b_k) is in essence not shared across iterations, as in (7.39) and Fig. 7.8.

While many neural networks are trained well with random initializations, it has been discovered that sometimes poor initializations can still hamper the effectiveness of first-order methods [71]. On the other hand, it is much easier to initialize our proposed models in the right regime. We first estimate the dictionary D using the standard K-SVD algorithm [72], and then inexactly solve (7.29) for up to K ($K = 2$) iterations, via the ADMM algorithm in Sect. 7.2.2, with the values of Lagrange multiplier p_t recorded for each iteration. Benefiting from the analytical correspondence relationships in (7.39), it is then straightforward to obtain high-quality initializations for W, S_k and b_k ($k = 1, 2$). As a result, we could achieve a steadily decreasing curve of training errors, without performing common tricks such as annealing the learning rate, which are found to be indispensable if random initialization is applied.

Datasets. The **CIFAR10** dataset [73] contains 60K labeled images of 10 different classes. The images are represented using 384-dimensional GIST descriptors [74]. Following the classical setting in [67], we used a training set of 200 images for each class, and a disjoint query set of 100 images per class. The remaining 59K images are treated as database.

NUS-WIDE [75] is a dataset containing 270K annotated images from Flickr. Every image is associated with one or more of the 81 different concepts, and is described using a 500-dimensional bag-of-features. In training and evaluation, we followed the protocol of [76]: two images were considered as neighbors if they share at least one common concept (only 21 most frequent concepts are considered). We use 100K pairs of images for training, and a query set of 100 images per concept in testing.

Comparison Methods. We compare the proposed deep ℓ_∞ Siamese network to six state-of-the-art hashing methods:

- Four representative "shallow" hashing methods: kernelized supervised hashing (KSH) [64], anchor graph hashing (AGH) [76] (we compare with its two alternative forms: AGH1 and AGH2; see the original paper), parameter-sensitive hashing (PSH) [77], and LDA Hash (LH) [78].[5]
- Two latest "deep" hashing methods: neural-network hashing (NNH) [65], and sparse neural-network hashing (SNNH) [67].

---

[5] Most of the results are collected from the comparison experiments in [67], under the same settings.

Table 7.7 Comparison of NNH, SNNH, and the proposed deep ℓ_∞ Siamese network

	encoder type	neuron type	structural prior on hashing codes
NNH	generic	tanh	/
SNNH	LISTA	SHeLU	sparse
Proposed	deep ℓ_∞	BLU	nearly antipodal & quantization-robust

Comparing the two "deep" competitors to the deep ℓ_∞ Siamese network, the only difference among the three is the type of encoder adopted in each's twin columns, as listed in Table 7.7. We reimplement the encoder parts of NNH and SNNH, with three hidden layers (i.e., two unfolded stages for LISTA), so that all three deep hashing models have the same depth.[6] Recalling that the input $y \in R^n$ and the hash code $x \in R^N$, we immediately see from (7.39) that $W \in R^{n \times N}$, $S_k \in R^{N \times N}$, and $b_k \in R^N$. We carefully ensure that both NNHash and SparseHash have all their weight layers of the same dimensionality with ours[7] for a fair comparison.

We adopt the following classical criteria for evaluation: (i) *precision and recall* (PR) for different Hamming radii, and the *F1 score* as their harmonic average; (ii) *mean average precision* (MAP) [79]. Besides, for NUS-WIDE, as computing mAP is slow over this large dataset, we follow the convention of [67] to compute the *mean precision* (MP) of top-5K returned neighbors (MP@5K), as well as report mAP of top-10 results (mAP@10).

We have not compared with convolutional network–based hashing methods [68–70], since it is difficult to ensure their models have the same parameter capacity as our fully-connected model in controlled experiments. We also do not include triplet loss–based methods, e.g., as in [70], into comparison because they will require three parallel encoder columns.

Results and Analysis. The performance of different methods on two datasets are compared in Tables 7.8 and 7.9. Our proposed method ranks top in almost all cases, in terms of mAP/MP and precision. Even under the Hamming radius of 0, our precision result is as high as 33.30% ($N = 64$) for CIFAR10, and 89.49% ($N = 256$) for NUS-WIDE. The proposed method also maintains the second best in most cases, in terms of recall, inferior only to SNNH. In particular, when the hashing code dimensionality is low, e.g., when $N = 48$ for CIFAR10, the proposed method outperforms all other

[6] The performance is thus improved than reported in their original papers using two hidden layers, although with extra complexity.

[7] Both the deep ℓ_∞ encoder and the LISTA network will introduce the diagonal layers, while the generic feed-forward networks will not. Besides, neither LISTA nor generic feed-forward networks contain layer-wise biases. Yet, since either a diagonal layer or a bias contains only N free parameters, the total amount is ignorable.

Table 7.8 Performance (%) of different hashing methods on the CIFAR10 dataset, with different code lengths N

Method	N	mAP	Hamming radius ≤ 2			Hamming radius $= 0$		
			Prec.	Recall	F1	Prec.	Recall	F1
KSH	48	31.10	18.22	0.86	1.64	5.39	5.6×10^{-2}	0.11
	64	32.49	10.86	0.13	0.26	2.49	9.6×10^{-3}	1.9×10^{-2}
AGH1	48	14.55	15.95	2.8×10^{-2}	5.6×10^{-2}	4.88	2.2×10^{-3}	4.4×10^{-3}
	64	14.22	6.50	4.1×10^{-3}	8.1×10^{-3}	3.06	1.2×10^{-3}	2.4×10^{-3}
AGH2	48	15.34	17.43	7.1×10^{-2}	3.6×10^{-2}	5.44	3.5×10^{-3}	6.9×10^{-3}
	64	14.99	7.63	7.2×10^{-3}	1.4×10^{-2}	3.61	1.4×10^{-3}	2.7×10^{-3}
PSH	48	15.78	9.92	6.6×10^{-3}	1.3×10^{-2}	0.30	5.1×10^{-5}	1.0×10^{-4}
	64	17.18	1.52	3.0×10^{-4}	6.1×10^{-4}	1.0×10^{-3}	1.69×10^{-5}	3.3×10^{-5}
LH	48	13.13	3.0×10^{-3}	1.0×10^{-4}	5.1×10^{-5}	1.0×10^{-3}	1.7×10^{-5}	3.4×10^{-5}
	64	13.07	1.0×10^{-3}	1.7×10^{-5}	3.3×10^{-5}	0.00	0.00	0.00
NNH	48	31.21	34.87	1.81	3.44	10.02	9.4×10^{-2}	0.19
	64	35.24	23.23	0.29	0.57	5.89	1.4×10^{-2}	2.8×10^{-2}
SNNH	48	26.67	32.03	12.10	17.56	19.95	**0.96**	**1.83**
	64	27.25	30.01	**36.68**	33.01	30.25	**9.8**	**14.90**
Proposed	48	**31.48**	**36.89**	**12.47**	**18.41**	**24.82**	0.94	1.82
	64	**36.76**	38.67	30.28	**33.96**	**33.30**	8.9	14.05

Table 7.9 Performance (%) of different hashing methods on the NUS-WIDE dataset, with different code lengths N

Method	N	mAP@10	MP@5K	Hamming radius ≤ 2			Hamming radius $= 0$		
				Prec.	Recall	F1	Prec.	Recall	F1
KSH	64	72.85	42.74	83.80	6.1×10^{-3}	1.2×10^{-2}	84.21	1.7×10^{-3}	3.3×10^{-3}
	256	73.73	45.35	84.24	1.4×10^{-3}	2.9×10^{-3}	84.24	1.4×10^{-3}	2.9×10^{-3}
AGH1	64	69.48	47.28	69.43	0.11	0.22	73.35	3.9×10^{-2}	7.9×10^{-2}
	256	73.86	46.68	75.90	1.5×10^{-2}	2.9×10^{-2}	81.64	3.6×10^{-3}	7.1×10^{-3}
AGH2	64	68.90	47.27	68.73	0.14	0.28	72.82	5.2×10^{-2}	0.10
	256	73.00	47.65	74.90	5.3×10^{-2}	0.11	80.45	1.1×10^{-2}	2.2×10^{-2}
PSH	64	72.17	44.79	60.06	0.12	0.24	81.73	1.1×10^{-2}	2.2×10^{-2}
	256	73.52	47.13	84.18	1.8×10^{-3}	3.5×10^{-3}	84.24	1.5×10^{-3}	2.9×10^{-3}
LH	64	71.33	41.69	**84.26**	1.4×10^{-3}	2.9×10^{-3}	84.24	1.4×10^{-3}	2.9×10^{-3}
	256	70.73	39.02	84.24	1.4×10^{-3}	2.9×10^{-3}	84.24	1.4×10^{-3}	2.9×10^{-3}
NNH	64	76.39	59.76	75.51	1.59	3.11	81.24	0.10	0.20
	256	78.31	61.21	83.46	5.8×10^{-2}	0.11	83.94	4.9×10^{-3}	9.8×10^{-3}
SNNH	64	74.87	56.82	72.32	**1.99**	**3.87**	81.98	**0.37**	**0.73**
	256	74.73	59.37	80.98	**0.10**	**0.19**	82.85	**0.98**	**1.94**
Proposed	64	**79.89**	**63.04**	79.95	1.72	3.38	**86.23**	0.30	0.60
	256	**80.02**	**65.62**	**84.63**	7.2×10^{-2}	0.15	**89.49**	0.57	1.13

competitors with a significant margin. It demonstrates the competitiveness of the proposed method in generating both compact and accurate hashing codes, achieving more precise retrieval results at lower computation and storage costs.

The next observation is that, compared to the strongest competitor SNNH, the recall rates of our method seem less compelling. We plot the precision and recall curves of the three best performers (NNH, SNNH, deep l_∞), with regard to the bit length of hashing codes N, within the Hamming radius of 2. Fig. 7.12 demonstrates that our method consistently outperforms both SNNH and NNH in precision. On the other hand, SNNH gains advantages in recall over the proposed method, although the margin appears vanishing as N grows.

Although it seems to be a reasonable performance tradeoff, we are curious about the behavior difference between SNNH and the proposed method. We are again reminded that they only differ in the encoder architecture, i.e., one with LISTA while the other using the deep l_∞ encoder. We thus plot the learned representations and binary hashing codes of one CIFAR image, using NNH, SNNH, and the proposed method, in Fig. 7.11. By comparing the three pairs, one could see that the quantization from (A) to (B) (also (C) to (D)) suffer visible distortion and information loss. Contrary to them, the output of the deep l_∞ encoder has a much smaller quantization error, as it naturally resembles an antipodal signal. Therefore, it suffers minimal information loss during the quantization step.

In view of those, we conclude the following points towards the different behaviors, between SNNH and deep l_∞ encoder:

- Both deep l_∞ encoder and SNNH outperform NNH, by introducing structure into the binary hashing codes.
- The deep l_∞ encoder generates nearly antipodal outputs that are robust to quantization errors. Therefore, it excels in preserving information against hierarchical information extraction as well as quantization. This explains why our method reaches the highest precisions, and performs especially well when N is small.
- SNNH exploits sparsity as a prior on hashing codes. It confines and shrinks the solution space, as many small entries in the SNNH outputs will be suppressed down to zero. That is also evidenced by Table 2 in [67], i.e., the number of unique hashing codes in SNNH results is one order smaller than that of NNH.
- The sparsity prior improves the recall rate, since its obtained hashing codes can be clustered more compactly in high-dimensional space, with lower intra-cluster variations. But it also runs the risk of losing too much information, during the hierarchical sparsifying process. In that case, the inter-cluster variations might also be compromised, which causes the decrease in precision.

Further, it seems that the sparsity and l_∞ structure priors could be complementary. We will explore it as future work.

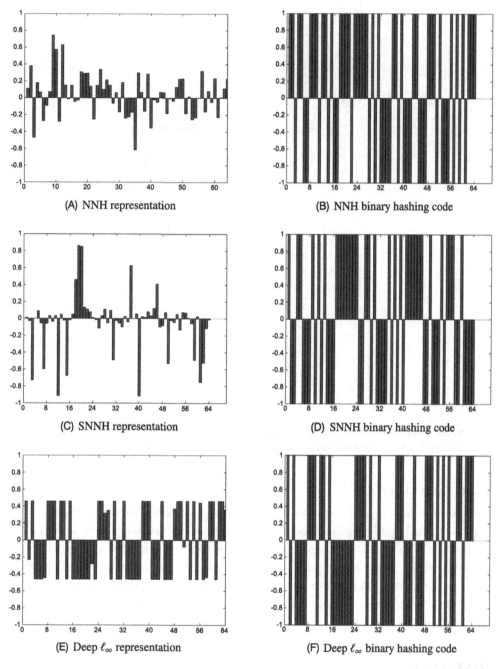

Figure 7.11 The learned representations and binary hashing codes of one test image from CIFAR10 through: (A)–(B) NNH; (C)–(D) SNNH; (E)–(F) proposed.

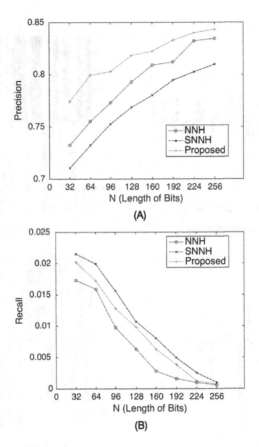

Figure 7.12 The comparison of three deep hashing methods on NUS-WIDE: (A) precision curve; (B) recall curve, both w.r.t. the hashing code length *N*, within the Hamming radius of 2.

7.2.6 Conclusion

This section investigates how to import the quantization-robust property of an ℓ_∞-constrained minimization model to a specially-designed deep model. It is done by first deriving an ADMM algorithm, which is then reformulated as a feed-forward neural network. We introduce the Siamese architecture concatenated with a pairwise loss, for the application purpose of hashing. We analyze in depth the performance and behaviors of the proposed model against its competitors, and hope it will evoke more interest from the community.

REFERENCES

[1] Wang S, Ding Z, Fu Y. Marginalized denoising dictionary learning with locality constraint. IEEE Transactions on Image Processing 2017.

[2] Yang J, Yu K, Gong Y, Huang T. Linear spatial pyramid matching using sparse coding for image classification. In: CVPR. IEEE; 2009. p. 1794–801.

[3] Rodriguez F, Sapiro G. Sparse representations for image classification: learning discriminative and reconstructive non-parametric dictionaries. tech. rep., DTIC Document; 2008.

[4] Elhamifar E, Vidal R. Robust classification using structured sparse representation. In: CVPR. IEEE; 2011. p. 1873–9.

[5] Jafari MG, Plumbley MD. Fast dictionary learning for sparse representations of speech signals. Selected Topics in Signal Processing, IEEE Journal of 2011;5(5):1025–31.

[6] Pique-Regi R, Monso-Varona J, Ortega A, Seeger RC, Triche TJ, Asgharzadeh S. Sparse representation and Bayesian detection of genome copy number alterations from microarray data. Bioinformatics 2008;24(3):309–18.

[7] Wright J, Yang AY, Ganesh A, Sastry SS, Ma Y. Robust face recognition via sparse representation. TPAMI 2009;31(2):210–27.

[8] Aharon M, Elad M, Bruckstein A. K-SVD: an algorithm for designing overcomplete dictionaries for sparse representation. TSP 2006;54(11):4311–22.

[9] Zhang Q, Li B. Discriminative K-SVD for dictionary learning in face recognition. In: CVPR. IEEE; 2010. p. 2691–8.

[10] Jiang Z, Lin Z, Davis LS. Learning a discriminative dictionary for sparse coding via label consistent K-SVD. In: CVPR. IEEE; 2011. p. 1697–704.

[11] Yang M, Zhang D, Feng X. Fisher discrimination dictionary learning for sparse representation. In: ICCV. IEEE; 2011. p. 543–50.

[12] Liu G, Lin Z, Yu Y. Robust subspace segmentation by low-rank representation. In: ICML; 2010. p. 663–70.

[13] Shen X, Wu Y. A unified approach to salient object detection via low rank matrix recovery. In: CVPR. IEEE; 2012. p. 853–60.

[14] Ding Z, Fu Y. Low-rank common subspace for multi-view learning. In: ICDM. IEEE; 2014. p. 110–9.

[15] Liu G, Lin Z, Yan S, Sun J, Yu Y, Ma Y. Robust recovery of subspace structures by low-rank representation. TPAMI 2013;35(1):171–84.

[16] Zhang C, Liu J, Tian Q, Xu C, Lu H, Ma S. Image classification by non-negative sparse coding, low-rank and sparse decomposition. In: CVPR. IEEE; 2011. p. 1673–80.

[17] Zhang D, Liu P, Zhang K, Zhang H, Wang Q, Jinga X. Class relatedness oriented-discriminative dictionary learning for multiclass image classification. Pattern Recognition 2015.

[18] Bengio Y. Learning deep architectures for AI. Foundations and Trends in Machine Learning 2009;2(1):1–127.

[19] Vincent P, Larochelle H, Lajoie I, Bengio Y, Manzagol PA. Stacked denoising autoencoders: learning useful representations in a deep network with a local denoising criterion. JMLR 2010;11:3371–408.

[20] Chen M, Xu Z, Weinberger K, Sha F. Marginalized denoising autoencoders for domain adaptation. arXiv preprint arXiv:1206.4683, 2012.

[21] Chen Y, Guo X. Learning non-negative locality-constrained linear coding for human action recognition. In: VCIP. IEEE; 2013. p. 1–6.

[22] Shabou A, LeBorgne H. Locality-constrained and spatially regularized coding for scene categorization. In: CVPR. IEEE; 2012. p. 3618–25.

[23] Liang Y, Song M, Bu J, Chen C. Colorization for gray scale facial image by locality-constrained linear coding. Journal of Signal Processing Systems 2014;74(1):59–67.

[24] Lin Z, Chen M, Ma Y. The augmented Lagrange multiplier method for exact recovery of corrupted low-rank matrices. arXiv preprint arXiv:1009.5055, 2010.

[25] Ma L, Wang C, Xiao B, Zhou W. Sparse representation for face recognition based on discriminative low-rank dictionary learning. In: CVPR. IEEE; 2012. p. 2586–93.

[26] Jiang X, Lai J. Sparse and dense hybrid representation via dictionary decomposition for face recognition. TPAMI 2015;37(5):1067–79.

[27] Wang J, Yang J, Yu K, Lv F, Huang T, Gong Y. Locality-constrained linear coding for image classification. In: CVPR. IEEE; 2010. p. 3360–7.

[28] Wang S, Fu Y. Locality-constrained discriminative learning and coding. In: CVPRW; 2015. p. 17–24.

[29] Boureau Y-Lan, LeCun Yann, et al. Sparse feature learning for deep belief networks. In: NIPS; 2008. p. 1185–92.

[30] Vincent P, Larochelle H, Bengio Y, Manzagol PA. Extracting and composing robust features with denoising autoencoders. In: ICML. ACM; 2008. p. 1096–103.

[31] Candès EJ, Li X, Ma Y, Wright J. Robust principal component analysis? Journal of the ACM (JACM) 2011;58(3):11.

[32] Yu K, Zhang T, Gong Y. Nonlinear learning using local coordinate coding. In: NIPS; 2009. p. 2223–31.

[33] Bertsekas DP. Constrained optimization and Lagrange multiplier methods. Computer Science and Applied Mathematics, vol. 1. Boston: Academic Press; 1982.

[34] Yang J, Yin W, Zhang Y, Wang Y. A fast algorithm for edge-preserving variational multichannel image restoration. SIAM Journal on Imaging Sciences 2009;2(2):569–92.

[35] Zhang G, Jiang Z, Davis LS. Online semi-supervised discriminative dictionary learning for sparse representation. In: ACCV. Springer; 2013. p. 259–73.

[36] Samaria FS, Harter AC. Parameterisation of a stochastic model for human face identification. In: Proceedings of the second IEEE workshop on applications of computer vision. IEEE; 1994. p. 138–42.

[37] Lee KC, Ho J, Kriegman D. Acquiring linear subspaces for face recognition under variable lighting. TPAMI 2005;27(5):684–98.

[38] Sim T, Baker S, Bsat M. The CMU pose, illumination, and expression database. TPAMI 2003;25(12):1615–8.

[39] Nayar S, Nene SA, Murase H. Columbia object image library (COIL 100). Tech Rep CUCS-006-96. Department of Comp Science, Columbia University; 1996.

[40] LeCun Y, Bottou L, Bengio Y, Haffner P. Gradient-based learning applied to document recognition. Proceedings of the IEEE 1998;86(11):2278–324.

[41] Belhumeur PN, Hespanha JP, Kriegman D. Eigenfaces vs. fisherfaces: recognition using class specific linear projection. TPAMI 1997;19(7):711–20.

[42] Naseem I, Togneri R, Bennamoun M. Linear regression for face recognition. TPAMI 2010;32(11):2106–12.

[43] Li L, Li S, Fu Y. Learning low-rank and discriminative dictionary for image classification. IVC 2014.

[44] Gu S, Zhang L, Zuo W, Feng X. Projective dictionary pair learning for pattern classification. In: NIPS; 2014. p. 793–801.

[45] Lyubarskii Y, Vershynin R. Uncertainty principles and vector quantization. Information Theory, IEEE Transactions on 2010.

[46] Studer C, Goldstein T, Yin W, Baraniuk RG. Democratic representations. arXiv preprint arXiv:1401.3420, 2014.

[47] Fuchs JJ. Spread representations. In: ASILOMAR. IEEE; 2011. p. 814–7.

[48] Gregor K, LeCun Y. Learning fast approximations of sparse coding. In: ICML; 2010. p. 399–406.

[49] Stark PB, Parker RL. Bounded-variable least-squares: an algorithm and applications. Computational Statistics 1995;10:129–41.

[50] Lawson CL, Hanson RJ. Solving least squares problems, vol. 161. SIAM; 1974.

[51] Adlers M. Sparse least squares problems with box constraints. Citeseer; 1998.

[52] Wang Z, Yang Y, Chang S, Li J, Fong S, Huang TS. A joint optimization framework of sparse coding and discriminative clustering. In: IJCAI; 2015.

[53] Bertsekas DP. Nonlinear programming. Belmont: Athena Scientific; 1999.

[54] Krizhevsky A, Sutskever I, Hinton GE. ImageNet classification with deep convolutional neural networks. In: NIPS; 2012.

[55] Wang Z, Ling Q, Huang T. Learning deep l0 encoders. AAAI 2016.

[56] Sprechmann P, Litman R, Yakar TB, Bronstein AM, Sapiro G. Supervised sparse analysis and synthesis operators. In: NIPS; 2013. p. 908–16.

[57] Sprechmann P, Bronstein A, Sapiro G. Learning efficient sparse and low rank models. TPAMI 2015.

[58] Wang Z, Chang S, Zhou J, Wang M, Huang TS. Learning a task-specific deep architecture for clustering. SDM 2016.

[59] Wang Z, Chang S, Liu D, Ling Q, Huang TS. D3: deep dual-domain based fast restoration of jpeg-compressed images. In: IEEE CVPR; 2016.

[60] Gionis A, Indyk P, Motwani R, et al. Similarity search in high dimensions via hashing. In: VLDB, vol. 99; 1999. p. 518–29.

[61] Gong Y, Lazebnik S. Iterative quantization: a procrustean approach to learning binary codes. In: CVPR. IEEE; 2011.

[62] Weiss Y, Torralba A, Fergus R. Spectral hashing. In: NIPS; 2009.

[63] Kulis B, Darrell T. Learning to hash with binary reconstructive embeddings. In: NIPS; 2009. p. 1042–50.

[64] Liu W, Wang J, Ji R, Jiang YG, Chang SF. Supervised hashing with kernels. In: CVPR. IEEE; 2012. p. 2074–81.

[65] Masci J, Migliore D, Bronstein MM, Schmidhuber J. Descriptor learning for omnidirectional image matching. In: Registration and recognition in images and videos. Springer; 2014. p. 49–62.

[66] Hadsell R, Chopra S, LeCun Y. Dimensionality reduction by learning an invariant mapping. In: CVPR. IEEE; 2006.

[67] Masci J, Bronstein AM, Bronstein MM, Sprechmann P, Sapiro G. Sparse similarity-preserving hashing. arXiv preprint arXiv:1312.5479, 2013.

[68] Xia R, Pan Y, Lai H, Liu C, Yan S. Supervised hashing for image retrieval via image representation learning. In: AAAI; 2014.

[69] Li WJ, Wang S, Kang WC. Feature learning based deep supervised hashing with pairwise labels. arXiv:1511.03855, 2015.

[70] Lai H, Pan Y, Liu Y, Yan S. Simultaneous feature learning and hash coding with deep neural networks. CVPR 2015.

[71] Sutskever I, Martens J, Dahl G, Hinton G. On the importance of initialization and momentum in deep learning. In: ICML; 2013. p. 1139–47.

[72] Elad M, Aharon M. Image denoising via sparse and redundant representations over learned dictionaries. TIP 2006;15(12):3736–45.

[73] Krizhevsky A, Hinton G. Learning multiple layers of features from tiny images. 2009.

[74] Oliva A, Torralba A. Modeling the shape of the scene: a holistic representation of the spatial envelope. IJCV 2001.

[75] Chua TS, Tang J, Hong R, Li H, Luo Z, Zheng Y. NUS-WIDE: a real-world web image database from National University of Singapore. In: ACM CIVR. ACM; 2009. p. 48.

[76] Liu W, Wang J, Kumar S, Chang SF. Hashing with graphs. In: ICML; 2011.

[77] Shakhnarovich G, Viola P, Darrell T. Fast pose estimation with parameter-sensitive hashing. In: ICCV. IEEE; 2003.

[78] Strecha C, Bronstein AM, Bronstein MM, Fua P. LDAHash: improved matching with smaller descriptors. TPAMI 2012;34(1):66–78.

[79] Müller H, Müller W, Squire DM, Marchand-Maillet S, Pun T. Performance evaluation in content-based image retrieval: overview and proposals. PRL 2001.

CHAPTER 8

Action Recognition

Yu Kong*, Yun Fu†
*B. Thomas Golisano College of Computing and Information Sciences, Rochester Institute of Technology, Rochester, NY, United States
†Department of Electrical and Computer Engineering and College of Computer and Information Science (Affiliated), Northeastern University, Boston, MA, United States

Contents

8.1.	Deeply Learned View-Invariant Features for Cross-View Action Recognition	183
	8.1.1 Introduction	183
	8.1.2 Related Work	185
	8.1.3 Deeply Learned View-Invariant Features	186
	8.1.4 Experiments	191
8.2.	Hybrid Neural Network for Action Recognition from Depth Cameras	198
	8.2.1 Introduction	198
	8.2.2 Related Work	199
	8.2.3 Hybrid Convolutional-Recursive Neural Networks	201
	8.2.4 Experiments	206
8.3.	Summary	209
	References	210

8.1. DEEPLY LEARNED VIEW-INVARIANT FEATURES FOR CROSS-VIEW ACTION RECOGNITION[1]

8.1.1 Introduction

Human action data are universal and are of interest to machine learning [1,2] and computer vision communities [3,4]. In general, action data can be captured from multiple views, such as multiple sensor views and various camera views, etc.; see Fig. 8.1. Classification on such action data in *cross-view* scenario is challenging since the raw data are captured by various sensor devices at different physical locations, and may look completely different. For example, in Fig. 8.1B, an action observed from side view and that observed from top view are visually different. Thus, it is less discriminative to use extracted features in one view than classifying actions in another view.

---

[1] ©2017 IEEE. Reprinted, with permission, from Yu Kong, Zhengming Ding, Jun Li, and Yun Fu. "Deeply learned view-invariant features for cross-view action recognition." IEEE Transactions on Image Processing (2017).

Deep Learning Through Sparse and Low-Rank Modeling
DOI: 10.1016/B978-0-12-813659-1.00008-1
183

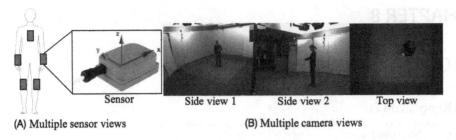

Figure 8.1 Examples of multiview scenarios: (A) multisensor-view, where multiple sensors (orange rectangles) are attached to torso, arms and legs, and human action data are recorded by these sensors; (B) multicamera-view, where human actions are recorded by multiple cameras at various viewpoints. (For interpretation of the colors in the figure(s), the reader is referred to the web version of this chapter.)

A line of work has been studied to build view–invariant representations for action recognition [5–9], where an action video is converted to a time series of frames. Approaches [5,6] take advantage of a so-called self-similarity matrix (SSM) descriptor to summarize actions in multiple views and have demonstrated their robustness in cross-view scenarios. Information shared between views is learned and transferred to each of the views in [7–9]. Their authors made an assumption that samples in different views contribute equally to the shared features. However, this assumption would be invalid as the cues in one view might be extraordinarily different from other views (e.g., the top view in Fig. 8.1B) and should have lower contribution to the shared features compared to other views. Furthermore, they do not constrain information sharing between action categories. This may produce similar features for videos in different classes but are captured from the same view, which would undoubtedly make classifiers confused.

We describe a deep network that classifies cross-view actions using the learned view–invariant features. A *sample-affinity matrix* (SAM) is introduced to measure the similarities between video samples in different camera views. This allows us to accurately balance information transfer between views and facilitates learning more informative shared features for cross-view action classification. The SAM structure controls information transfer among samples in different classes, which enables us to extract distinctive features in each class. Besides the shared features, private features are also learned to capture motion information exclusively existing in each view that cannot be modeled using shared features. We learn discriminative view–invariant information from shared and private features separately by encouraging incoherence between them. Label information and stacking multiple layers of features are used to further boost the performance of the network. This feature learning problem is formulated in a marginalized autoencoder framework (see Fig. 8.2) [10], particularly developed for learning view-invariant features. Specifically, cross-view shared features are summarized by one autoencoder, and private feature particularly for one view is learned using a group of autoencoders.

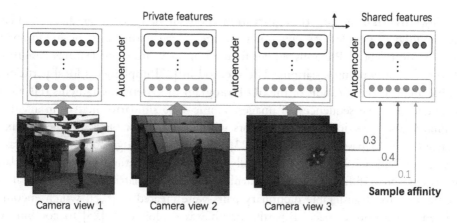

Figure 8.2 Overview of the method described for cross-view action recognition.

We obtain incoherence between the two types of features by encouraging the orthogonality between mapping matrices in the two categories of autoencoders. A Laplacian graph is built to encourage samples in the same action categories to have similar shared and private features. We stack multiple layers of features and learn them in a layerwise fashion. We evaluate our approach on two multi-view datasets, and show that our approach significantly outperforms state-of-the-art approaches.

8.1.2 Related Work

The aim of **multiview learning** methods is to find mutual agreement between two distinct views of data. Researchers made several attempts to learn more expressive and discriminative features from low-level observations [11–13,2,14–16]. Cotraining approach [17] finds consistent relationships between a pair of data points across different views by training multiple learning algorithms for each view. Canonical correlation analysis (CCA) was also used in [18] to learn a common space between multiple views. Wang et al. [19] proposed a method which learns two projection matrices to map multimodal data onto a common feature space, in which cross-modal data matching can be executed. Incomplete view problem was discussed in [20]. Its authors presumed that a shared subspace generated different views. A generalized multiview analysis (GMA) method was introduced in [21], and was proved to be a supervised extension of CCA. Liu et al. [13] took advantage of matrix factorization in multiview clustering. Their method leverages factors representing clustering structures gained from multiple views toward a common consensus. A collective matrix factorization (CMF) method was explored in [12], which obtains correlations among relational feature matrices. Ding et al. [16] proposed a low-rank constrained matrix factorization model, which works perfectly in the multiview learning scenario even if the view information of test data is unknown.

View-invariant action recognition methods are designed to predict action labels given multiview samples. As viewpoint changes, large within-class pose and appearance variation appear. Previous studies focused on view-invariant features designs that are robust to viewpoint variations. The method in [22] implements local partitioning and hierarchical classification of the 3D Histogram of Oriented Gradients (HOG) descriptor to produce sequences of images. Frame-wise similarity matrix in a video is computed, and view-invariant descriptors within a log-polar block on the matrix are extracted in SSM-based approaches [5,23]. Sharing knowledge among views was reviewed in [24,25,8,7,9,26–28]. Specifically, MRM-Lasso method in [9] captures latent corrections across different views by learning a low-rank matrix consisting of pattern-specific weights. Transferable dictionary pairs were created in [8,7], which encourage a shared sparse feature space. Bipartite graph was exploited in [25] to combine two view-dependent vocabularies into visual-word clusters called bilingual-words in order to bridge the semantic gap across view-dependent vocabularies.

8.1.3 Deeply Learned View-Invariant Features

The goal of this work is to extract view-invariant features that allow us to train the classification model on one (or multiple) view(s), and examine on the other view.

8.1.3.1 Sample-Affinity Matrix (SAM)

We introduce SAM to measure the similarity among pairs of video samples in multiple views. Suppose that we are given training videos of V views, $\{X^v, \mathbf{y}^v\}_{v=1}^V$. The data of the vth view X^v consist of N action videos, $X^v = [\mathbf{x}_1^v, \ldots, \mathbf{x}_N^v] \in \mathbb{R}^{d \times N}$ with corresponding labels $\mathbf{y}^v = [y_1^v, \ldots, y_N^v]$. SAM $Z \in \mathbb{R}^{VN \times VN}$ is interpreted as a block diagonal matrix

$$
Z = \mathrm{diag}(Z_1, \ldots, Z_N), \quad Z_i = \begin{pmatrix} 0 & z_i^{12} & \cdots & z_i^{1V} \\ z_i^{21} & 0 & \cdots & z_i^{2V} \\ \vdots & \vdots & \ddots & \vdots \\ z_i^{V1} & z_i^{V2} & \cdots & 0 \end{pmatrix},
$$

where $\mathrm{diag}(\cdot)$ creates a diagonal matrix, and $z_i^{\mu v}$ is the distance between two views in the ith sample computed by $z_i^{\mu v} = \exp(\|\mathbf{x}_i^v - \mathbf{x}_i^\mu\|^2 / 2c)$ parameterized by c.

Essentially, SAM Z captures within-class between-view information and between-class within-view information. Block Z_i in Z characterizes appearance variations in different views within one class. This explains how an action varies if view changes. Such information makes it possible to transfer information among views and build robust cross-view features. Additionally, since the off-diagonal blocks in SAM Z are zeros, this restricts information sharing among classes in the same view. Consequently, the features from different classes but in the same view are encouraged to be distinct. This enables us to differentiate various action categories if they appear similarly in some views.

8.1.3.2 Preliminaries on Autoencoders

Our approach is based on a popular deep learning approach, called autoencoder (AE) [29,10,30]. AE maps the raw inputs x to hidden units h using an "encoder" $f_1(\cdot)$, $h = f_1(x)$, and then maps the hidden units to outputs using a "decoder" $f_2(\cdot)$, $o = f_2(h)$. The objective of learning AE is to encourage similar or identical input–output pairs, where the reconstruction loss is minimized after decoding, $\min \sum_{i=1}^{N} \| x_i - f_2(f_1(x_i)) \|^2$. Here, N is the number of training samples. In this way, the neurons in the hidden layer are good representations for the inputs as the reconstruction process captures the intrinsic structure of the input data.

As opposed to the two-level encoding and decoding in AE, marginalized stacked denoising autoencoder [10] (mSDA) reconstructs the corrupted inputs with a single mapping W, $\min \sum_{i=1}^{N} \| x_i - W\tilde{x}_i \|^2$, where $\tilde{x}_i$ is the corrupted version of x_i obtained by setting each feature to 0 with a probability p. mSDA performs m passes over the training set, each time with different corruptions. This essentially performs a dropout regularization on the mSDA [31]. By setting $m \to \infty$, mSDA effectively computes the transformation matrix W that is robust to noise using infinitely many copies of noisy data. mSDA is stackable and can be calculated in closed-form.

8.1.3.3 Single-Layer Feature Learning

The single-layer feature learner described in this subsection builds on mSDA. We attempt to learn both discriminative shared features between multiple views and private features particularly owned by one view for cross-view action classification. Considering large motion variations in different views, we incorporate SAM Z in learning shared features to balance information transfer between views so as to build more robust features.

We use the following objective function to learn shared features and private features:

$$\min_{W,\{G^v\}} \mathcal{Q}, \quad \mathcal{Q} = \| W\tilde{X} - XZ \|_F^2 + \sum_v \Big[\alpha \| G^v \tilde{X}^v - X^v \|_F^2$$
$$+ \beta \| W^{\mathrm{T}} G^v \|_F^2 + \gamma \operatorname{Tr}(P^v X^v L X^{v\mathrm{T}} P^{v\mathrm{T}}) \Big], \tag{8.1}$$

where W is the mapping matrix for learning shared features, $\{G^v\}_{v=1}^{V}$ is a group of mapping matrices for learning private features particularly for each view, and $P^v = (W; G^v)$. The above objective function contains 4 terms: $\psi = \| W\tilde{X} - XZ \|_F^2$ learns shared features between views, which essentially reconstructs an action data from one view with the data from all views; $\phi_v = \| G^v \tilde{X}^v - X^v \|_F^2$ learns view-specific private features that are complementary to the shared features; $r_{1v} = \| W^{\mathrm{T}} G^v \|_F^2$ and $r_{2v} = \operatorname{Tr}(P^v X^v L X^{v\mathrm{T}} P^{v\mathrm{T}})$ are model regularizers. Here, r_{1v} reduces redundancy between two mapping matrices, and r_{2v} encourages the shared and private features of the same class and the same view to be similar, while α, β, γ are parameters balancing the importance of these components. Further details about these terms are discussed in the following.

Note that in cross-view action recognition, data from all the views are available in training to learn shared and private features. Data from some views are unavailable only in testing.

Shared Features. Humans can perceive an action from one view and envision what the action will look like if we observe from other views. This is possibly because we have studied similar actions before from multiple views. This inspires us to reconstruct an action data from one view (target view) using the action data from all the views (source view). In this way, information shared between views can be summarized and transferred to the target view.

We define the discrepancy between the data of the vth target view and the data of all the V source views as

$$\psi = \sum_{i=1}^{N} \sum_{v=1}^{V} \| W\tilde{x}_i^v - \sum_{u} x_i^u z_i^{uv} \|^2 = \| W\tilde{X} - XZ \|_F^2, \tag{8.2}$$

where z_i^{uv} is a weight measuring the contributions of the uth view action in the reconstruction of the sample x_i^v of the vth view, $W \in \mathbb{R}^{d \times d}$ is a single linear mapping for the corrupted input $\tilde{x}_i^v$ of all the views, $Z \in \mathbb{R}^{VN \times VN}$ is a sample-affinity matrix encoding all the weights $\{z_i^{uv}\}$. Matrices $X, \tilde{X} \in \mathbb{R}^{d \times VN}$ denote the input training matrix and the corresponding corrupted version of X, respectively [10]. The corruption essentially performs a dropout regularization on the model [31].

The SAM Z here allows us to precisely balance information transfer among views and assists learn more discriminative shared features. Instead of using equal weights [7, 8], we reconstruct the ith training sample of the vth view based on the samples from all V views with different contributions. As shown in Fig. 8.3, a sample of side view (source 1) will be more identical to the one also from side view (target view) than the one from top view (source 2). Thus, more weight should be given to source 1 in order to learn more descriptive shared features for the target view. Note that SAM Z limits information sharing across samples (off-diagonal blocks are zeros) since it cannot capture view-invariant information for cross-view action recognition.

Private Features. Besides the information shared across views, there is still some remaining discriminative information that exclusively exists in each view. In order to utilize such information and make it robust to viewpoint variations, we adopt the robust feature learning in [10], and learn view-specific private features for the samples in the vth view using a mapping matrix $G^v \in \mathbb{R}^{d \times d}$,

$$\phi_v = \sum_{i=1}^{N} \| G^v \tilde{x}_i^v - x_i^v \|^2 = \| G^v \tilde{X}^v - X^v \|_F^2. \tag{8.3}$$

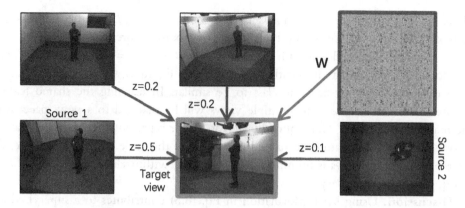

Figure 8.3 Learning shared features using weighted samples.

Here, $\tilde{X}^v$ is the corrupted version of the feature matrix X^v of the vth view. We will learn V mapping matrices $\{G^v\}_{v=1}^{V}$ given corresponding inputs of different views.

It should be noted that using Eq. (8.3) may also capture some redundant shared information from the vth view. In this work, we reduce such redundancy by encouraging the incoherence between the view-shared mapping matrix W and view-specific mapping matrix G^v,

$$r_{1v} = \| W^{\mathrm{T}} G^v \|_F^2. \tag{8.4}$$

The incoherence between W and $\{G^v\}$ enables our approach to independently exploit the discriminative information included in the view-specific features and view-shared features.

Label Information. Large motion and posture variations may appear in action data captured from various views. Therefore, the shared and private features extracted using Eqs. (8.2) and (8.3) may not be discriminative enough for actions classification with large variations. We enforce the shared and private features of the same class and same view to be similar to address the issue. A within-class within-view variance is defined to regularize the learning of the view-shared mapping matrix W and view-specific mapping matrix G^v as

$$
\begin{aligned}
r_{2v} &= \sum_{i=1}^{N} \sum_{j=1}^{N} \left[\| W\boldsymbol{x}_i^v - W\boldsymbol{x}_j^v \|^2 + \| G^v\boldsymbol{x}_i^v - G^v\boldsymbol{x}_j^v \|^2 \right] \\
&= \mathrm{Tr}(WX^v L X^{v\mathrm{T}} W^{\mathrm{T}}) + \mathrm{Tr}(G^v X^v L X^{v\mathrm{T}} G^{v\mathrm{T}}) \\
&= \mathrm{Tr}(P^v X^v L X^{v\mathrm{T}} P^{v\mathrm{T}}).
\end{aligned}
\tag{8.5}
$$

Here, $L \in \mathbb{R}^{N \times N}$ is the label-view Laplacian matrix, $L = D - A$, D is the diagonal degree matrix with $D_{(i,i)} = \sum_{j=1}^{N} a_{(i,j)}$, A is the adjacent matrix that represents the label relationships of training videos. The (i,j)th element $a_{(i,j)}$ in A is 1 if $y_i = y_j$ and 0 otherwise.

Note that since we have implicitly used this idea in Eq. (8.2), we do not need features from different views in the same class to be similar. In learning the shared feature, features of the same class from multiple views will be mapped to a new space using the mapping matrix W. Consequently, we can better represent the projected features of one sample by the features from multiple views of the same sample. Therefore, the discrepancy among views is minimized, and thus the within-class cross-view variance in Eq. (8.5) is not necessary.

Discussion. Using label information in Eq. (8.5) contributes to a supervised approach. We can also replace this term with an unsupervised one by making $\gamma = 0$. We refer to the **unsupervised** approach as **Ours-1** and the **supervised** approach as **Ours-2** in the following discussions.

8.1.3.4 Learning

We develop a coordinate descent algorithm to solve the optimization problem in Eq. (8.1) and optimize parameters W and $\{G^v\}_{v=1}^{V}$. More specifically, in each step, one parameter matrix is updated by fixing the others, and computing the derivative of Q w.r.t. to the parameter and setting it to 0.

Update W. Parameters $\{G^v\}_{v=1}^{V}$ are fixed in updating W, which can be updated by setting the derivative $\frac{\partial Q}{\partial W} = 0$, yielding

$$W = \left[\sum_v (\beta G^v G^{vT} + \gamma X^v L X^{vT} + I) \right]^{-1}$$
$$\cdot (XZ\tilde{X}^T)[\tilde{X}\tilde{X}^T + I]^{-1}. \tag{8.6}$$

It should be noted that $XZ\tilde{X}^T$ and $\tilde{X}\tilde{X}^T$ are computed by repeating the corruption $m \to \infty$ times. By the weak law of large numbers [10], $XZ\tilde{X}^T$ and $\tilde{X}\tilde{X}^T$ can be computed by their expectations $E_p(XZ\tilde{X}^T)$ and $E_p(\tilde{X}\tilde{X}^T)$ with the corruption probability p, respectively.

Update G^v. Fixing W and $\{G^u\}_{u=1,u \neq v}^{V}$, parameter G^v is updated by setting the derivative $\frac{\partial Q}{\partial G^v} = 0$, giving

$$G^v = \left(\beta WW^T + \gamma X^v L X^{vT} + I \right)^{-1}$$
$$\cdot (\alpha X^v \tilde{X}^{vT})[\alpha \tilde{X}^v \tilde{X}^{vT} + I]^{-1}. \tag{8.7}$$

Similar to the procedure of updating W, $X^v \tilde{X}^{vT}$ and $\tilde{X}^v \tilde{X}^{vT}$ are computed by their expectations with corruption probability p.

Convergence. Our learning algorithm iteratively updates W and $\{G^v\}_{v=1}^V$. The problem in Eq. (8.1) can be divided into $V + 1$ subproblems, each of which is a convex problem with respect to one variable. Therefore, by solving the subproblems alternatively, the learning algorithm is guaranteed to find an optimal solution to each subproblem. Finally, the algorithm will converge to a local solution.

8.1.3.5 Deep Architecture

Inspired by the deep architecture in [10,32], we also design a deep model by stacking multiple layers of feature learners described in Sect. 8.1.3.3. A nonlinear feature mapping is performed layer by layer. More specifically, a nonlinear squashing function $\sigma(\cdot)$ is applied on the output of one layer, $H_w = \sigma(WX)$ and $H_g^v = \sigma(G^v X^v)$, resulting in a series of hidden feature matrices.

A layer-wise training scheme is used in this work to train the networks $\{W_k\}_{k=1}^K$, $\{G_k^v\}_{k=1,v=1}^{K,V}$ with K layers. Specifically, the outputs of the fth layer H_{kw} and H_{kg}^v are used as the input to the $(k+1)$th layer. The mapping matrices W_{k+1} and $\{G_{k+1}^v\}_{v=1}^V$ are then trained using these inputs. For the first layer, the inputs H_{0w} and H_{0g}^v are the raw features X and X^v, respectively. More details are shown in Algorithm 8.1.

Algorithm 8.1: Learning deep sequential context networks.

1: **Input:** $\{(\boldsymbol{x}_i^v, y_i)\}_{i=1,v=1}^{N,V}$.

2: **Output:** $\{W_k\}_{k=1}^K$, $\{G_k^v\}_{k=1,v=1}^{K,V}$.

3: **for** Layer $k = 1$ to K **do**

4:    Input $H_{(k-1)w}$ for learning W_k.

5:    Input $H_{(k-1)g}^v$ for learning G_k^v.

6:    **while** not converged **do**

7:       Update W_k using Eq. (8.6);

8:       Update $\{G_k^v\}_{v=1}^V$ using Eq. (8.7);

9:    **end while**

10:    Compute H_{kw} by: $H_{kw} = \sigma(W_k H_{(k-1)w})$.

11:    Compute $\{H_{kg}^v\}_{v=1}^V$ by: $H_{kg}^v = \sigma(G_k^v H_{(k-1)g}^v)$.

12: **end for**

8.1.4 Experiments

We evaluate Ours-1 and Ours-2 approaches on two multiview datasets: multiview IXMAS dataset [33], and the Daily and Sports Activities (DSA) dataset [1], both of which have been popularly used in [1,24,25,8,7,9].

We consider two cross-view classification scenarios in this work, many-to-one and one-to-one. The former trains on $V - 1$ views and tests on the remaining view, while

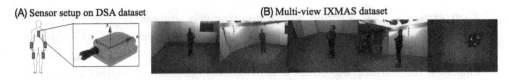

(A) Sensor setup on DSA dataset   (B) Multi-view IXMAS dataset

Figure 8.4 Examples of multi-view problem settings: (A) multiple sensor views in the Daily and Sports Activities (DSA) dataset, and (B) multiple camera views in the IXMAS.

the latter trains on one view and tests on the other views. For the vth view that is used for testing, we simply set the corresponding X^v used in training to $\mathbf{0}$ in Eq. (8.1) during training. Intersection kernel support vector machine (IKSVM) with parameter $C = 1$ is adopted as the classifier. Default parameters are $\alpha = 1, \beta = 1, \gamma = 0, K = 1, p = 0$ for Ours-1 approach, and $\alpha = 1, \beta = 1, \gamma = 1, K = 1, p = 0$ for Ours-2 approach unless specified. The default number of layers is set to 1 for efficiency consideration.

IXMAS is a multicamera-view video dataset, where each view corresponds to a camera view (see Fig. 8.4B). The IXMAS dataset consists of 12 actions performed by 10 actors. An action was recorded by 4 side view cameras and 1 top view camera. Each actor repeated one action 3 times.

We adopt the *bag-of-words* model in [34]. An action video is described by a set of detected local spatiotemporal trajectory-based and global frame-based descriptors [35]. A k-means clustering method is employed to quantize these descriptors and build so-called *video words*. Consequently, a video can be represented by a histogram of the video words detected in the video, which is essentially a feature vector. An action captured by V camera views is represented by V feature vectors, each of which is a feature representation for one camera view.

DSA is a multisensor-view dataset comprising 19 daily and sports activities (e.g., sitting, playing basketball, and running on a treadmill with a speed of 8 km/h), each performed by 8 subjects in their own style for 5 minutes. Five Xsens MTx sensor units are used on the torso, arms, and legs (Fig. 8.4A), resulting in a 5-view data representation. Sensor units are calibrated to acquire data at 25 Hz sampling frequency. The 5-min signals are divided into 5-second segments so that $480(= 60 \text{ seconds} \times 8 \text{ subjects})$ signal segments are obtained for each activity. One 5-second segment is used as an action time series in this work.

We follow [1] to preprocess the raw action data in a 5-s window, and represent the data as a 234-dimensional feature vector. Specifically, the raw action data is represented as a 125×9 matrix, where 125 is the number of sampling points ($125 = 25 \text{ Hz} \times 5 \text{ s}$), and 9 is the number of values (the x, y, z axes' acceleration, the x, y, z axes' rate of turn, and the x, y, z axes' Earth's magnetic field) obtained on one sensor. We first compute the minimum and maximum values, the mean, skewness, and kurtosis on the data matrix. The resulting features are concatenated and generate a 45-dimensional (5 features $\times$ 9 axes) feature vector. Then, we compute discrete Fourier transform on the raw data matrix, and select the maximum 5 Fourier peaks. This yields a 45-dimensional (5 peaks $\times$ 9 axes) feature vector. The 45 frequency values that correspond to these Fourier peaks are also extracted, resulting in a 45-dimensional (5 frequency $\times$ 9 axes) as well. Afterwards, 11 autocorrelation samples are computed for each of the 9 axes, resulting in a 99-dimensional (11 samples $\times$ 9 axes) features. The three types of features are concatenated and generate a 234-dimensional feature vector, representing the human motion captured by one sensor in a 5-second window. A human action captured by V sensors is represented by V feature vectors, each of which corresponds to a sensor view.

8.1.4.1 IXMAS Dataset

Dense trajectory and histogram of oriented optical flow [35] are extracted from videos. A dictionary of size 2000 is built for each type of features using k-means. We use the bag-of-words model to encode these features, and represent each video as a feature vector.

We adopt the same leave-one-action-class-out training scheme in [25,7,8] for fair comparison. At each time, one action class is used for testing. In order to evaluate the effectiveness of the information transfer in our approaches, all the videos in this action are excluded from the feature learning procedure including k-means and our approaches. Note that these videos can be seen in training the action classifiers. We evaluate both the unsupervised approach (**Ours-1**) and the supervised approach (**Ours-2**).

One-to-One Cross-view Action Recognition

This experiment trains on data from one camera view (training view), and tests the on data from the other view (test view). We only use the learned shared features and discard the private features in this experiment as the private features learned on one view does not capture too much information of the other view.

We compare Ours-2 approach with [36,7,8] and report recognition results in Table 8.1. Ours-2 achieves the best performance in 18 out of 20 combinations, significantly better than all the compared approaches. It should be noted that Ours-2 achieves 100% in 16 cases, demonstrating the effectiveness of the learned shared features. Thanks to the abundant discriminative information from the learned shared features and label information, our approach is robust to viewpoint variations and can achieve high performance in cross-view recognition.

We also compare Ours-1 approach with [25,7,8,24,37], and report comparison results in Table 8.2. Our approach achieves the best performance in 19 out of 20 combinations. In some cases, our approach outperforms the comparison approaches by a large margin, for example, C4 → C0 (C4 is the training view and C0 is the test view), C4 → C1, and C1 → C3. The overall performance of Ours-1 is slightly worse than Ours-2 due to the removal of the label information.

Table 8.1 One-to-one cross-view recognition results of various supervised approaches on IXMAS dataset. Each row corresponds to a training view (from view C0 to view C4) and each column is a test view (also from view C0 to view C4). The results in brackets are the recognition accuracies of [36,7,8] and our supervised approach, respectively

	C0	C1	C2	C3	C4
C0	NA	(79, 98.8, 98.5, **100**)	(79, 99.1, **99.7**, **99.7**)	(68, 99.4, 99.7, **100**)	(76, 92.7, 99.7, **100**)
C1	(72, 98.8, **100**, **100**)	NA	(74, **99.7**, 97.0, **99.7**)	(70, 92.7, 89.7, **100**)	(66, 90.6, **100**, 99.7)
C2	(71, 99.4, 99.1, **100**)	(82, 96.4, 99.3, **100**)	NA	(76, 97.3, **100**, **100**)	(72, 95.5, 99.7, **100**)
C3	(75, 98.2, 90.0, **100**)	(75, 97.6, 99.7, **100**)	(73, **99.7**, 98.2, 99.4)	NA	(76, 90.0, 96.4, **100**)
C4	(80, 85.8, 99.7, **100**)	(77, 81.5, 98.3, **100**)	(73, 93.3, 97.0, **100**)	(72, 83.9, 98.9, **100**)	NA
Ave.	(74, 95.5, 97.2, **100**)	(77, 93.6, 98.3, **100**)	(76, 98.0, 98.7, **99.7**)	(73, 93.3, 97.0, **100**)	(72, 92.4, 98.9, **99.9**)

Table 8.2 One-to-one cross-view recognition results of various unsupervised approaches on IXMAS dataset. Each row corresponds to a training view (from view C0 to view C4) and each column is a test view (also from view C0 to view C4). The results in brackets are the recognition accuracies of [25,7,8,24,37] and our unsupervised approach, respectively

	C0	C1	C2	C3	C4
C0	NA	(79.9, 96.7, 99.1, 92.7, 94.8, **99.7**)	(76.8, 97.9, 90.9, 84.2, 69.1, **99.7**)	(76.8, 97.6, 88.7, 83.9, 83.9, **98.9**)	(74.8, 84.9, 95.5, 44.2, 39.1, **99.4**)
C1	(81.2, 97.3, 97.8, 95.5, 90.6, **100**)	NA	(75.8, 96.4, 91.2, 77.6, 79.7, **99.7**)	(78.0, 89.7, 78.4, 86.1, 79.1, **99.4**)	(70.4, 81.2, 88.4, 40.9, 30.6, **99.7**)
C2	(79.6, 92.1, 99.4, 82.4, 72.1, **100**)	(76.6, 89.7, 97.6, 79.4, 86.1, **99.7**)	NA	(79.8, 94.9, 91.2, 85.8, 77.3, **100**)	(72.8, 89.1, **100**, 71.5, 62.7, 99.7)
C3	(73.0, 97.0, 87.6, 82.4, 82.4, **100**)	(74.1, 94.2, 98.2, 80.9, 79.7, **100**)	(74.0, 96.7, 99.4, 82.7, 70.9, **100**)	NA	(66.9, 83.9, 95.4, 44.2, 37.9, **100**)
C4	(82.0, 83.0, 87.3, 57.1, 48.8, **99.7**)	(68.3, 70.6, 87.8, 48.5, 40.9, **100**)	(74.0, 89.7, 92.1, 78.8, 70.3, **100**)	(71.1, 83.7, 90.0, 51.2, 49.4, **100**)	NA
Ave	(79.0, 94.4, 93.0, 79.4, 74.5, **99.9**)	(74.7, 87.8, 95.6, 75.4, 75.4, **99.9**)	(75.2, 95.1, 93.4, 80.8, 72.5, **99.9**)	(76.4, 91.2, 87.1, 76.8, 72.4, **99.9**)	(71.2, 84.8, 95.1, 50.2, 42.6, **99.7**)

Table 8.3 Many-to-one cross-view action recognition results on IXMAS dataset. Each column corresponds to a test view

Methods	C0	C1	C2	C3	C4
Junejo et al. [5]	74.8	74.5	74.8	70.6	61.2
Liu and Shah [38]	76.7	73.3	72.0	73.0	N/A
Weinland et al. [22]	86.7	89.9	86.4	87.6	66.4
Liu et al. [25]	86.6	81.1	80.1	83.6	82.8
Zheng et al. [7]	98.5	99.1	99.1	100	90.3
Zheng and Jiang [8]-1	97.0	99.7	97.2	98.0	97.3
Zheng and Jiang [8]-2	99.7	99.7	98.8	99.4	99.1
Yan et al. [6]	91.2	87.7	82.1	81.5	79.1
No-SAM	95.3	93.9	95.3	93.1	94.7
No-private	98.6	98.1	98.3	99.4	100
No-incoherence	98.3	97.5	98.9	98.1	100
Ours-1 (unsupervised)	100	99.7	100	100	99.4
Ours-2 (supervised)	100	100	100	100	100

Many-to-One Cross-view Action Recognition

In this experiment, one view is used as test view and all the other views are used as training views. We evaluate the performance of our approaches in this experiment, which use both the learned shared and private features.

Our unsupervised (Ours-1) and supervised (Ours-2) approaches are compared with existing approaches [38,5,22,25,7,8,6]. The importance of SAM Z in Eq. (8.2), the incoherence in Eq. (8.4), and the private features in Ours-2 model are also evaluated.

Table 8.3 shows that our supervised approach (Ours-2) achieves an impressive 100% recognition accuracy in all the 5 cases, and Ours-1 achieves an overall accuracy of 99.8%. Ours-1 and Ours-2 achieve superior overall performance over all the other comparison approaches, demonstrating the benefit of using both shared and private features in this work. Our approaches use the sample-affinity matrix to measure the similarities between video samples across camera views. Consequently, the learned shared features accurately characterize the commonness across views. In addition, the redundancy is reduced between shared and private features, making the learned private features more informative for classification. Although the two methods in [8] exploit private features as well, they do not measure different contributions of samples in learning the shared dictionary, making the shared information less discriminative.

Ours-2 outperforms No-SAM approach, suggesting the effectiveness of SAM Z. Without SAM Z, No-SAM treats samples across views equally, and thus cannot accurately weigh the importance of samples in different views. The importance of the private features can be clearly seen from the performance gap between Ours-2 and No-private approach. Without private features, the No-private approach only uses shared

Table 8.4 Many-to-one cross-view action classification results on DSA dataset. Each column corresponds to a test view. V0–V4 are sensor views on torso, arms, and legs

Methods	Overall	V0	V1	V2	V3	V4
IKSVM	54.6	36.5	53.4	63.4	60.1	59.7
DRRL [39]	55.4	35.5	56.7	62.1	61.7	60.9
mSDA [10]	56.1	34.4	57.7	62.8	61.5	64.1
No-SAM	55.4	35.1	57.0	60.7	62.2	62.2
No-private	55.4	35.1	57.0	60.7	62.2	62.1
No-incoherence	55.4	35.1	56.9	60.7	62.2	62.2
Ours-1	**57.1**	**35.7**	**57.4**	**64.4**	**64.2**	**63.9**
Ours-2	**58.0**	**36.1**	**58.9**	**65.8**	**64.2**	**65.2**

features for classification, which are not discriminative enough if some informative motion patterns exclusively exist in one view and are not sharable across views. The performance variation between Ours-2 and the No-incoherence method suggests the benefit of encouraging the incoherence in Eq. (8.4). Using Eq. (8.4) allows us to reduce the redundancy between shared and private features, and helps extract discriminative information in each of them. Ours-2 slightly outperforms Ours-1 in this experiment, indicating the effectiveness of using label information in Eq. (8.5).

8.1.4.2 Daily and Sports Activities Data Set

Many-to-One Cross-view Action Classification

In this experiment, data from 4 sensors are used for training (36,480 time series) and the data from the remaining 1 sensor (9,120 time series) are used for testing. This process is repeated 5 times and the average results are reported.

Our unsupervised (**Ours-1**) and supervised (**Ours-2**) approaches are compared with mSDA [10], DRRL [39] and IKSVM. The importance of SAM Z in Eq. (8.2), the incoherence in Eq. (8.4), and the private features in Ours-2 model are also evaluated. We remove Z in Eq. (8.2) and the incoherence component in Eq. (8.4) from the supervised model, respectively, and obtain the "No-SAM", and the "No-incoherence" model. We also remove the learning of parameter $\{G^v\}_{v=1}^{V}$ from the supervised model and obtain the "No-private" model. Comparison results are shown in Table 8.4.

Ours-2 achieves superior performance over all the other comparison methods in all the 5 cases with an overall recognition accuracy of 58.0%. Ours-2 outperforms Ours-1 by 0.9% in overall classification result due to the use of label information. Note that cross-view classification on DSA dataset is challenging as the sensors on different body parts are weakly correlated. The sensor on torso (V0) has the weakest correlations with the other four sensors on arms and legs. Therefore, results of all the approaches on V0 are the lowest performance compared to sensors V1–V4. Ours-1 and Ours-2 achieve superior overall performance over the comparison approaches IKSVM and mSDA due to the use of both shared and private features. IKSVM and mSDA do not discover shared and private features, and thus cannot use correlations between views and exclusive information in each view for classification. To better balance the information transfer between views, Ours-1 and Ours-2 use the sample-affinity matrix to measure the similarities between video samples across camera views. Thus, the learned shared features accurately characterize the commonness across views. Though the overall improvement of Ours-1 and Ours-2 over mSDA is 1% and 1.9%, Ours-1 and Ours-2 correctly classify 456 and 866 more sequences than mSDA in this experiment, respectively.

The performance gap between Ours-2 and the No-SAM approach suggests the effectiveness of SAM Z. Without SAM Z, No-SAM treats samples across views equally, and thus cannot accurately weigh the importance of samples in different views. Ours-2 outperforms No-private approach, suggesting the importance of the private features in learning discriminative features for multiview classification. Without private features, No-private approach only uses shared features for classification, which are not discriminative enough if some informative motion patterns exclusively exist in one view and are not sharable across views. Ours-2 achieves superior performance over No-incoherence method, indicating the benefit of encouraging the incoherence in Eq. (8.4). Using Eq. (8.4) allows us to reduce the redundancy between shared and private features, and helps extract discriminative information in each of them. Ours-2 slightly outperforms Ours-1, indicating the effectiveness of using label information in Eq. (8.5).

8.2. HYBRID NEURAL NETWORK FOR ACTION RECOGNITION FROM DEPTH CAMERAS

8.2.1 Introduction

Using depth cameras for action recognition is receiving increasing interest in computer vision community due to the recent advent of the cost-effective Kinect. Depth sensors provide several advantages over typical visible light cameras. Firstly, 3D structural information can be easily captured, which helps simplify the intra-class motion variation. Secondly, depth information provides useful cues for background subtraction and occlusion detection. Thirdly, depth data are generally not affected by the lighting variations, and thus it is a robust information in different lighting conditions.

Unfortunately, improving the recognition performance via depth data is not an easy task. One reason is that depth data are noisy, and may have spatial and temporal discontinuities when undefined depth points exist. Existing methods resort to mining discriminative actionlets from noisy data [40], exploiting a sampling scheme [41], or developing depth spatiotemporal interest point detectors [42,43], in order to overcome the problem of noisy depth data. However, these methods directly use low-level features, which may not be expressive enough for discriminating depth videos. Another problem is that depth information alone is not discriminative enough, as most of the body parts in different actions have similar depth values. It is desirable to extract useful information from depth data, e.g., surface normals in the 4D space [44] and 3D human shapes [45], and then use additional cues effectively to improve the performance, e.g., joint data [40,46]. It should be noted that most of the existing approaches for depth action videos only capture low-order context, such as hand–arm and foot–leg. High-order context such as head–arm–leg and torso–arm–leg is not considered. In addition, all these methods depend on hand-crafted, problem-dependent features, whose importance for the recognition task is rarely known. This is extremely noticeable since intra-class action data are generally highly varied, and inter-class action data usually appear similar.

In this chapter, we describe a hybrid convolutional-recursive neural network (HCRNN), a cascade of 3D convolutional neural network (3D-CNN) and 3D recursive neural network (3D-RNN), to learn high-order, compositional features for recognizing RGB-D action videos. The hierarchical nature of HCRNN helps us abstract low-level features to yield powerful features for action recognition. HCRNN models the relationships between local neighboring body parts and allows deformable body parts in actions. This makes our model robust to pose variations and geometry changes in intra-class RGB-D action data. In addition, HCRNN captures high-order body part context information in RGB-D action data, which is significantly important for learning actions with large pose variations [47,48]. A new 3D convolution is performed on spatiotemporal 3D patches, thereby capturing rich motion information in adjacent frames and reducing noise. We organize all the components in HCRNN in different layers, and train HCRNN in an unsupervised fashion without network tuning. More importantly, we demonstrate that high quality features can be learned by 3D-RNNs, even with random weights.

The goal of HCRNN is to learn discriminative features from RGB-D videos. As the flowchart illustrated in Fig. 8.5, HCRNN starts with raw RGB-D videos, and first extract features from each of the RGB and depth modalities separately. The two separate modalities, RGB and depth data, are then fed into 3D convolutional neural networks (3D-CNN), and are convolved with K filters, respectively. 3D-CNN outputs translational invariant low-level features, a matrix of filter responses. These features are then given to the 3D-RNN to learn compositional high-order features. To improve feature discriminability, multiple 3D-RNNs are jointly employed to learn the features. The final feature vectors learned by all the 3D-RNNs from all the modalities are combined into a single feature vector, which is the action representation for the input RGB-D video. The softmax classifier is applied to recognize the RGB-D action video.

8.2.2 Related Work

RGBD Action Recognition. In depth videos, depth images generally contain undefined depth points, causing spatial and temporal discontinuities. This is an obstacle

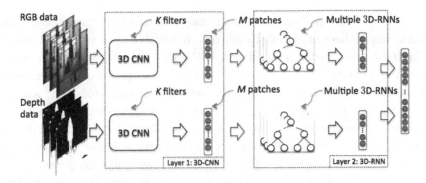

Figure 8.5 Architecture of the our hybrid convolutional-recursive neural network (HCRNN) model. Given a RGB-D video, HCRNN learns a discriminative feature vector from both RGB and depth data. We use 3D-CNN to learn features of local neighbor body parts, and 3D-RNN to learn compositional features hierarchically.

for using informative depth information. For example, popular spatio-temporal interest point (STIP) detectors [34,49] can not be applied to depth videos directly, as they will falsely fire on those discontinuous black regions [44]. To overcome this problem, counterparts of the popular spatio-temporal interest point (STIP) detectors [34,49] for depth videos have been proposed in [42,43]. Depth STIP [42], a filtering method, was introduced to detect interest points from RGB-D videos with noise reduction.

To obtain useful information from noisy depth videos, [40] proposed to select informative joints that are most relevant to the recognition task. Consequently, an action can be represented by subsets of joints (actionlets), and is learned by the multiple-kernel SVM, where each kernel corresponds to an actionlet. In [44], the histogram of oriented 4D surface normals (HON4D) are computed to effectively exploit geometric changing structure of actions in depth videos. Li et al. [45] projected depth maps onto 2D planes and sampled a set of points along the contours of the projections. Then, the points are clustered to obtain salient postures. GMM is further used to model the postures, and an action graph is applied for inference. Holistic features [14] and human pose (joint) information are also used for action recognition from RGB-D videos [46,50,51].

Hollywood 3D dataset, a new 3D action dataset, was released in [43], and evaluated using both conventional STIP detectors and their extensions for depth videos. Results show that those new STIP detectors for depth videos can effectively take advantage of depth data and suppress false detections caused by the spatial and temporal discontinuities in depth images.

Applications Using Deep Models. In recent years, feature learning using deep models have been successfully applied in object recognition [52–54] and detection [55, 56], scene understanding [57,58], face recognition and action recognition [59–61].

Feature learning methods for object recognition are generally composed of a filter bank, a nonlinear transformation, and some pooling layers. To evaluate their influences, [52] built several hierarchies by different combinations of those components, and reported their performance on object recognition and handwritten digits recognition datasets. 3D object recognition task were also addressed in [53,54]. In [55], mutual visibility relationship in pedestrian detection is modeled by summarizing human body part interaction in a hierarchical way.

Deep models have achieved promising results in conventional action recognition tasks. Convolutional neural network [59] was applied to extract features from both the spatial and temporal dimensions by performing 3D convolutions. An unsupervised gate RBM model [61] was proposed for action recognition. Le et al. [60] combined independent subspace analysis with deep learning techniques to build features robust to local translation. All these methods are designed for color videos. In this chapter, we introduce a deep architecture for recognizing actions from RGB-D videos.

8.2.3 Hybrid Convolutional-Recursive Neural Networks

We describe the hybrid convolutional-recursive neural networks (HCRNN) to learn high-order compositional features for action recognition from depth camera. The HCRNN consists of two components, the 3D-CNN model and the 3D-RNN model. The 3D-CNN model is utilized to generate low-level, translationally invariant features, and the 3D-RNN model is employed to compose high-order features that can be used to classify actions. Architecture is shown in Fig. 8.5, which is a cascade of 3D-CNN and 3D-RNN.

8.2.3.1 Architecture Overview

Our method takes an RGB-D video v as input and outputs the corresponding action label. Our HCRNN is employed to find a transform h that maps the RGB-D video into a feature vector x, $x = h(v)$. Feature vector x is then fed into a classifier, and the action label y is obtained. We treat an RGB-D video as multichannel data, and extract gray, gradient, optical flow and depth data from RGB and depth modalities. HCRNN is applied to each of these channels.

3D-CNN. The lower part of our HCRNN is a 3D-CNN model that extracts features hierarchically. The 3D-CNN (Fig. 8.6) in this work has five stages: 3D convolution (Sect. 8.2.3.2), absolute rectification, local normalization, average pooling and subsampling. We sample N 3D patches of size (s_r, s_c, s_t) (height, width, frame) from each channel with stride size s_p. The 3D-CNN g takes these patches as the input, and outputs a K-dimensional vector u for each patch, $g : \mathbb{R}^S \to \mathbb{R}^K$. Here, K is the number of learned filters and $S = s_r \times s_c \times s_t$.

Rich motion and geometry change information is captured using the 3D convolution in 3D-CNN. Each video of size (height, width, frame) d_I is convolved with K

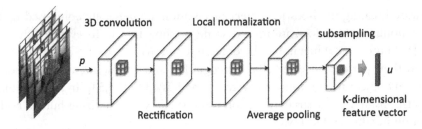

Figure 8.6 Graphical illustration of 3D-CNN. Given a 3D video patch p, 3D-CNN g performs five stages of computations: 3D convolution, rectification, local normalization, average pooling and subsampling. Then the K-dimensional feature vector u is obtained, $u = g(p)$.

filters of size d_p, resulting in K filter responses of dimensionality N (N is the number of patches extracted from one channel of a video). Then, absolute rectification is performed which applies absolute value function to all the components of the filter responses. This step is followed by local contrast normalization (LCN). LCN module performs local subtractive and divisive normalization, enforcing a sort of local competition between adjacent features in a feature map, and between features at the same spatiotemporal location in different feature maps. To improve the robustness of features to small distortions, we add average pooling and subsampling modules to 3D-CNN. Patch features whose locations are within a small spatiotemporal neighborhood are averaged and pooled to generate one parent feature of dimensionality K.

In this chapter, we augment gray and depth feature maps (channels) with gradient-x, gradient-y, optflow-x, and optflow-y as in [59]. The gradient-x and gradient-y feature maps are computed by computing the gradient along the horizontal and vertical directions, respectively. The optflow-x and optflow-y feature maps are computed by running the optical flow algorithm and separating the horizontal and vertical flow field data.

3D-RNN. The 3D-RNN model is to hierarchically learn compositional features. The graphical illustration of 3D-RNN is shown in Fig. 8.7. It merges a spatiotemporal block of patch feature vectors and generate a parent feature vector. The input for 3D-RNN is a $K \times M$ matrix, where M is the number of feature vectors generated by 3D-CNN ($M \neq N$ since we apply subsampling in 3D-CNN). The output of 3D-RNN is a K-dimensional vector, which is the feature for one channel of the video. We adopt a tree-structured 3D-RNN with J layers. At each layer, child vectors, whose spatiotemporal locations are within a 3D block, are merged into one parent vector. This procedure continues to generate one parent vector in the top layer. We concatenate feature vectors of all the channels generated by 3D-RNN, and derive a KC-dimensional vector as the action feature given C channels of data.

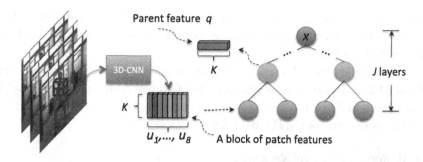

Figure 8.7 Graphical illustration of 3D-RNN. $u_1, \ldots, u_8$ are a block of patch features generated by 3D-CNN. 3D-RNN takes these features as inputs and produces a parent feature vector q. 3D-RNN recursively merges adjacent feature vectors, and generates feature vector x for one channel of data.

8.2.3.2 3D Convolutional Neural Networks

We use 3D-CNN to extract features from RGB-D action videos. 2D-CNNs have been successfully used in 2D images, such as object recognition [52–54] and scene understanding [57,58]. In these methods, 2D convolution are adopted to extract features from local neighborhood on the feature map. Then an addictive bias is added and a sigmoid function is used for feature mapping. However, in action recognition, a 3D convolution is desired as successive frames in videos encode rich motion information. In this work, we develop a new 3D convolution operation for RGB-D videos.

The 3D convolution in the 3D-CNN is achieved by convolving a filter on 3D patches extracted from RGB-D videos. Consequently, local spatiotemporal motion and structure information can be well captured in the convolution layer. It also captures motion and structure relationships of body parts in a local neighborhood, e.g., arm–hand and leg–foot. Suppose p is a 3D spatiotemporal patch randomly extracted from a video. We apply a nonlinear mapping to map p onto the feature map in the next layer,

$$g_k(p) = \max(\bar{d}_k - \|p - z_k\|_2, 0), \tag{8.8}$$

where $\bar{d}_k = \frac{1}{K}\sum_k \|x - z_k\|_2$ is the averaged distance of the sample p to all the filters, z_k is the kth filter, and K is the number of the learned filters. Note that the convolution in Eq. (8.8) is different from [59,62], but philosophically similar; $\bar{d}_k$ can be considered as the bias, $\|p - z_k\|_2$ is similar to the convolution operation, which is a similarity measure between the filter z_k and the patch p. The $\max(\cdot)$ function is a nonlinear mapping function as the sigmoid function. The filters in Eq. (8.8) are easily trained in unsupervised fashion (Sect. 8.2.3.5) by running k-means [63].

After 3D convolution, rectification, local contrast normalization and averaged down sampling are applied as in object recognition but they are performed on 3D patches.

3D-CNN generates a list of K-dimensional vectors $u_i^{rct} \in U$ ($i = 1, \ldots, M$), where r, c, and t are the locations of the vector in row, column and temporal dimensions, respectively, U is the set of vectors generated by 3D-CNN. Each vector in U is the feature for a patch. All these patch features are then given as inputs to 3D-RNN to compose high-order features.

Our 3D-CNN extracts discriminative motion and geometry changes information from RGB-D data. It also captures relationships of human body parts in local neighborhood, and allows body part to be deformable in actions. Therefore, the learned features are robust to pose variations in RGB-D data.

8.2.3.3 3D Recursive Neural Networks

The idea of recursive neural networks is to learn hierarchical features by applying the same neural network recursively in a tree structure. In our case, 3D-RNN can be regarded as combining convolution and pooling over 3D patches into one efficient, hierarchical operation.

We use balanced fixed-tree structure of 3D-RNN. Compared with previous RNN approaches, the tree structure offers high speed operation, and making use of parallelization. In our tree-structure 3D-RNN, each leaf node is a K-dimensional vector, which is an output of the 3D-CNN. At each layer, the 3D-RNN merges adjacent vectors into one vector. As this process repeats, high-order relationships and long-range dependencies of body parts can be encoded in the learned feature.

3D-RNN takes a list of K-dimensional vectors $u_i^{rct} \in U$ generated by 3D-CNN ($i = 1, \ldots, M$) as inputs, and recursively merges a block of vectors into one parent vector $q \in \mathbb{R}^K$. We define a 3D block of size $b_r \times b_c \times b_t$ as a list of adjacent vectors to be merged. For example, if $b_r = b_c = b_t = 2$, then $B = 8$ vectors will be merged. We define the merging function as

$$q = f\left(W \begin{bmatrix} u_1 \\ \vdots \\ u_B \end{bmatrix} \right). \tag{8.9}$$

Here, W is the parameter of size $K \times BK$ ($B = b_r \times b_c \times b_t$), $f(\cdot)$ is a nonlinear function (e.g., $\tanh(\cdot)$). Bias term is omitted here as it does not affect performance.

3D-RNN is a tree with multiple layers, where the jth layer composes high-order features over the $(j-1)$th layer. In the jth layer, all the blocks of vectors in the $(j-1)$th are merged into one parent vector using the same weight W in Eq. (8.9). This process is repeated until only one parent vector x remains. Fig. 8.7 shows an example of a pooled CNN output of size $K \times 2 \times 2 \times 2$ and an RNN tree structure with blocks of 8 children, $u_1, \ldots, u_8$.

We apply 3D-RNN to C channel data separately, and obtain C parent vectors from 3D-RNN x_c, $c = 1, \ldots, C$. Each parent vector x_c in 3D-RNN is a K-dimensional vector, computed from one channel data of an RGB-D video. The vectors from all the channels are then concatenated into a long vector to encode rich motion and structure information for the RGB-D video. Finally, this feature is fed into the softmax classifier for action classification.

The feature learned by 3D-RNN captures high-order relationships of body parts and encodes long-range dependencies of body parts. Therefore, human actions can be well represented and can be classified accurately.

8.2.3.4 Multiple 3D-RNNs

3D-RNN abstracts high-order features using the same weight W in a recursive way. The weight W, randomly learned, expresses the knowledge of which vector is more important in the parent vector for the classification task. It may not be accurate due to the randomness of W.

This problem can be solved by using multiple 3D-RNNs. Similar to [54], we use multiple 3D-RNNs with different random weights. Consequently, different importance of adjacent vectors can be well captured by different weights, and high quality feature vectors can then be produced. We concatenate vectors generated by multiple 3D-RNNs to feed into the softmax classifier.

8.2.3.5 Model Learning

Unsupervised Learning of 3D-CNN Filters. CNN models can be learned using supervised or unsupervised approaches [64,59]. Since convolution operates on millions of 3D patches, using back propagation and fine tuning the entire networks may not be practical or efficient. Instead, we train our 3D-CNN model using an unsupervised approach.

Inspired by [63], we learn 3D-CNN filters in an unsupervised way by clustering random 3D patches. We treat multichannel data (gray, gradient, optical flow, and depth) as separated feature maps, and randomly generate spatiotemporal 3D patches from each channel. The extracted 3D patches are then normalized and whitened. Finally, these 3D patches are clustered to build K cluster centers z_k, $k = 1, \ldots, K$, which are used in the 3D convolution (Eq. (8.8)).

Random Weights for 3D-RNN. Recent work [54] shows that RNNs with random weights can also generate features with high discriminability. We follow [54] to learn 3D-RNNs with random weights W. We show that learning RNNs with random weights provides an efficient, yet powerful model for action recognition from depth camera.

8.2.3.6 Classification

As described in Sect. 8.2.3.4, features generated by multiple 3D-RNNs will be concatenated to produce the feature vector x for the depth video. We train a multiclass

(A) MSR Action 3D dataset   (B) MSR Gesture 3D dataset

Figure 8.8 Example frames from three RGB-D action datasets.

softmax classifier to classify the depth action x,

$$f(x, y) = \frac{\exp(\boldsymbol{\theta}_y^{\mathrm{T}} x_i)}{\sum_{l \in y} \exp(\boldsymbol{\theta}_l^{\mathrm{T}} x_i)}, \tag{8.10}$$

where $\boldsymbol{\theta}_y$ is the parameter for class y. The prediction is performed by taking the $\arg\max$ of the vector whose lth element is $f(x, l)$, $y^* = \arg\max_l f(x, l)$. The multiclass cross entropy loss function is defined in learning the model parameter $\boldsymbol{\theta}$ for all the classes. The model parameter $\boldsymbol{\theta}$ is learned using the limited-memory variable-metric gradient ascent (BFGS) method.

8.2.4 Experiments

We evaluate our HCRNN model on two popular 3D action datasets, MSR-Gesture3D Dataset [65] and MSR-Action3D dataset [40]. Example frames of these datasets are shown in Fig. 8.8. We use gray, depth, gradient-x, gradient-y, optflow-x, and optflow-y feature maps for all the datasets.

8.2.4.1 MSR-Gesture3D Dataset

MSR-Gesture3D dataset is a hand gesture dataset containing 336 depth sequences captured by a depth camera. There are 12 categories of hand gestures in the dataset: "bathroom", "blue", "finish", "green", "hungry", "milk", "past", "pig", "store", "where", "j", and "z". This is a challenging dataset due to the self-occlusion issue and visually similarity. Our HCRNN takes an input video of size $80 \times 80 \times 18$. The number of filters in 3D-CNN is set to 256 and the number of 3D-RNNs is set to 16. The kernel size (filter size) in 3D-CNN is $6 \times 6 \times 4$, and the receptive filed size in 3D-RNN is $2 \times 2 \times 2$. As in [65], only depth frames are used in the experiments. The leave-one-out cross validation is employed to in evaluation.

Fig. 8.9 shows the confusion matrix of HCRNN on the MSR-Gesture3D dataset. Our method achieves 93.75% accuracy in classifying hand gestures. Our method misses some of the examples in "ASL Past" and "ASL Store", "ASL Finish" and "ASL Past", and "ASL Blue" and "ASL J" due to their visual similarities. As shown in Fig. 8.9, our method can recognize visually similar hand gestures as the HCRNN discovers discriminative features and abstracts expressive high-order features for the task. HCRNN

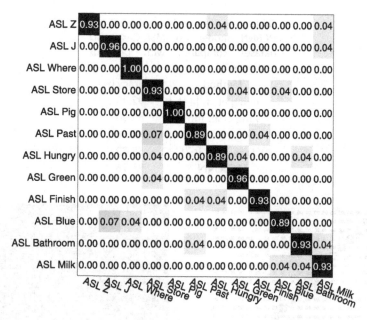

Figure 8.9 Confusion matrix of HCRNN on the MSR-Gesture3D dataset. Our method achieves 93.75% recognition accuracy.

confuses some of examples shown in Fig. 8.9 due to the self-occlusion and intra-class motion variations.

We compare our HCRNN with [44,66,65,67] on the MSR-Gesture3D dataset. Methods in [44,66,65] are particularly designed for depth sequences and [67] proposed HoG3D descriptor which was originally designed for color sequences. Compared with these methods that are based on hand-crafted features, the HCRNN learns features from data. Results in Table 8.5 indicate that our method outperforms all these comparison methods. Our method learns features from data, which better represent intra-class variations and inter-class similarities, and thus achieves better performance. In addition, HCRNN encodes high-order context information of body part, and allows deformable body parts. These two benefits help improve the expressiveness of the learned features.

8.2.4.2 MSR-Action3D Dataset

MSR-Action3D dataset [40] consists of 20 classes of human actions: "bend", "draw circle", "draw tick", "draw x", "forward kick", "forward punch", "golf swing", "hammer", "hand catch", "hand clap", "high arm wave", "high throw", "horizontal arm wave", "jogging", "pick up & throw", "side boxing", "side kick", "tennis serve", "tennis swing", and "two hand wave". A total of 567 depth videos are contained in the dataset which are captured using a depth camera.

Table 8.5 The performance of our HCRNN model on MSR-Gesture3D dataset compared with previous methods

Method	Accuracy (%)
Oreifej et al. [44]	92.45
Yang et al. [66]	89.20
Jiang et al. [65]	88.50
Klaser et al. [67]	85.23
HCRNN	**93.75**

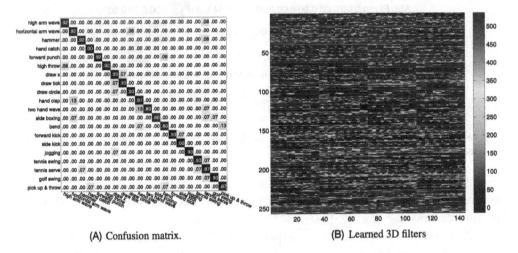

(A) Confusion matrix.   (B) Learned 3D filters

Figure 8.10 (A) Confusion matrix and (B) learned 3D filters of HCRNN on the MSR-Action3D dataset. Our method achieves 90.07% recognition accuracy.

In this dataset, the background is preprocessed to remove the discontinuities caused by the undefined depth regions. However, it is still challenging since many actions are visually very similar. The same training/testing splits in [44] is adopted in this experiment, i.e., the videos of the first five subjects (295 videos) are used for training and the remaining 272 videos are for testing. Our HCRNN takes an input video of size $120 \times 160 \times 30$. The number of filters in 3D-CNN is set to 256, and the number of 3D-RNNs is set to 32. The kernel size (filter size) in 3D-CNN is $6 \times 6 \times 4$, and the receptive filed size in 3D-RNN is $2 \times 2 \times 2$.

The confusion matrix of HCRNN is displayed in Fig. 8.10A. Our method achieves 90.07% recognition accuracy on the MSR–Action3D dataset. Confusions mostly occur between visually similar actions, e.g., "horizontal hand wave" and "clap", "hammer" and "tennis serve", and "draw x" and "draw tick". The learned filters used in the ex-

Table 8.6 The performance of our HCRNN model on MSRAction3D dataset

(A) Comparison results		(B) Performance with different n_r	
Data	Accuracy (%)	Number of RNNs	Accuracy (%)
RGGP [14]	89.30	1	40.44
Xia and Aggarwal [42]	89.30	2	57.72
Oreifej et al. [44]	88.89	4	63.24
Jiang et al. [40]	88.20	8	73.90
Jiang et al. [65]	86.50	16	83.09
Yang et al. [66]	85.52	32	90.07
Klaser et al. [67]	81.43	64	80.88
Vieira et al. [68]	78.20	128	68.01
Dollar [34]	72.40		
Laptev [49]	69.57		
HCRNN	**90.07**		

periment are also illustrated in Fig. 8.10B. Our filter learning method discovers various representative patterns, which can be used to accurately describe local 3D patches.

We compare with methods particularly designed for depth sequences [40,65,66, 68], as well as conventional action recognition methods that use spatiotemporal interest point detectors [34,49,67]. Results in Table 8.6A show that our method outperforms all the comparison methods. Our method achieves 90.07% recognition accuracy, which is higher than that of the state-of-the-art methods [42,14]. It should be noted that our method does not utilize the skeleton tracker, and yet outperforms the skeleton-based method [40]. Table 8.6B shows the recognition of HCRFF with different number of 3D-RNNs. The HCRNN achieves the best performance at $n_r = 32$. HCRNN obtains the worse performance with $n_r = 1$. With more 3D-RNNs, HCRNN achieves higher accuracy until 32 3D-RNNs are used. Then, with more 3D-RNNs, its performance degrades due to the overfitting problem.

8.3. SUMMARY

This chapter studies the problem of action recognition using two different types of data, multi-view data and RGB-D data. In the first scenario, action data are captured by multiple cameras, and thus the appearance of the human subject looks significantly different in different camera view, making action recognition more challenging. To address this problem, we have proposed feature learning approaches for learning view-invariant features. The proposed approaches utilize both shared and private features to accurately characterize human actions with large viewpoint and appearance variations. The sample affinity matrix is introduced in this chapter to compute sample similarities across views. The matrix is elegantly embedded in the learning of shared features in order to accurately weigh the contribution of each sample to the shared features, and balance information transfer. Extensive experiments on the IXMAS and DSA datasets show that our approaches outperform state-of-the-art approaches in cross-view action classification.

Actions can also be captured by RGB-D sensors such as Kinect since there are cost-effective. Action data captured by a Kinect sensor have multiple data channels including RGB, depth, and skeleton. However, it is challenging to use all of them for recognition as they are in different feature space. To address this problem, a new 3D convolutional recursive deep neural network (3DCRNN) is proposed for action recognition from RGB-D cameras. The architecture of the network consists of a 3D-CNN layer and a 3D-RNN layer. The 3D-CNN layer learns low-level translationally invariant features, which are then given as the input to the 3D-RNN. The 3D-RNN combines convolution and pooling into an efficient and hierarchical operation, and learns high-order compositional features. Results on two datasets show that the proposed method achieves state-of-the-art performance.

REFERENCES

[1] Altun K, Barshan B, Tunçel O. Comparative study on classifying human activities with miniature inertial and magnetic sensors. Pattern Recognition 2010;43(10):3605–20.

[2] Grabocka J, Nanopoulos A, Schmidt-Thieme L. Classification of sparse time series via supervised matrix factorization. In: AAAI; 2012.

[3] Kong Y, Fu Y. Bilinear heterogeneous information machine for RGB-D action recognition. In: IEEE conference on computer vision and pattern recognition; 2015.

[4] Kong Y, Fu Y. Max-margin action prediction machine. IEEE Transactions on Pattern Analysis and Machine Intelligence 2016;38(9):1844–58.

[5] Junejo I, Dexter E, Laptev I, Perez P. Cross-view action recognition from temporal self-similarities. In: ECCV; 2008.

[6] Yan Y, Ricci E, Subramanian R, Liu G, Sebe N. Multitask linear discriminant analysis for view invariant action recognition. IEEE Transactions on Image Processing 2014;23(12):5599–611.

[7] Zheng J, Jiang Z, Philips PJ, Chellappa R. Cross-view action recognition via a transferable dictionary pair. In: BMVC; 2012.

[8] Zheng J, Jiang Z. Learning view-invariant sparse representation for cross-view action recognition. In: ICCV; 2013.

[9] Yang W, Gao Y, Shi Y, Cao L. MRM-Lasso: a sparse multiview feature selection method via low-rank analysis. IEEE Transactions on Neural Networks and Learning Systems 2015;26(11):2801–15.

[10] Chen M, Xu Z, Weinberger KQ, Sha F. Marginalized denoising autoencoders for domain adaptation. In: ICML; 2012.

[11] Ding G, Guo Y, Zhou J. Collective matrix factorization hashing for multimodal data. In: CVPR; 2014.

[12] Singh AP, Gordon GJ. Relational learning via collective matrix factorization. In: KDD; 2008.

[13] Liu J, Wang C, Gao J, Han J. Multi-view clustering via joint nonnegative matrix factorization. In: SDM; 2013.

[14] Liu L, Shao L. Learning discriminative representations from rgb-d video data. In: IJCAI; 2013.

[15] Argyriou A, Evgeniou T, Pontil M. Convex multi-task feature learning. IJCV 2008;73(3):243–72.

[16] Ding Z, Fu Y. Low-rank common subspace for multi-view learning. In: IEEE international conference on data mining. IEEE; 2014. p. 110–9.

[17] Kumar A, Daume H. A co-training approach for multi-view spectral clustering. In: ICML; 2011.

[18] Zhang W, Zhang K, Gu P, Xue X. Multi-view embedding learning for incompletely labeled data. In: IJCAI; 2013.

[19] Wang K, He R, Wang W, Wang L, Tan T. Learning coupled feature spaces for cross-modal matching. In: ICCV; 2013.

[20] Xu C, Tao D, Xu C. Multi-view learning with incomplete views. IEEE Transactions on Image Processing 2015;24(12).

[21] Sharma A, Kumar A, Daume H, Jacobs DW. Generalized multiview analysis: a discriminative latent space. In: CVPR; 2012.

[22] Weinland D, Özuysal M, Fua P. Making action recognition robust to occlusions and viewpoint changes. In: ECCV; 2010.

[23] Junejo IN, Dexter E, Laptev I, Pérez P. View-independent action recognition from temporal self-similarities. IEEE Transactions on Pattern Analysis and Machine Intelligence 2011;33(1):172–85.

[24] Rahmani H, Mian A. Learning a non-linear knowledge transfer model for cross-view action recognition. In: CVPR; 2015.

[25] Liu J, Shah M, Kuipers B, Savarese S. Cross-view action recognition via view knowledge transfer. In: CVPR; 2011.

[26] Li B, Campus OI, Sznaier M. Cross-view activity recognition using Hankelets. In: CVPR; 2012.

[27] Li R, Zickler T. Discriminative virtual views for cross-view action recognition. In: CVPR; 2012.

[28] Zhang Z, Wang C, Xiao B, Zhou W, Liu S, Shi C. Cross-view action recognition via continuous virtual path. In: CVPR; 2013.

[29] Hinton GE, Salakhutdinov RR. Reducing the dimensionality of data with neural networks. Science 2006;313(5786):504–7.

[30] Li J, Zhang T, Luo W, Yang J, Yuan X, Zhang J. Sparseness analysis in the pretraining of deep neural networks. IEEE Transactions on Neural Networks and Learning Systems 2016. https://doi.org/10.1109/TNNLS.2016.2541681.

[31] Chen M, Weinberger K, Sha F, Bengio Y. Marginalized denoising auto-encoders for nonlinear representations. In: ICML; 2014.

[32] Vincent P, Larochelle H, Lajoie I, Bengio Y, Manzagol PA. Stacked denoising autoencoders: learning useful representations in a deep network with a local denoising criterion. JMLR 2010;11:3371–408.

[33] Weinland D, Ronfard R, Boyer E. Free viewpoint action recognition using motion history volumes. Computer Vision and Image Understanding 2006;104(2–3).

[34] Dollar P, Rabaud V, Cottrell G, Belongie S. Behavior recognition via sparse spatio-temporal features. In: VS-PETS; 2005.

[35] Wang H, Kläser A, Schmid C, Liu CL. Dense trajectories and motion boundary descriptors for action recognition. IJCV 2013;103(1):60–79.

[36] Farhadi A, Tabrizi MK, Endres I, Forsyth DA. A latent model of discriminative aspect. In: ICCV; 2009.

[37] Gupta A, Martinez J, Little JJ, Woodham RJ. 3d pose from motion for cross-view action recognition via non-linear circulant temporal encoding. In: CVPR; 2014.

[38] Liu J, Shah M. Learning human actions via information maximization. In: CVPR; 2008.

[39] Kong Y, Fu Y. Discriminative relational representation learning for rgb-d action recognition. IEEE Transactions on Image Processing 2016;25(6).

[40] Wang J, Liu Z, Wu Y, Yuan J. Mining actionlet ensemble for action recognition with depth cameras. In: CVPR; 2012.

[41] Wang Y, Mori G. A discriminative latent model of object classes and attributes. In: ECCV; 2010.

[42] Xia L, Aggarwal J. Spatio-temporal depth cuboid similarity feature for activity recognition using depth camera. In: CVPR; 2013.

[43] Hadfield S, Bowden R. Hollywood 3d: recognizing actions in 3d natural scenes. In: CVPR; 2013.

[44] Oreifej O, Liu Z. HON4D: histogram of oriented 4D normals for activity recognition from depth sequences. In: CVPR; 2013. p. 716–23.

[45] Li W, Zhang Z, Liu Z. Action recognition based on a bag of 3d points. In: CVPR workshop; 2010.

[46] Rahmani H, Mahmood A, Mian A, Huynh D. Real time action recognition using histograms of depth gradients and random decision forests. In: WACV; 2013.

[47] Kong Y, Fu Y, Jia Y. Learning human interaction by interactive phrases. In: ECCV; 2012.

[48] Lan T, Wang Y, Yang W, Robinovitch SN, Mori G. Discriminative latent models for recognizing contextual group activities. PAMI 2012;34(8):1549–62.

[49] Laptev I. On space–time interest points. IJCV 2005;64(2):107–23.

[50] Koppula HS, Saxena A. Learning spatio-temporal structure from RGB-D videos for human activity detection and anticipation. In: ICML; 2013.

[51] Luo J, Wang W, Qi H. Group sparsity and geometry constrained dictionary learning for action recognition from depth maps. In: ICCV; 2013.

[52] Jarrett K, Kavukcuoglu K, Ranzato M, LeCun Y. What is the best multi-stage architecture for object recognition? In: ICCV; 2009.

[53] Nair V, Hinton GE. 3d object recognition with deep belief nets. In: NIPS; 2009.

[54] Socher R, Huval B, Bhat B, Manning CD, Ng AY. Convolutional-recursive deep learning for 3d object classification. In: NIPS; 2012.

[55] Ouyang W. Modeling mutual visibility relationship in pedestrian detection. In: CVPR; 2013.

[56] Szegedy C, Toshev A, Erhan D. Deep neural networks for object detection. In: NIPS; 2013.

[57] Farabet C, Couprie C, Najman L, LeCun Y. Learning hierarchical features for scene labeling. PAMI 2013.

[58] Socher R, Lim CCY, Ng AY, Manning CD. Parsing natural scenes and natural language with recursive neural networks. In: ICML; 2011.

[59] Ji S, Xu W, Yang M, Yu K. 3d convolutional neural networks for human action recognition. IEEE Transactions on Pattern Analysis and Machine Intelligence 2013;35(1):221–31.

[60] Le QV, Zou WY, Yeung SY, Ng AY. Learning hierarchical invariant spatio-temporal features for action recognition with independent subspace analysis. In: CVPR; 2011.

[61] Taylor GW, Fergus R, LeCun Y, Bregler C. Convolutional learning of spatio-temporal features. In: ECCV; 2010.

[62] LeCun Y, Bottou L, Bengio Y, Haffner P. Gradient-based learning applied to document recognition. In: Proceedings of the IEEE; 1998.

[63] Coates A, Lee H, Ng AY. An analysis of single-layer networks in unsupervised feature learning. In: AISTATS; 2011.

[64] Ranzato M, Huang FJ, Boureau YL, LeCun Y. Unsupervised learning of invariant feature hierarchies with applications to object recognition. In: CVPR; 2007.

[65] Wang J, Liu Z, Chorowski J, Chen Z, Wu Y. Robust 3d action recognition with random occupancy patterns. In: ECCV; 2012.

[66] Yang X, Zhang C, Tian Y. Recognizing actions using depth motion maps-based histograms of oriented gradients. In: ACM multimedia. ISBN 978-1-4503-1089-5, 2012.

[67] Klaser A, Marszalek M, Schmid C. A spatio-temporal descriptor based on 3d-gradients. In: BMVC; 2008.

[68] Vieira AW, Nascimento ER, Oliveira GL, Liu Z, Campos MFM. STOP: space–time occupancy patterns for 3D action recognition from depth map sequences. In: 17th Iberoamerican congress on pattern recognition (CIARP); 2012.

CHAPTER 9

Style Recognition and Kinship Understanding

Shuhui Jiang*, Ming Shao†, Caiming Xiong‡, Yun Fu§
*Department of Electrical and Computer Engineering, Northeastern University, Boston, MA, United States
†Computer and Information Science, University of Massachusetts Dartmouth, Dartmouth, MA, United States
‡Salesforce Research, Palo Alto, CA, United States
§Department of Electrical and Computer Engineering and College of Computer and Information Science
(Affiliated), Northeastern University, Boston, MA, United States

Contents

9.1. Style Classification by Deep Learning		213
9.1.1 Background		213
9.1.2 Preliminary Knowledge of Stacked Autoencoder (SAE)		217
9.1.3 Style Centralizing Autoencoder		217
9.1.4 Consensus Style Centralizing Autoencoder		221
9.1.5 Experiments		226
9.2. Visual Kinship Understanding		230
9.2.1 Background		230
9.2.2 Related Work		232
9.2.3 Family Faces		233
9.2.4 Regularized Parallel Autoencoders		234
9.2.5 Experimental Results		239
9.3. Research Challenges and Future Works		246
References		246

9.1. STYLE CLASSIFICATION BY DEEP LEARNING[1]

9.1.1 Background

Style classification has attracted increasing attention from both researchers and artists in many fields. Style classification is related, but essentially different from most existing classification tasks. For example, in current online clothing shopping website, usually the items are categorized to skirt, dress, suit, etc. However, one clothing category may consist of diverse fashion styles. For example, a "suit" could be either *casual* or *renascent* fashion styles and a dress could be either *romantic* and *elegant* fashion styles. Style classification may help people identify style classes and generate relationships between styles.

---

[1] ©2017 IEEE. Reprinted, with permission, from Jiang, Shuhui, Ming Shao, Chengcheng Jia, and Yun Fu. "Learning consensus representation for weak style classification." IEEE Transactions on Pattern Analysis and Machine Intelligence (2017).

Deep Learning Through Sparse and Low-Rank Modeling
DOI: 10.1016/B978-0-12-813659-1.00009-3

Therefore, learning robust and discriminative feature representation for style classification becomes an interesting and challenging research topic. Most style classification methods mainly focus on extracting discriminative local patches or patterns based on low-level features. Some recent works of the fashion, manga, and architecture style classification based on low-level feature representation are described as follows:

Fashion style classification. "Fashion and AI" is becoming a hot research topic recently, e.g., in clothing parsing [2], retrieval [3], recognition [4] and generation [5]. Bossard et al. densely extracted feature descriptors like HOG in the bounding box of the upper body followed by a bag-of-words model [6]. In Hipster Wars [7], Kiapour et al. proposed an online game to collect a fashion dataset. A style descriptor was formed by accumulating visual features like color and texture. Then they applied mean-std pooling and concatenated all the pooled features as the final style descriptor, followed by a linear SVM for classification.

Manga style classification. Chu et al. [8] paved the way for manga style classification, which classifies whether the manga is targeting young boys (i.e., shonen) or young girls (i.e., shojo). They designed both explicit (e.g., the density of line segments) and implicit (e.g., included angles between lines) feature descriptors, and concatenated these descriptors as feature representation.

Architecture style classification. Goel et al. focused on architectural style classification (e.g., baroque and gothic) [9]. They mined characteristic features with the semantic utility from the low-level features with various scales. Van et al. created correspondences across images by a generalized spatial pyramid matching scheme [10]. They assumed that images within a category share a similar style defined by attributes such as colorfulness and lighting. Xu et al. adopted deformable part-based models (DPM) to capture the morphological characteristics of basic architectural components [11].

However, style is usually reflected by the high-level abstract concepts. These works may fail to extract some mid/high level features for style presentation. Furthermore, they haven't discussed the *spread out* phenomenon of style images, which is observed in [12,1]. Fig. 9.1 illustrates this "spread out" phenomenon. We take fashion style classification of "Goth" and "Preppy" as an example. Representative images in the center of each class are assigned *strong style* level l_3. It is easy to distinguish "Goth" and "Preppy" images in strong style. Images which are less representative and distant to the center are assigned as lower style level l_1. They are named as *weak style* images. Week style images within one style could be visually diverse, and images in two classes could be visually similar (shown in red frames). The spread out natural makes the weak style images easily get misclassified with other classes. To better illustrate the spread out idea, in Fig. 9.2, two feature descriptors of manga data consisting of shojo and shonen classes [8] are visualized. PCA is conducted to reduce the dimension of feature descriptors into two for visualization. We could see that strong style data points (e.g., in blue) are with high density and well separated; however, weak style data points (e.g., in magenta) are spread out and hard to be separated.

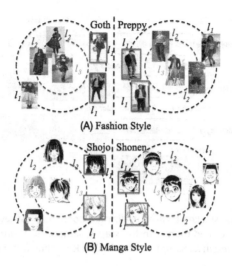

Figure 9.1 Illustration of *weak style* phenomenon in fashion and manga style images in (A) and (B), respectively. Style images are usually "spread out". Images in the center of each style circle are representatives of this style and defined as *strong style*. They are easy to be distinguished from other styles. Images far from the center are less similar to strong style images, and easy to be misclassified. Images in red frames on the boundary are weak style images from two different classes, but they seem visually similar. We denote by l_1 to l_3 the style levels from weakest to strongest. (For interpretation of the colors in the figures, the reader is referred to the web version of this chapter.)

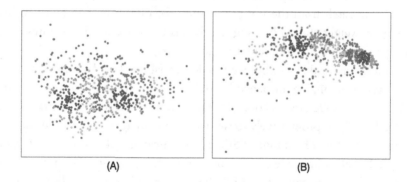

Figure 9.2 Data visualization of the "spread out" phenomenon in "shoji" and "shonen" classes of manga style. PCA is conducted to reduce the dimension of feature descriptors of "line strength" and "included an angle between lines" into 2D for visualization in (A) and (B), respectively. In both (A) and (B), five colors, blue, green, red, cyan, and magenta, are used to present the data points in different style levels from the strongest to the weakest. We could see that strong style data points are dense and weak style data points are diffuse.

Furthermore, as described above, usually all the feature descriptors are concatenated together to form the style descriptor. It means that all the feature descriptors are treated equally important. However, for different styles, the importance of different feature

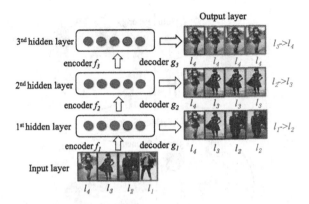

Figure 9.3 Illustration of the stacked style centralizing autoencoder (SSCAE). Example images of each level l are presented with colored frames. In the step k, samples in l_k are replaced by l_{k+1} samples' (red) nearest neighbors found in l_k. Samples in the higher level than l_k are not changed (blue).

descriptors maybe different. For example, color may be more important than other feature descriptors when describing "Goth" style, which is usually in black color. Thus, adaptively allocating weights for different feature descriptors becomes another challenge in style classification. To address this challenge, a "consensus" idea is introduced in [12, 1] to jointly learn weights for different visual features in representation learning. For example, if one patch from the image is critical for discrimination (e.g., eye patch for face), a higher weight should be assigned to this patch for all the features, meaning a consistency of weights across different feature descriptors.

In the following of this section, we would describe a deep learning solution named consensus style centralizing autoencoder (CSCAE) for robust style feature extraction, especially for weak style classification [1]. First, we describe a style centralizing autoencoder (SCAE), which progressively draws weak style images back to the class center of one feature descriptor. The inputs of SCAE are concatenated low-level features from all the local patches of an image (e.g., eyes, nose, mouth patches in a face image). As shown in Fig. 9.3, for each autoencoder (AE), the corresponding output feature is the same type of feature in the same class but one style level stronger than the input feature. Only the neighbor samples are pulled together towards the center of this class, since weak style images could be very diverse even within one class. Then the progressive steps slowly mitigate the weak style distinction, and ensures the smoothness of the model. In addition, to approach the lack of consensus issue among different kinds of visual features, the weights are jointly learned for them by a CSCAE through rank-constrained group sparsity autoencoder (RCGSAE), based on the consensus idea. We show the evaluation of both deep learning and non–deep learning methods on three applications: fashion style classification, manga style classification and architecture style classification.

Deep learning structures such as autoencoders (AEs) have been exploited to learn discriminative feature representation [13–15]. Conventional AEs [13] include two parts: (1) encoder and (2) decoder. An encoder $f(\cdot)$ attempts to map the input feature $x_i \in \mathbb{R}^D$ to the hidden layer representation $z_i \in \mathbb{R}^d$,

$$z_i = f(x_i) = \sigma(W_1 \times x_i + b_1), \tag{9.1}$$

where $W_1 \in \mathbb{R}^{d \times D}$ is a linear transform, $b_1 \in \mathbb{R}^d$ is the bias, and σ is the nonlinear activation (e.g., sigmoid function). The decoder $g(\cdot)$ manages to map the hidden representation z_i back to the input feature x_i, namely,

$$x_i = g(z_i) = \sigma(W_2 \times z_i + b_2), \tag{9.2}$$

where $W_2 \in \mathbb{R}^{D \times d}$ is a linear transform, $b_2 \in \mathbb{R}^D$ is the bias.

To optimize the model parameters W_1, b_1, W_2 and b_2, the least squared error problem is formulated as

$$\min_{\substack{W_1, b_1 \\ W_2, b_2}} \frac{1}{2N} \sum_{i=1}^{N} \left\| x_i - g(f(x_i)) \right\|^2 + \lambda R(W_1, W_2), \tag{9.3}$$

where N is the number of data points, $R(W_1, W_2) = (\|W_1\|_F^2 + \|W_2\|_F^2)$ works as a regularizer, $\|\cdot\|_F^2$ is the Frobenius norm, and λ is the weight decay parameter to suppress arbitrarily large weights.

9.1.2 Preliminary Knowledge of Stacked Autoencoder (SAE)

A stacked autoencoder (SAE) [16,17] stacks multiple AEs to form a deep structure. It feeds the hidden layer of the kth AE as the input feature to the $(k+1)$th layer. However, in the weak style classification problem, the performance of AE or SAE degrades due to the "spread out" phenomenon. The reason is that the conventional AE or SAE runs in an unsupervised fashion to learn mid/high-level feature representation, meaning there is no guidance to lead images in the same class close and images in the different classes far away to each other. This is very similar to the conventional PCA (SAE can be seen as a multilayer nonlinear PCA). In Fig. 9.2 where data are illustrated after PCA, weak-style classes represented by cyan and magenta are diffused and overlap with other classes, from which we can see that the mid/high-level feature representation by AE or SAE will suffer from the "spread out" phenomenon. A style centralizing autoencoder (SCAE) is introduced addressing these issues [1,12].

9.1.3 Style Centralizing Autoencoder

Local visual features are applied as the input for SCAE. Assume that there are N images from N_c style classes, and x_i ($i \in \{1, \ldots, N\}$) is the feature representation of the ith image.

First, each image is divided into several patches (e.g., eyes, nose and mouth patches in a face image). Then visual features (e.g., HoG, RGB, Gabor) are extracted from each patch. For each feature descriptor (e.g., HoG), the extracted features from all the patches are concatenated as one part of input features for SCAE. By concatenating all different visual features, we obtain the final input features for SCAE. In addition, each image is assigned a style level label. Intuitively, representative images of each style are usually assigned the strong style level, while less representative images are assigned the weak style level. We use L distinct style levels denoted as $\{l_1, l_2, \ldots, l_k, \ldots, l_L\}$ from the weakest to the strongest.

9.1.3.1 One Layer Basic SCAE

Different from the conventional AE taking identical input and output features, the input and output of SCAE are different. Illustration of the full pipeline of SCAE can be found in Fig. 9.3. Suppose that we have $L = 4$ style levels, and the inputs of the SCAE in the first layer are the features of images in the ascent order of style level, namely, X_1, X_2, X_3 and X_4. For example, X_2 is the set of features of images in style level l_2. Let $X^{(k)}$ be the input feature of the kth step, where $x_i^{(k)} \in X^{(k)}$ is the feature of the ith sample with the hidden representation learning from the $(k-1)$th step.

SCAE handles the following mappings:

$$\{X_k^{(k)}, X_{k+1}^{(k)}, \ldots, X_L^{(k)}\} \rightarrow \{X_{k+1}^{(k)}, X_{k+1}^{(k)}, \ldots, X_L^{(k)}\}, \tag{9.4}$$

where only X_k is pulled towards stronger style level l_{k+1}, and others keep the same style level before and after the kth step. In this way, the weak style level will be gradually pulled towards the strong style level, i.e., centralization, till $k = L - 1$. Thus, the $L - 1$ stacked AEs embody the stacked SCAE. Note that the mappings between X_k and X_{k+1} are still unclear. To keep the style level transition smooth, for each output feature $x \in X_{k+1}$, the nearest neighbor in X_k in the same style class is applied as the corresponding input to learn SCAE. The whole process is shown in Fig. 9.3.

9.1.3.2 Stacked SCAE (SSCAE)

After introducing the one layer basic SCAE, we explain how to build the Stacked SCAE (SSCAE). Suppose that we have L style levels, in the kth step and style class c, the corresponding input for the output $x_{i,\xi+1}^{(k,c)}$ is given by

$$\tilde{x}_{i,\xi}^{(c)} = \begin{cases} x_{j,\xi}^{(k,c)} \in u(x_{i,\xi+1}^{(k,c)}), & \text{if } \xi = k, \\ x_{i,\xi}^{(k,c)}, & \text{if } \xi = k+1, \ldots, L, \end{cases} \tag{9.5}$$

where $u(x_{i,\xi+1}^{(k,c)})$ is the set of nearest neighbors of $x_{i,\xi+1}^{(k,c)}$ in the ξth layer.

As there are N_c style classes, in each layer, we first separately learn parameters and hidden layer features $Z^{(k,c)}$ of SCAE of each class, and then combine all the $Z^{(k,c)}$ together as $Z^{(k)}$. Mathematically, SSCAE can be formulated as

$$\min_{\substack{W_1^{(k,c)}, b_1^{(k,c)} \\ W_2^{(k,c)}, b_2^{(k,c)}}} \sum_{\substack{i,j \\ x_{j,k}^{(k,c)} \in u(x_{i,k+1}^{(k,c)})}} \left\| x_{i,k+1}^{(k,c)} - g(f(x_{j,k}^{(k,c)})) \right\|^2 +$$

$$\sum_{\xi=k+1}^{L} \sum_{i} \left\| x_{i,\xi}^{(k,c)} - g(f(x_{i,\xi}^{(k,c)})) \right\|^2 + \lambda R(W_1^{(k,c)}, W_2^{(k,c)}). \quad (9.6)$$

The problem above can be solved in a similar way as the conventional AE by back propagation algorithms [18]. Similarly, the deep structure can be built in a layer-wise way, which is outlined in Algorithm 9.1.

Algorithm 9.1: Stacked Style Centralizing Autoencoder.

1 **INPUT**: Style feature X including weak style feature.
2 **OUTPUT**: Style centralizing feature $Z^{(k)}$, model parameters: $W_m^{(k,c)}$, $b_m^{(k,c)}$, $k \in [1, L-1]$, $c \in [1, N_c]$, $m \in \{1, 2\}$.
   1: Initialize $Z^{(0)} = X$.
   2: **for** $k = 1, 2, \ldots, L\text{-}1$ **do**
   3:    $X^{(k)} = Z^{(k-1)}$.
   4:    **for** $c = 1, \ldots, N_c$ **do**
   5:       Calculate $W_m^{(k,c)}$, $b_m^{(k,c)}$ by Eq. (9.6).
   6:       Calculate $Z^{(k,c)}$ by Eq. (9.1).
   7:    **end for**
   8:    Combine all $Z^{(k,c)}$ into $Z^{(k)}$.
   9: **end for**

9.1.3.3 Visualization of Encoded Feature in SCAE

Fig. 9.4 shows the visualization of encoded features in the progressive step $k = 1, 2, 3$ in manga style classification. Similar to Fig. 9.2, PCA is employed to reduce the dimensionality of the descriptors. The low-level input feature is "density of line segments" in [8]. In all the subfigures, a dot represents a sample. In the right sub-figures, colors are used to distinguish *different styles*, while in left subfigures, colors are used to distinguish *different style levels*.

From Fig. 9.4, we could see that at step $k = 1$, in the right subfigures, the "shoji" samples (in red) and "shonen" samples (in blue) have overlaps. In the left subfigures, samples in the strong style level (in blue) are separated from each other. However, sam-

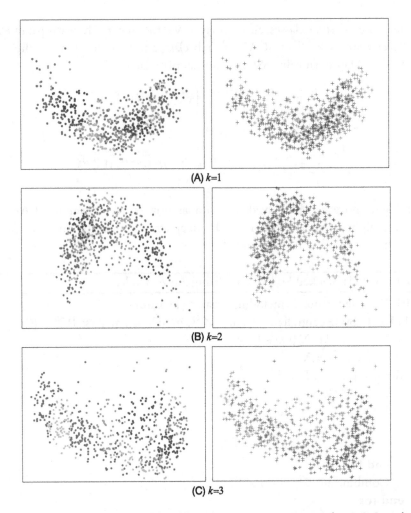

Figure 9.4 Visualization of the encoded features in SCAE in progressive step $k = 1, 2, 3$ on the Manga dataset. Meanings of different colors for different style levels are: Blue, level 5; Green, level 4; Red, level 3; Cyan, level 2; Magenta, level 1. Note that PCA is applied for dimensionality reduction before visualization.

ples in weak style levels overlap with each other. For example, it is hard to separate the samples in cyan in two styles. During progressive steps, samples in different styles gradually separate due to the style centralization. At progressive step $k = 2, 3$, in the right subfigures we could see the samples in red and blue gradually become separable. In the left subfigures, we could see that during style centralizing, the weak style samples, shown in green, red, cyan and magenta move closely to the locations of blue samples in two different styles. Since the blue samples represent the strong style level

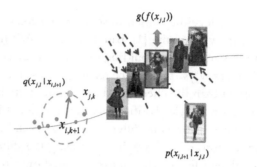

$g(f(x_{j,l}))$

$q(x_{j,l} \mid x_{i,l+1})$

$x_{j,k}$

$x_{i,k+1}$

$p(x_{i,l+1} \mid x_{j,l})$

Figure 9.5 Manifold learning perspective of SCAE with fashion images in "Goth" style.

and could easily be separated, the centralization process makes the weak style samples more distinguishable.

9.1.3.4 Geometric Interpretation of SCAE

Recalling the mappings presented in Eq. (9.4), if we consider X_k as a corrupted version of X_{k+1}, SCAE can be recognized as a denoising autoencoder (DAE) [16] that uses partially corrupted feature as the input and the clean noise free features as the output to learn robust representations [16,19]. Thus, inspired by the geometric perspective under the manifold assumption [20], we may offer a geometric interpretation for the SCAE, analogical to that of DAE [16].

Fig. 9.5 illustrates the manifold learning perspective of SCAE where images of Goth fashion style are shown as the examples. Suppose that the higher level l_{k+1} Goth style images lie close to a low dimensional manifold. The weak style examples are more likely being far away from the manifold than the higher level ones. Note that $x_{j,k}$ is the corrupted version of $x_{i,k+1}$ by the operator $q(X_k|X_{k+1})$, and therefore lies far away from the manifold. In SCAE, $q(X_k|X_{k+1})$ manages to find the nearest neighbor of $x_{i,k+1}$ in level l_k to obtain the corrupted the version of $x_{i,k+1}$ as $x_{j,k}$ in the same category. During the centralizing training, similar to DAE, SCAE learns the stochastic operator $p(X_{k+1}|X_k)$ that maps the lower style level samples X_k back to a higher level. Successful centralization implies that the operator $p(\cdot)$ is able to map spread-out weak style data back to the strong style data which are close to the manifold.

9.1.4 Consensus Style Centralizing Autoencoder

Given multiple low-level visual descriptors (e.g., HOG, RGB, Gabor), usually low-level feature based methods treat them equally and concatenate them to formulate the final representation [7,8]. Thus, they may fail to consider the correlation between different kinds of feature descriptors, i.e., consensus [21,22].

In this section, we introduce the consensus style centralizing autoencoder (CSCAE) with low-rank group sparsity constraint [1], based on the SCAE. Intuitively, the weights of two features from the same patch should be similar, as they encode the same visual information, but in different ways. Taking face recognition as an example, the eyes patch should be more important than the cheek patch, as demonstrated by many face recognition works. Thus, for manga style classification, given different kinds of features used by different SCAEs, the eyes patches in different SCAEs should be equally important. To that end, a consensus constraint through minimizing the differences of weights of the same patch from different feature descriptors is added onto the SCAE to form the CSCAE.

In the following of this section, we firstly introduce the low-rank constraint and group sparsity constraint for achieving the consensus idea. Then we introduce the rank-constrained group sparsity autoencoder (RCGSAE) and its solution. Finally, CSCAE is introduced based on the RCGSAE.

9.1.4.1 Low-Rank Constraint on the Model

Low-rank constraint has been widely used for discovering underlying structure and data recovery [23,24]. It has been suggested to extract the salient features despite of noises in the following formulations for latent low-rank representation (LaLRR) [24]. As described above, similar weights are expected across different feature descriptors for the same patch. This will give rise to an interesting phenomenon: if we concatenate all the weight matrices of different feature descriptors together denoted as W, the rank of W should be low. The main reason is the similar values across different columns in W. Thus, this idea is pursued through a low-rank matrix constraint on the concatenated weight matrix W.

To that end, we introduce a rank-constrained autoencoder model [1] to pursue the low-rankness of the weight matrix W in the following formula:

$$\min_{W,E} \|W\|_* + \lambda \|E\|_{2,1}, \text{ s.t. } X = W\tilde{X} + E, \tag{9.7}$$

where $\tilde{X}$ is the input feature and X is the output feature of AE, similar to SCAE; $\|\cdot\|_*$ is the nuclear norm of a matrix used as the convex surrogate of the original rank constraint, $\|E\|_{2,1}$ is the matrix $\ell_{2,1}$ norm for characterizing the sparse noise E, and λ is a balancing parameter. Intuitively, the residual term E encourages sparsity, as both W and $W\tilde{X}$ are low-rank matrices. This is also well explored by many low-rank recover/representation works [24,23,14,22]. Equation (9.7) can also be considered as a special form of the work [24] by only considering features in the column space.

Fig. 9.6A illustrates this phenomenon. Each row represents a specific feature of one patch, and different colors indicate different kinds of features. In addition, each column represents all features of one sample. It can be seen that the concatenated features for

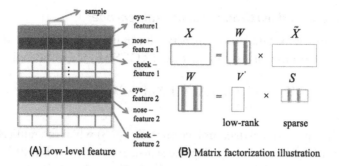

(A) Low-level feature   (B) Matrix factorization illustration

Figure 9.6 Illustration of the consensus style centralizing autoencoder (CSCAE). The illustration of the low-rank group sparsity structure of low-level features is shown in (A). The illustration of the matrix factorization in the solution of the model is shown in (B).

one sample are formulated by first stacking different features of the same patch, and then stacking different patches. Note that for simplicity, one cell is used to represent one kind of features of one patch. We could see that ideally, both eye-feature1 and eye-feature2 should have the highest weights, and nose-feature1 and nose-feature2 have the second highest weights. Meanwhile, the less important patches cheek-feature1 and cheek-feature2 should be given lower weights to suppress noises. In brief, based on the definition of the matrix rank, the consensus constraint among different features induces the low-rank structure of matrix W identified in Fig. 9.6A.

9.1.4.2 Group Sparsity Constraint on the Model

To further consider the regularizers introduced in Eq. (9.3) under the new rank constraint autoencoder framework, we introduce an additional group sparsity constraint on W. The reasons are three-fold. First, like conventional regularizers in neural networks, it helps avoid the arbitrarily large magnitude in W. Second, it enforces the row-wise selection on W to ensure a better consensus effect together with the low-rank constraint. Third, it helps find the most discriminative representation.

Mathematically, we can achieve this by adding a matrix $\ell_{2,1}$ norm $\|W\|_{2,1} = \sum_{i=1}^{D} \|W(i)\|_2$ which is equal to the sum of the Euclidean norms of all columns of W. The ℓ_2-norm constraint is applied to each group separately (i.e., each column of W). It ensures that all elements in the same column are either approaching zero or nonzero at the same time. The ℓ_1 norm guarantees that only a few columns are nonzero. Fig. 9.6A also illustrates how the group sparsity works. If the entry in the jth row and ith column of X indicates an unimportant patch, all the entries from the jth row are also less important, and vice-versa. As discussed above, these patches should have been assigned very low or zero weights to suppress the noise.

9.1.4.3 Rank-Constrained Group Sparsity Autoencoder

Considering both rank and group sparsity constraints, the objective function of the rank–constrained group sparsity autoencoder (RCGSAE) is formulated as

$$\hat{W}_r = \underset{\text{rank}(W) \leq r}{\arg\min} \{||X - W\tilde{X}||_F^2 + 2\lambda ||W||_{2,1}\}, \qquad (9.8)$$

where $\hat{W}_r$ is the optimized matrix projection in Eq. (9.8) when the $\text{rank}(W) \leq r$; λ is the balancing parameter. Note that we skip the sparse error term in Eq. (9.7) for simplicity. Clearly, for $r = D$, we have no rank constraint in Eq. (9.8) which degrades to a group sparsity problem, while for $\lambda = 0$, we obtain the reduced-rank regression estimator. Thus, an appropriate rank r and λ will balance the two parts to yield better performance.

Low-rank and group sparsity constraints not only minimize the difference of patch weights among different descriptors, but also assign the weights of unimportant weights to be all zero. In this way, the influence of the noise of unimportant patches are decreased and we are able to find the most discriminative representation.

9.1.4.4 Efficient Solutions for RCGSAE

Here, we introduce how to solve the objective function of RCGSAE. It should be noted that the problem defined in Eq. (9.8) is nonconvex and has no closed-form solutions for W. Thus, an iterative algorithm is used to solve this in a fast manner. As shown in Fig. 9.6B, W is factorized into $W = V'S$, where V is an $r \times D$ orthogonal matrix, V' is the inverse of V and S is an $r \times D$ matrix with the group sparse constraint [25]. Then the optimization problem of W in Eq. (9.8) turns out to be

$$(\hat{S}, \hat{V}) = \underset{S \in \mathbb{R}^{r \times D}, V \in \mathbb{R}^{r \times D}}{\arg\min} \{||X - V'S\tilde{X}||_F^2 + 2\lambda ||S||_{2,1}\}. \qquad (9.9)$$

The details of the algorithm are outlined in Algorithm 9.2. In addition, the following theorem presents a convergence analysis for Algorithm 9.2 and ensures that the algorithm converges well regardless of the initial point.

Theorem 9.1. *Given λ and an arbitrary starting point $V_{r,\lambda}^{(0)} \in \mathbb{O}^{r \times D}$, let $(S_{r,\lambda}^{(j)}, V_{r,\lambda}^{(j)})$ $(j = 1, 2, \dots)$ be the sequence of iterates generated by Algorithm 9.2. Then, any accumulation point of $(S_{r,\lambda}^{(j)}, V_{r,\lambda}^{(j)})$ is a coordinate-wise minimum point (and a stationary point) of F and $F(S_{r,\lambda}^{(j)}, V_{r,\lambda}^{(j)})$ converges monotonically to $F(S_{r,\lambda}^*, V_{r,\lambda}^*)$ for some coordinate-wise minimum point $(S_{r,\lambda}^*, V_{r,\lambda}^*)$.*

The proof is given in Appendix A.7 of [25].

9.1.4.5 Progressive CSCAE

After introducing RCGSAE, we introduce progressive CSCAE, which stacks multiple RCGSAEs in a progressive way. In the kth step, it will increase the style level of $\tilde{X}$

Algorithm 9.2: Solutions for RCGSAE.

---

1 **INPUT**: Original style feature X, corrupted style feature $\tilde{X}$, $1 \leq r \leq N \wedge D$, $\lambda \geq 0$, $V_{r,\lambda}^{(0)} \in \mathbb{O}^{n \times r}$, $j \leftarrow 0$, converged $\leftarrow$ FALSE

2 **OUTPUT**: Model parameters, W.

  1: **while** not converged **do**

  2:    (a) $S_{r,\lambda}^{(j+1)} \leftarrow \arg\min_S \frac{1}{2} ||V_{r,\lambda}^{(j)} X - S\tilde{X}||_F^2 + \lambda ||S||_{2,1}$.

  3:    (b) Let $Q \leftarrow S_{r,\lambda}^{(j+1)} \tilde{X} X' \in \mathbb{R}^{n \times r}$ and perform SVD, $Q = U_w D_w V_w'$; X' is the inverse of X.

  4:    (c) $V_{r,\lambda}^{(j+1)} \leftarrow U_m V_w'$.

  5:    (d) $W_{r,\lambda}^{(j+1)} \leftarrow (V_{r,\lambda}^{(j+1)})' S_{r,\lambda}^{(j+1)}$.

  6:    (e) converged $\leftarrow |F(W_{r,\lambda}^{(j+1)}; \lambda) - F(W_{r,\lambda}^{(j)}; \lambda)| < \varepsilon$

  7:    (f) $j \leftarrow j + 1$

  8: **end while**

  9: $\hat{W}_{r,\lambda} = W_{r,\lambda}^{(j+1)}$, $\hat{S}_{r,\lambda} = S_{r,\lambda}^{(j+1)}$, $\hat{V}_{r,\lambda} = V_{r,\lambda}^{(j+1)}$.

Algorithm 9.3: Progressive CSCAE.

---

1 **INPUT**: Original style feature X. The number of style levels L.

2 **OUTPUT**: Style centralizing feature $h^{(k)}$, projection matrices $W^{(k)}$, $k \in [1, L-1]$.

  1: Initialize $h^{(0)} = X$.

  2: **for** $k = 1, 2, \ldots, L-1$ **do**

  3:    $X^{(k)} = h^{(k-1)}$.

  4:    Calculate $\tilde{X}^{(k)}$ by Eq. (9.5).

  5:    Calculate $W^{(k)}$ by Algorithm 9.2.

  6:    Calculate encoded feature by $h^{(k)} = \tanh(W^{(k)} X^{(k)})$.

  7: **end for**

---

from k to $k+1$, and meanwhile keep the consensus of different features. As shown in Algorithm 9.3, the input of the CSCAE is the style feature X. The output of the algorithm is the encoded feature $h^{(k)}$ and the projection matrix $W^{(k)}$ in the kth step, $k \in [1, L-1]$.

To initialize, we set $h^{(0)} = X$. For step k, the encoded feature $h^{(k-1)}$ is regarded as the input. First, we calculate the output for $X^{(k)}$ as described in Eq. (9.5). Second, we optimize W by CSCAE using Algorithm 9.2. The learned W achieves the properties of both group sparsity and low-rankness. Afterwards, we calculate the new features $W\tilde{X}$ followed by a nonlinear function for normalization. Following the suggestion in [26],

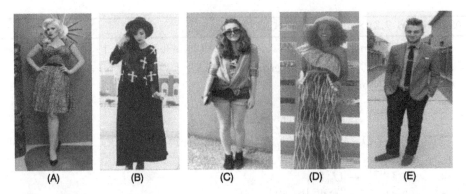

Figure 9.7 Examples for 5 categories in the Hipster Wars dataset: (A) bohemian, (B) hipster, (C) goth, (D) pinup, and (E) preppy.

we use tanh($\cdot$) to achieve the nonlinearity performance. The encoded feature $h^{(k)}$ is regarded as the input in the next step $k + 1$. After $(L - 1)$ steps, we obtain $(L - 1)$ sets of weight matrices and the corresponding encoded features.

9.1.5 Experiments

In this section, we discuss the performance of several low-level based and deep learning based feature representation methods for fashion, manga and architecture style classification tasks.

9.1.5.1 Dataset

Fashion style classification dataset. Kiapour et al. collected a fashion style dataset named Hipster Wars [7] including 1,893 images of 5 fashion styles, as shown in Fig. 9.7. They also launched an online style comparison game to collect human judgments and provide style level information for each image.

Pose estimation is applied to extract key boxes of the human body [27]. Seven dense features are extracted for each box following [7]: RGB color value, LAB color value, HSI color value, Gabor, MR8 texture response [28], HOG descriptor, and the probability of pixels belonging to skin categories.

Manga style classification dataset. Chu et al. collected a shonen (boy targeting) and shojo (girl targeting) manga dataset, which includes 240 panels [8]. Six computational features, including angle between lines, line orientation, density of line segments, orientation of nearby lines, number of nearby lines with similar orientation and line strength, are calculated. Example shojo and shonen style panels are shown in Fig. 9.8.

Since Manga dataset does not provide manually labeled style level information, an automatic way to calculate the style level is applied. First, a mean-shift clustering is applied to find the peak of the density of the images for each style based on the line

Figure 9.8 Examples for shojo style and shonen style in the Manga dataset. The first row is shoji style and the second row is shonen style.

strength feature. Line strength feature is most discriminative among 6 features measured by p-values. Images at the peak of the density are regarded the most representative ones. Then images are ranked according to the distances between the most centralized images, and evenly divided from lowest to highest distance as five style levels.

Architecture style classification dataset. Xu et al. collected an architecture style dataset containing 5000 images [11]. It is the largest publicly available dataset for architectural style classification. The category definition is according to "Architecture_by_style" of Wikimedia.[2] Example images of ten classes are shown in Fig. 9.9. As there is no manually labeled style level information, a similar strategy to that used in Manga dataset is applied to generate the style level information.

9.1.5.2 Compared Methods

Here we briefly describe several low-level feature representation based methods on fashion [7,29,6], manga [8] and architecture [11] style classification tasks. Then we describe general deep learning methods and deep learning methods for style classification tasks.

Low-level based methods:

Kiapour et al. [7] applied mean-std pooling for 7 dense low-level features from clothing images, and then concatenated them as the input to the classifier and named the concatenated features as the style descriptor.

Yamaguchi et al. [29] approached clothing parsing via retrieval, and considered robust style feature for retrieving similar style. They concatenated the pooling features similar to [7], followed by PCA for dimension reduction.

Bossard et al. [6] focused on apparel classification with style. For style feature representation, they first learned a codebook through k-means clustering based on low-level

---

[2] https://commons.wikimedia.org/wiki/Category:Architecture_by_style.

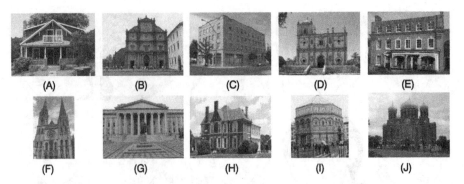

Figure 9.9 Examples for 10 categories in the Architecture Style dataset: (A) American craftsman, (B) Baroque architecture, (C) Chicago school architecture, (D) Colonial architecture, (E) Georgian architecture, (F) Gothic architecture, (G) Greek Revival architecture, (H) Queen Anne architecture, (I) Romanesque architecture, and (J) Russian Revival architecture.

features. Then the bag-of-words features were further processed by spatial pyramids and max-pooling.

Chu and Chao [8] designed 6 computational features derived from line segments to describe drawing styles. Then they concatenated 6 features with equal weights.

Xu et al. [11] adopted the deformable part-based models (DPM) to capture the morphological characteristics of basic architectural components, where DPM describes an image by a multiscale HOG feature pyramid.

MultiFea [12,1]. The baseline in [11] only employed the HOG feature, but CSCAE employed multiple features. Another low-level feature based method using multiple features is generated for fair comparisons. Six low-level features are chosen according to SUN dataset,[3] including HoG, GIST, DSIFT, LAB, LBP, and tinny image. First, PCA dimension reduction is applied to each feature. Then, the normalized features are concatenated together.

Deep learning based methods:

AE [13]. A conventional autoencoder (AE) [13] is applied for learning mid/high-level features. The inputs of AE are the concatenated low-level features.

DAE [26]. Marginalized stacked denoising autoencoder (mSDA) [26] is a widely applied version of denoising autoencoder (DAE). Both SCAE and DAE share the spirit of "noise". The inputs of the DAE are corrupted image features. As in [15], [19] and [26], the corruption rate of the dropout noise is learned by cross-validation. Other settings such as the number of stacked layers and the layer size are the same as SCAE.

SCAE [12,1]. Style centralizing autoencoder (SCAE) [12,1] is applied for learning mid/high-level features. The inputs of SCAE are the concatenated various kinds of low-level feature descriptors, i.e., an early fusion for SCAE.

---

[3] http://vision.cs.princeton.edu/projects/2010/SUN/.

Table 9.1 Performances (%) of fashion style classification on Hipster Wars dataset. The best and second best results under each setting are shown in bold font and underline

Performance	$L=5$	$L=4$	$L=3$	$L=2$	$L=1$
Kiapour et al. [7]:	77.73	62.86	53.34	37.74	34.61
Yamaguchi et al. [29]:	75.75	62.42	50.53	35.36	33.36
Bossard et al. [6]:	76.36	62.43	52.68	34.64	33.42
AE [13]	83.76	75.73	60.33	44.42	39.62
DAE [26]	83.89	73.58	58.83	46.87	38.33
SCAE [12,1]	84.37	72.15	59.47	48.32	38.41
CAE [1]	87.55	76.34	63.55	50.06	41.33
CSCAE [1]	**90.31**	**78.42**	**64.35**	**54.72**	**45.31**

CAE [1]. To demonstrate the roles of "progressive style centralizing" in CSCAE, a consensus autoencoder (CAE) is generated as another baseline. CAE is similar to CSCAE except that in each progressive step, the input and output features are exactly the same. **CSCAE [1].** This method contains the full pipeline of consensus style centralizing autoencoder (CSCAE) in [1].

In all the classification tasks, cross-validation is applied with a 9 : 1 training-to-test ratio. *SVM classifier* is applied in Hipster Wars and Manga datasets by following the settings in [7,8], while *nearest neighbor classifier* (NN) is applied on Architecture datasets. For all deep learning baselines, same number of layers are used.

9.1.5.3 Experimental Results

Results on fashion style classification

Table 9.1 shows the accuracy (%) of low-level and deep learning based methods under different style levels $L = 1, \ldots, 5$. First, from Table 9.1 we can see that deep learning based methods (bottom 5 in the table) in general perform better than low-level based methods (top 3 in the table). CSCAE and CAE achieve the best and second best performance under all the style levels. When comparing DAE with SCAE, we could see that SCAE outperforms DAE, which shows the effectiveness of the style centralizing strategy compared with general denoising strategy for style classification. When comparing CSCAE with CAE, we see that CSCAE outperforms CAE in all the settings, which also due to the style centralizing learning strategy.

Results on manga style classification

Table 9.2 shows the accuracy (%) of the deep learning based methods (bottom 5 in the table) and low-level feature based method (top in the table) under five style levels on Manga dataset. CSCAE and CAE achieve the highest and second highest performance under all the style levels. Compared to fashion and architecture images, CSCAE works especially well for the face images. We think that the face structure and different weights

Table 9.2 Performance (%) of manga style classification

Performance	$L=5$	$L=4$	$L=3$	$L=2$	$L=1$
LineBased [8]	83.21	71.35	68.62	64.79	60.07
AE [13]	83.61	72.52	69.32	65.18	61.28
DAE [26]	83.67	72.75	69.32	65.86	62.86
SCAE [12,1]	83.75	73.43	69.32	65.42	63.60
CAE [1]	<u>85.35</u>	<u>76.45</u>	<u>72.57</u>	<u>67.85</u>	<u>65.79</u>
CSCAE [1]	**90.70**	**80.96**	**77.97**	**77.63**	**79.90**

Table 9.3 Performance (%) of architecture style classification

Performance	$L=5$	$L=4$	$L=3$	$L=2$	$L=1$
Xu et al. [8]	40.32	35.96	32.65	33.32	31.34
MultiFea	52.78	53.00	50.29	49.93	46.79
AE [13]	58.72	56.32	52.32	52.32	48.31
DAE [26]	58.55	56.99	53.34	52.39	50.33
SCAE [12,1]	<u>59.61</u>	57.00	53.27	<u>54.28</u>	51.76
CAE [1]	59.54	<u>58.66</u>	<u>54.55</u>	53.46	<u>51.88</u>
CSCAE [1]	**60.37**	**59.41**	**55.12**	**54.74**	**54.68**

of patches (e.g., the weights of eye patch should be higher than cheek patch) work especially well with the low-rankness and group sparsity assumptions.

Results on architecture style classification

Table 9.3 shows the classification accuracy on the architecture style dataset. First, comparing the method in [8] with MultiFea, we learn that the additional low-level features do contribute to the performance. Second, all the deep learning methods (bottom 5 in the table) achieve better performance than low-level features based methods (top 2 in the table).

9.2. VISUAL KINSHIP UNDERSTANDING

9.2.1 Background

Kinship analysis and parsing have been popular research topics in psychology and biology community [30] for a long time, since kin relations build up the most fundamental social connections. However, verifying the kin relationship between people is not an easy task since there is not instant yet economic way to precisely verify. Although with the modern technology, advanced tools like DNA paternity test is able to provide top-level accuracy, its high-cost in terms of both time and money prevents it being an off-the-shelf verification tool. Recently, kinship verification has attracted substantial attention from computer vision and artificial intelligence society [31–42]. Inspired by the

Figure 9.10 Illustration of kinship verification and family membership recognition problems.

truth that children inherit gene from parents, these works exploit facial appearance as well as social context in the photo to predict the kin relationship. With confirmed relationship, applications like building family tree, seeking missing child, and family photos retrieval in real-world become promising.

Most of the state-of-the-art works concentrate on the pair-wise kinship verification, meaning given a pair of facial images, the algorithm determines if the two people have the kin relation or not. In general, the relationship is restricted to "parent–child". In a more general case, kinship could include siblings as well, and a higher-order relationship, i.e., many-to-one verification, should be considered in the real-world applications. A typical scenario is, given a single test image and a family album, we determine if the test is from this family or not, which is first discussed in [37]. Apparently, this problem is more general than one-to-one verification and the latter one is actually a special case. In this part, we formally name it as "family membership recognition" (FMR). The problem illustration can be found in Fig. 9.10.

In addition to the conventional kinship verification problem, in FMR, we encounter new challenges: (1) How to define familial features in terms of family rather than individual; (2) How to effectively extract familial features from input images. In this part, we proposed a low-rank regularized family faces guided parallel autoencoders (rPAE) for FMR problem, and the proposed method solves challenges through an integrated framework. In addition, rPAE can easily adapt to conventional kinship verification problem, and significantly boost the performance.

To that end, we first propose a novel concept called "family faces" which are constructed by family mean-face and its component-wise nearest neighbors in the training set. Second, we design a novel structure called parallel autoencoders whose outputs are guided by multiple family faces. To be concrete, the inputs of each encoder are all facial images from different families, while the outputs are corresponding family faces. In this way, we have better chance to capture the inherited family facial feature under the assumption that they are from different family members. Finally, to guarantee learned autoencoders yield common feature space, we impose a low-rank constraint on the model parameters in the first layer which enables us to learn parallel autoencoders in a unified framework, rather than one-by-one. Extensive experiments conducted on KFW [36] and Family101 databases [37] demonstrate the proposed model is effective in solving both kinship verification and family membership recognition problems compared with the state-of-the-art methods.

9.2.2 Related Work

There are three lines in the related work: (1) familial features, (2) autoencoder, (3) low-rank matrix analysis.

Facial appearance and its representation [43,44] have been considered as two of the most important familial features in kinship verification and FRM problems. Among the existing work [31–33,35,34,36–38,41,42], local features [45–47] and components based strategy provide better performance [31–33]. These facial components carry explicit semantics such as eyes, nose, mouth, cheek, forehead, jaw, brows, by which people could empirically determine the kin relationship. The final feature vector usually concatenates all the local descriptors extracted from these components for feature assembling. In addition, metric learning has been widely discussed in kinship verification to yield a high-level familial representation [36,39,40].

Although above hand-craft features [45–47] empirically provide superior performance, they are recently chased by learning based feature representation [48–52]. Among them, autoencoders [50–52] are able to generate honestly reflected features through hidden layers. Moreover, autoencoder based learning method has the flexibility to stack more than one encoder to build a deep structure [53] which has been empirically proved effective in visual recognition [54,17]. In this part, different from the traditional autoencoder that uses identical data for both input and output, we tune the model and enforce the output to be multiple family faces. Therefore, we have a parallel structure for a group of autoencoders and each of them runs in the supervised fashion guided by the family faces. It should be noted most recently gated autoencoders have been applied to kinship verification and family member recognition problems, and achieved appealing results [55].

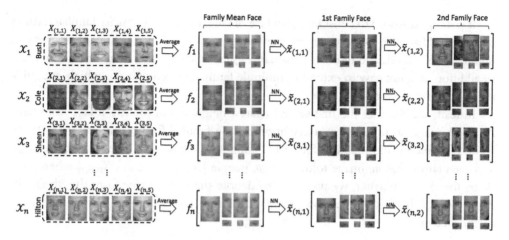

Figure 9.11 Illustration of building family faces. For each family album $\mathcal{X}_j$, we compute its family mean face by averaging all face images in the family folder. Then for each family, we find family mean face's *k*-nearest-neighbor within this family, in a component-wise way. Note in each family face, the single face image on the left is the nearest neighbor search result based on the whole face, while the facial components on the right are those results from a component-wise way. The second family face in the first row with red border demonstrates components of family face are not necessarily coming from the same face.

Low-rank matrix constraint has been widely discussed recently due to its successful application in data recovery [56], subspace segmentation [57], image segmentation [58], visual domain adaptation [59], and multitask learning [60]. The underlying assumption is low-rank constraint on the linear feature space is able to discover the principal component in spite of noises with arbitrarily large magnitude. This constraint can work on the reconstruction coefficients to recover the hidden subspace structure [57,59]. It is also preferred by multiple features/tasks that have intrinsic connections [58,60]. In this part, different from them, we encourage the model parameters of different autoencoders to be low-rank, which helps with discovering common feature space shared by different family faces. In addition, this low-rank constraint enables us to learn parallel autoencoders in a joint framework, rather than one by one.

9.2.3 Family Faces

Familial features have been discussed recently for kinship verification, and most of them are based on empirical observation that children inherit facial traits from their parents. Therefore, the foundation of this group of methods is pair-wise comparisons between children and parents in terms of facial components. The introduction of FMR breaks through the limitation of pair-wise comparison since the test needs to refer to facial traits of a whole family, rather than a single person. Therefore, formulating family familial features from a group of family photos becomes critical.

It is straightforward to consider mean face of family photos as concise familial features of a family. However, using mean face inevitably introduces blurring and artificial effects on appearance level, and discards personal features of family members on feature level. In addition, it is not easy to extend to multiple familial representations which can guide the parallel autoencoders.

Instead, we consider the first several nearest neighbors of the mean face from family facial images as the ideal representations of the familial feature and call it "family faces", which is illustrated in Fig. 9.11. Suppose we have n families $[\mathcal{X}_1, \mathcal{X}_2, \ldots, \mathcal{X}_n]$, and each family has m_i image folders for m_i people $[X_{(i,1)}, X_{(i,2)}, \ldots, X_{(i,m_i)}]$ where i indexes the family. Further, we use $x_{(i,j,k)}$ to denote the facial image from the ith family, jth person's kth photo. Therefore, the family mean face from the ith family can be formulated as

$$f_i = \frac{1}{m_i \times m_{ij}} \sum_{j,k} x_{(i,j,k)}, \tag{9.10}$$

where m_i and m_{ij} are numbers of people in family i, and number of images for the jth person in family i, and $x_{(i,j,k)}$ could be either raw images or visual descriptors. Therefore, the first family face for the ith family can be easily found by the nearest neighbor search and we denote it as $\tilde{x}_{(i,1)}$. To expand the size of family faces, more neighbors of the mean face can be added to the family faces by considering the second, third and kth neighbors, namely, $\tilde{x}_{(i,2)}, \tilde{x}_{(i,3)}, \ldots, \tilde{x}_{(i,k)}$.

Inspired by the previous works that facial components work better than holistic feature, we construct family faces in a component-wise way. The facial components can be defined by a few key points on the face. Consequently, each facial image in Eq. (9.10) is now replaced by certain facial component $x^c_{(i,j,k)}$, and f_i replaced by f^c_i, where c indexes this component. After the nearest neighbor search, we obtain the local family face $\tilde{x}^c_{(i,j)}$ for the ith family jth nearest neighbor and cth component. Finally, we assemble these local components into one feature vector, and still use $\tilde{x}_{(i,j)}$ to indicate this family face.

Interestingly, when we assemble these local family faces into an integrated one on the pixel level, we found that not all components come from the same image, or even the same person, as mentioned in Fig. 9.11. This is reasonable since mean face is essentially a virtual face and each local component from the same face may have different rankings in nearest neighbor search of the mean face. In addition, it supports the fact that children inherit genes from both parents and different people may carry different family traits.

9.2.4 Regularized Parallel Autoencoders

In this section, we detail the structure of parallel autoencoders (shown in Fig. 9.12) and explain how to utilize low-rank regularizer in building parallel autoencoders.

9.2.4.1 Problem Formulation

A typical objective function for autoencoder with N training samples is

$$\min_{W,b} \frac{1}{N} \sum_i L(W, b; x_i, y_i) + \lambda \Omega(W), \tag{9.11}$$

where $L(W, b; x, y)$ is the loss function, $\Omega(W)$ is the regularization term, W, b are model parameters, x_i, y_i are input and target value, and weight decay parameter λ balances the relative importance of the two terms. In this part, the loss function is implemented by the squared error between hypothesis and target values, namely

$$L(W, b; x_i, y_i) = \|h_{W,b}(x_i) - y_i\|^2, \tag{9.12}$$

and the regularization term can be written in square sum of all elements in the weight matrix of the first and second layers, equal to the norms $\|W^{(1)}\|_F^2$ and $\|W^{(2)}\|_F^2$, where $\|\cdot\|_F$ is the matrix Frobenius norm.

Suppose we have k family faces, then we can formulate the new loss function for the family faces guided autoencoders by

$$\frac{1}{N \times k} \sum_{i,j} \|h_{W_j,b_j}(x_i) - \tilde{x}_{(i,j)}\|^2, \tag{9.13}$$

where j indexes the family face, and k denotes the number of family faces in each family. Apparently, Eq. (9.13) includes k autoencoders, and these encoders can be learned at the same time. Therefore, we call it parallel autoencoders, which provide a wide structure rather than a deep one.

However, trivially combining these autoencoders will not necessarily boost the final performance since there are no connections between autoencoders. To guarantee all the encoders share common feature spaces, we introduce a low-rank regularization term to encourage the weight matrices assembled from k autoencoders in a matrix low-rank structure, which ensures the familial features generated by the hidden layers from k autoencoders to have common feature space. Combining with the parallel autoencoders, the proposed low-rank regularized objective function is written as

$$\min_{W_j,b_j} \frac{1}{N \times k} \sum_{i,j} \|h_{W_j,b_j}(x_i) - \tilde{x}_{(i,j)}\|^2 \tag{9.14}$$

$$+ \lambda_1 (\|\mathcal{W}^{(1)}\|_F^2 + \|\mathcal{W}^{(2)}\|_F^2) + \lambda_2 \|\mathcal{W}^{(1)}\|_*,$$

where $\mathcal{W}^{(1)} = [W_1^{(1)}, \ldots, W_k^{(1)}]$ is a column-wise matrices concatenation, $\mathcal{W}^{(2)} = [W_1^{(2)}, \ldots, W_k^{(2)}]$ has a similar structure, and $\|\cdot\|_*$ is the matrix nuclear norm, which is identified as convex surrogate of the original rank minimization problem. Solving

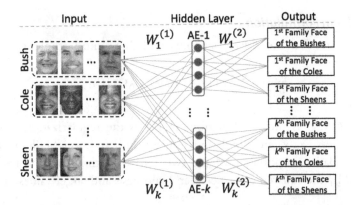

Figure 9.12 Illustration of the proposed low-rank regularized parallel autoencoders (rPAE) for family membership recognition. Here $W_1^{(1)}, \ldots, W_k^{(1)}$ and $W_1^{(2)}, \ldots, W_k^{(2)}$ are weight matrices of the input and hidden layers. Low-rank constraint is imposed on $\mathcal{W} = [W_1^{(1)}, \ldots, W_k^{(1)}]$ to achieve better representation for family membership features.

this problem is nontrivial since we introduce a nonsmooth term. Therefore, we cannot directly use gradient descent method to solve both $\mathcal{W}^{(1)}$ and $\mathcal{W}^{(2)}$ following the traditional way for the proposed autoencoders. Since in practice $\mathcal{W}^{(1)}$ and $\mathcal{W}^{(2)}$ are solved in an iterative way, which means one's update relies on another's update, we break down it into two subproblems and concentrate on $\mathcal{W}^{(1)}$ first.

9.2.4.2 Low-Rank Reframing

Although the nonsmooth property of low-rank prevents from solving it directly, we need to keep it since it helps in recovering the common feature space among different encoders. Recently, atomic decomposition [61] has been proposed to tackle large-scale low-rank regularized classification problem [60], where the matrix trace norm is converted to vector l_1-norm and therefore can be solved efficiently through coordinate descent algorithm. We borrow the idea of low-rank reframing and use it to solve our problem since it is fast and efficient in our problem.

Suppose there is an overcomplete and uncountable infinite dictionary of all possible "atoms", or rank-one matrices in our problem, that is denoted by a matrices set $\mathcal{M}$,

$$\mathcal{M} = \{uv^T | u \in \mathbb{R}^{d_1}, v \in \mathbb{R}^{d_2}, \|u\|_2 = \|v\|_2 = 1\}. \tag{9.15}$$

Note that $\mathcal{M}$ does not necessarily build a basis in $\mathbb{R}^{d_1 \times d_2}$. Further, we use $\mathcal{I}$ to represent the index set spanning the rank-1 matrix in the $\mathcal{M}$, namely,

$$\mathcal{M} = \{M_i \in \mathbb{R}^{d_1 \times d_2} | i \in \mathcal{I}\} = \{u_i v_i^T | i \in \mathcal{I}\}. \tag{9.16}$$

Next, we consider a vector $\theta \in \mathbb{R}^{\mathcal{I}}$ and its support $\text{supp}(\theta) = \{i, \theta_i \neq 0\}$, and further define a vector set as $\Theta = \{\theta \in \mathbb{R}^{\mathcal{I}} | \text{supp}(\theta) \text{ is finite}\}$. Then we have a decomposition for the matrix $\mathcal{W}^{(1)}$ onto atoms in $\mathcal{M}$,

$$\mathcal{W}^{(1)} = \sum_{i \in \text{supp}(\theta)} \theta_i M_i. \tag{9.17}$$

Actually, this is the atom decomposition of the matrix $\mathcal{W}^{(1)}$ onto a series of rank-1 matrices and, since the atom dictionary is overcomplete, this decomposition might not be unique. Through the formulation of this decomposition, we can reframe the original low-rank regularized problem proposed in Eq. (9.14) into the following one:

$$\min_{\theta \in \Theta^+} I(\theta) = \min_{\theta \in \Theta^+} \lambda_2 \sum_{i \in \text{supp}(\theta)} \theta_i + R(\mathcal{W}_\theta), \tag{9.18}$$

where $R(\mathcal{W}_\theta)$ represents the remaining parts in Eq. (9.14) other than low-rank regularizer. We can see that $\sum_{i \in \text{supp}(\theta)} \theta_i$ is essentially the vector l_1-norm for θ.

9.2.4.3 Solution

We describe how to use coordinate descent method to solve Eq. (9.18). The algorithm described here actually belongs to the family of atom descent algorithms, whose stopping criteria are given by ε-approximation optimality:

$$\begin{cases} \forall i \in \mathcal{I} : \frac{\partial R(\mathcal{W})}{\partial \theta_i} \geq -\lambda_2 - \varepsilon, \\ \forall i \in \text{supp}(\theta) : |\frac{\partial R(\mathcal{W})}{\partial \theta_i} + \lambda_2| \leq \varepsilon. \end{cases} \tag{9.19}$$

The basic flow of coordinate descent is similar to gradient descent, but at each iteration with θ_t,[4] we need to find the coordinate along which we can achieve the steepest descent while remaining in Θ^+. In our model, this is equal to pick $i \in \mathcal{I}$ with the largest $-\partial I(\theta_t)/\partial \theta_i$.

In practice, it is easy to find that the coordinate corresponding to the largest $-\partial I(\theta_t)/\partial \theta_i$ can be computed by the singular vector pair corresponding to the top value of the matrix $-\nabla R(\mathcal{W}_{\theta_t})$, namely,

$$\max_{\|u\|_2 = \|v\|_2 = 1} u^T \left(-\nabla R(\mathcal{W}_{\theta_t}) \right) v. \tag{9.20}$$

There are two possible cases after we find the descent direction: if coordinate $i \notin \text{supp}(\theta_t)$, then we only move in the positive direction; else we can move in either

---

[4] Note we slightly abuse θ_t here to indicate supports vector at iteration t, while θ_i stands for the ith element in the vector.

the positive or negative direction. To avoid aggressive update, we use a steep-enough direction (up to $\varepsilon/2$) instead in the practice.

In each iteration, after we find the steep-enough direction defined by $u_t v_t^T$, we need to tackle the step size of each update. Since we do not adopt the steepest direction in the algorithm, we will use a line-search to guarantee the objective value $I(\theta)$ is decreased in each iteration. So each update can be described as $\mathcal{W}_{t+1} = \mathcal{W}_t + \delta u_t v_t^T$, and $\theta_{t+1} = \theta_t + \delta e_t$, where e_t is an indicator vector, with the tth element being 1 in the tth iteration. The overall process can be found in Algorithm 9.4.

Algorithm 9.4: Coordinate descent algorithm for the reframed low-rank regularized problem.

Input: Weight decay and regularization parameters λ_1 and λ_2, initial point $\mathcal{W}_{\theta_0}$, convergence threshold ε.

Output: ε-optimal $\mathcal{W}_\theta$

1 **for** $t = 0$ **to** T **do**

2   Compute the top singular vector pairs u_t and v_t, and make sure $u^T(-\nabla R(\mathcal{W}_{\theta_t}))v \geq \|\nabla R(\mathcal{W}_t)\|_{\sigma,\infty}$

3      $g_t = \lambda_2 + < \nabla R(\mathcal{W}_t), u_t v_t^T >$

4   IF

         $g_t \leq -\varepsilon/2$, then $\mathcal{W}_{t+1} = \mathcal{W}_t + \delta u_t v_t^T$ where δ is learned by line-search, and $\theta_{t+1} = \theta_t + \delta e_t$.

5   ELSE

         If θ_t satisfies stopping condition, stop and return θ_t; otherwise solve θ_{t+1} by $\min_\theta \lambda_2 \sum_{i=1}^s \theta_i + R(\sum_i^s \theta_i u_i v_i^T)$, $\forall \theta_i \geq 0$, where s is the number of support of θ at iteration t.

6 **end**

Recall that the proposed regularized family faces guided parallel autoencoders in Eq. (9.14) should have been solved by gradient descent algorithm, however, the nonsmoothness property of the new regularized term makes it non-differential. Algorithm 9.4 actually tells us how to solve it with gradient algorithm even with low-rank term. To solve the original problem, we still need to run gradient descent algorithm on Eq. (9.14) with partial derivative solved by back-propagation. The difference is each time when we update $\mathcal{W}^{(1)}$, we need to run one iteration of Algorithm 9.4. The entire solution for Eq. (9.14) can be found in Algorithm 9.5.

Remarks. (1) We mainly focus on the solution of rPAE in this section and do not list the detailed solution of autoencoders, which involves "feed-forward" and "back-propagation" processes, and "nonlinear activation" function, since they can be easily checked from many relevant works. (2) We empirically set the model parameters as: $\lambda_1 = \lambda_2 = 0.1$, $T = 50$, and $\varepsilon = 0.01$ in our experiments, and achieve acceptable results.

Algorithm 9.5: Gradient descent algorithm for the regularized family faces guided parallel autoencoders.

Input: Weight decay and regularization parameters λ_1 and λ_2, initial point $\mathcal{W}_0^{(1)}$, and $\mathcal{W}_0^{(2)}$.

Output: $\mathcal{W}^{(1)}$ and $\mathcal{W}^{(2)}$

1 Initialization: random initialization for $\mathcal{W}^{(1)}$ and $\mathcal{W}^{(2)}$ with purpose of symmetry breaking.

2 **repeat**

3 $\quad \mathcal{W}_{t+1}^{(1)} = \mathcal{W}_t^{(1)} - \alpha(\frac{1}{N}\nabla\mathcal{W}_t^{(1)} + \lambda_1\mathcal{W}_t^{(1)})$

4 $\quad \mathcal{W}_{t+1}^{(2)} = \mathcal{W}_t^{(2)} - \alpha(\frac{1}{N}\nabla\mathcal{W}_t^{(2)} + \lambda_1\mathcal{W}_t^{(2)})$

$\quad$ where $\mathcal{W}_t^{(1)}$ is learned in Algorithm 9.4, and $\mathcal{W}_t^{(2)}$ is learned by back-propagation, α is the step size, and $\nabla\mathcal{W}_t$ is the gradient of the objective w.r.t. $\mathcal{W}$ at iteration t.

5 **until** $\mathcal{W}^{(1)}$ *and* $\mathcal{W}^{(2)}$ do not change by more than a predefined threshold;

There might be better settings for them, but we leave the space for discussions of other important model parameters. (3) Although rPAE is proposed for FMR problem, it can easily adapt to kinship verification problems by sampling a few subsets of the training data to guide the learning of parallel autoencoders. We will introduce the implementation details in the experiment section. (4) The deep structure can be trained in a layer-wise way which only involves the training of single hidden layer of rPAE.

9.2.5 Experimental Results

In this section, we demonstrate the effectiveness of the proposed method through two groups of experiments: (1) kinship verification and (2) family membership recognition. For kinship verification experiments, we use KFW database published in [36,40] while for face recognition through familial feature experiments, we use Family101 database published in [37].

9.2.5.1 Kinship Verification

Kinship Face in the Wild (KFW)[5] is a database of face images collected for studying the problem of kinship verification from unconstrained face images. It includes two parts, KFW-I and KFW-II, both of which include four detailed kin relations: father–son (F–S), father–daughter (F–D), mother–son (M–S), mother–daughter (M–D). In the KFW-I dataset, there are 156, 134, 116, and 127 pairs of kinship images for these four relations, while in KFW-II dataset, each relation contains 250 pairs of kinship images. The difference of these two datasets is that KFW-I uses facial images from different photo to build kinship pairs, but KFW-II uses facial images from the same photo. Therefore,

[5] http://www.kinfacew.com/.

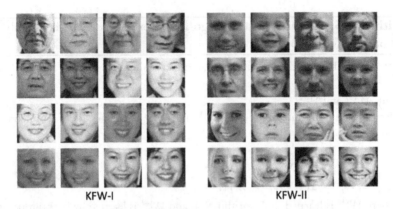

Figure 9.13 Sample faces from KFW-I and KFW-II datasets. Four rows illustrate four different kin relations: father–son, father–daughter, mother–son, and mother–daughter.

the variations of lighting and expressions may be more dramatic in KFW-I, leading to a relatively lower performance. All faces in the database have been manually aligned and cropped to 64 × 64 images to exclude background. Sample images can be found in Fig. 9.13. In the following KFW relevant experiments, we use the HOG features provided by KFW benchmark website as the input to rPAE.

Different from family faces guided rPAE in FMR, here we sample a few small sets from the training data and use each of them to build an autoencoder and all of them to build rPAE. For each specific relation, we randomly sample half of the images from the given positive pairs, and then put these samples back. We repeat this several times to obtain enough sets for the training of rPAE. The input of rPAE is [child, parent] and corresponding target value is [parent, child]. After we learn the rPAE, we encode the input feature through rPAE, and concatenate all of them to form the final feature vector which will be fed to a binary SVM classifier. We use the absolute value of vector difference from positive kinship pairs as the positive samples and that from negative kinship pairs as the negative samples. Note that we use LibSVM [63] with rbf kernel to train the binary model and use its probabilistic output to compute both ROC and AUC. The model parameters of SVM such as slack variable C and bandwidth σ is learned through grid search on the training data. We strictly follow the benchmark protocol of KFW, and report the `image restricted` experimental results with five-fold cross-validation.

There are several key factors in the model: the number of hidden units in each layer, the number of hidden layers, and the number of sampling to generate different autoencoders. To evaluate their impacts on our model, we experiment step by step to show their effects. We first use a single hidden layer autoencoder and experiment with four kin relations on KFW-II, with the number of hidden units changing from 200 to 1800, as shown in Fig. 9.14A. It can be seen from four relations that there exists

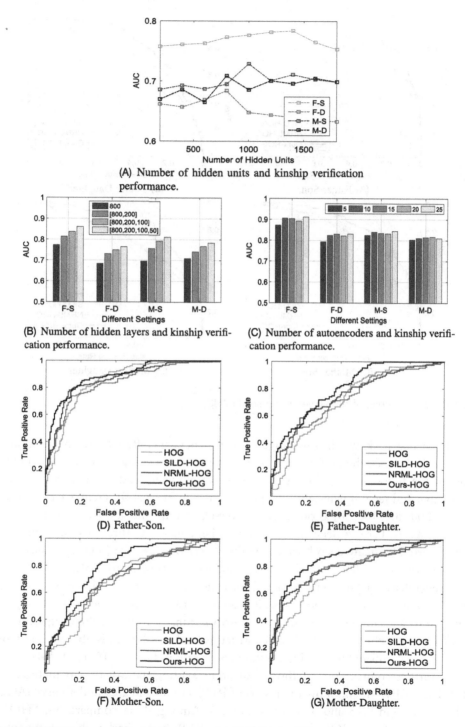

(A) Number of hidden units and kinship verification performance.

(B) Number of hidden layers and kinship verification performance.

(C) Number of autoencoders and kinship verification performance.

(D) Father-Son.

(E) Father-Daughter.

(F) Mother-Son.

(G) Mother-Daughter.

Figure 9.14 ROC curves of kinship verification on KFW-I.

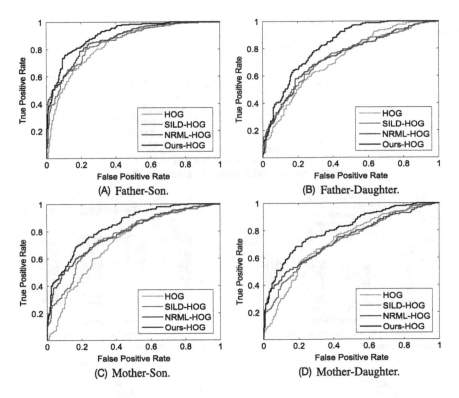

Figure 9.15 ROC curves of kinship verification on KFW-II.

a performance peak within this range, and in general it differs from one relation to another. To balance the performance and the time cost, we suggest to use 800 hidden units in the following experiments. Second, we gradually add the number of layers from 1 to 4 and see if it helps with the performance in Fig. 9.14B. Here, we empirically choose [800, 200, 100, 50] as our deepest layersize setting, and add layer one by one to see the impacts. Clearly, deep structure always benefits the feature learning, which has been reported by many deep learning works. Therefore, we take [800, 200, 100, 50] as our layersize setting. In addition, we also conduct experiments to analyze the number of autoencoder in rPAE in Fig. 9.14C, from which we can observe that more sampling could bring in performance boost, but it takes more time as well. Therefore, we suggest to use 10 parallel autoencoders in our framework. Finally, from the three experiments, we can also conclude that the parallel structure and regularizer is able to boost the performance compared to the single autoencoder case in Figs. 9.14A and 9.14B.

We further compare our method with other existing state-of-the-art methods on KFW-I and KFW-II, and show the results of ROC curves and area under curve (AUC) in Figs. 9.14 and 9.15 and Tables 9.4 and 9.5. Among these comparisons, "HOG" means we directly use HOG feature provided by KFW benchmark website as the input

Table 9.4 AUC of KFW-I dataset

Method	F–S	F–D	M–S	M–D	Average
HOG [47]	0.849	0.717	0.703	0.747	0.754
SILD [62]	0.838	0.730	0.708	0.797	0.769
NRML [40]	0.862	0.757	0.721	0.801	0.785
Ours	0.890	0.808	0.801	0.856	0.838

Table 9.5 AUC of KFW-II dataset

Method	F–S	F–D	M–S	M–D	Average
HOG [47]	0.833	0.723	0.721	0.723	0.750
SILD [62]	0.853	0.739	0.765	0.723	0.770
NRML [40]	0.871	0.740	0.784	0.738	0.783
Ours	0.906	0.823	0.840	0.811	0.845

Table 9.6 Comparisons with gated autoencoder (GAE) based methods [55]. Mean average precision (%) is reported for this experiment

KFW-I	F–S	F–D	M–S	M–D	Average
GAE [55]	76.4	72.5	71.9	77.3	74.5
Ours	87.7	78.9	81.1	86.4	83.5
KFW-II	**F–S**	**F–D**	**M–S**	**M–D**	**Average**
GAE [55]	83.9	76.7	83.4	84.8	82.2
Ours	90.5	82.0	82.6	78.5	83.4

to the binary SVM, to train and classify the kinship pairs. "SILD (image restricted)" [62] and "NRML (image unrestricted)" [36,40] mean that we first feed the original HOG features to the two comparisons and then use their output as the new features for the binary SVM. From the result, we can see that almost all the methods perform better than the direct use of HOG, which demonstrates that these methods work well on kinship verification problem. With the help of rPAE, our method performs better than the other two closely related methods in the four cases on both KFW-I and KFW-II. Finally, we also compare with the most recent work based on gated autoencoders (GAE) [55], which is shown in Table 9.6.

9.2.5.2 Family Membership Recognition

We conduct familial feature based face recognition experiments in this section, which includes two parts: (1) family membership recognition and (2) face identification through familial features. For both experiments, we use Family101 database published in [37]. Family 101 database has 101 different family trees, 206 nuclear families, and 607 individuals, including 14,816 images. Most of individual in the database are pub-

Table 9.7 Experimental results of family membership recognition on Family101 database. NN, SVM and Our method uses local Gabor as feature extraction methods

Method	Random	NN	SVM	Group Sparsity [37]	Ours
Accuracy (%)	4.00	15.59 ± 5.6	19.93 ± 6.94	20.94 ± 5.97	23.96 ± 5.78

lic figures. Since the number of family members in each family tree are different, and the image qualities are diverse, we select 25 different family trees, and ensure that each family tree contains at least five family members. In the preprocessing, we found that there are some dominant members in the family tree, i.e., an individual with significantly many images. Therefore, we restrict the number of each individual's images less than 50. In our test, we randomly select one family member as test and use rest family members for training (including both rPAE training and classifier training). We use nearest neighbor as the classifier, repeat this five times, and the average performance plus standard deviation is reported in Table 9.7.

In the following evaluation, we still follow the parameters setting of kinship verification discussed in the last section, and use local Gabor [33] as our input feature. That is, we first crop the image to 127×100 and then partition the faces into components by the four key points on the face: two eyes, nose tip, center of the mouth. Gabor features are extracted from each component in 8 directions and 5 magnitudes, and then concatenate to a long vector. We use PCA to reduce the vector length to 1000 for simplicity. From Table 9.7, we can see that family membership recognition is very challenging [37], and fewer results have been reported before. Most methods are slightly better than random guess. Nonetheless, our method performs best thanks to family face + rPAE framework. It is interesting to find that the standard deviation is relatively large in all experiments. The reason is when selected as the test, the current individual may not inherit too much character from the family members, while for others they may inherit more. Therefore, the accuracy fluctuates significantly during the five evaluations.

In addition, we showcase how the familial feature assists in general face recognition. If we consider all the training data in the FMR as references/gallery in the face recognition, then FMR can also be regarded as a face recognition problem. The only difference is in face recognition problem, references are other facial images of the test individual, while in FMR, references are facial images of family members'. From Table 9.7, we are inspired that FMR can boost the face recognition since FMR can identify people by auxiliary data. Therefore, in the following experiments, we compare face recognition (FR) results with/without auxiliary data. In Fig. 9.16, we can observe that familial features and family member's facial images are helpful in face recognition problem.

Finally, we illustrate some face recognition results with query images and returned nearest neighbor based on the proposed familial feature. In Fig. 9.17, we show three cases in rows 1–3. The first one is a failure case that returns incorrect facial images from other family. The second query image is correctly recognized because its nearest neigh-

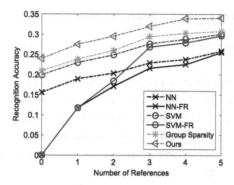

Figure 9.16 Family membership recognition accuracy and the number of references in the training set. Methods with "FR" mean they do not use family members' facial images as training data, and only use several references from the same individual for training. Note group sparsity [37] is the only state-of-the-art method for FMR so far.

Figure 9.17 The first column of images are query images while the 2nd to 6th columns are their first 5 nearest neighbor in the training set. The first row shows a failure case where the nearest neighbor is not a facial image from Babbar family. The second row is the success case because the nearest neighbor of query is from the same family. Last row is also a success case, but different from the previous one. Since we also include 2 images of the query in the training set, the nearest neighbor of the query turns out to be an image of the query itself, similar to conventional face recognition problem. Note we use blue border for the query, green for the correct family member, doted green for training images of the query, and red for the incorrect family member.

bor is from the same family as the query image. In the third case, we use doted frame to represent the training images from the query itself which follows conventional face recognition procedure. From Figs. 9.16 and 9.17, we can conclude that the proposed approach indeed can help with face recognition using family members.

9.3. RESEARCH CHALLENGES AND FUTURE WORKS

In this chapter, we described using deep learning for style recognition and kinship understanding.

For style classification, the style centralizing autoencoder (SCAE) progressively drew weak style images to the class center to increase the feature discrimination. The weights of different descriptors are automatically allocated due to the consensus constraints. We described a novel rank-constrained group sparsity autoencoder, and a corresponding fast solution to achieve competitive performance but saving half of the training time compared to nonlinear ones.

Currently, we are working on the scenario that each image only belongs to one single style. However, sometimes one image may be with multiple styles, such as the mix-and-match styles in fashion. In the future, we plan to explore multiple-style classification. Furthermore, the style classification applications described in this chapter are all vision based, in the future, we plan to explore audio and document style classification, e.g., music style classification. We believe that weak style phenomenon also exists in audio or document styles. In vision based application, we apply patch consensus of images to constrain the consensus of different features. It would be very interesting to find out the consensus rules for audio or document.

REFERENCES

[1] Jiang S, Shao M, Jia C, Fu Y. Learning consensus representation for weak style classification. IEEE Transactions on Pattern Analysis and Machine Intelligence 2017.

[2] Yamaguchi K, Kiapour MH, Ortiz LE, Berg TL. Parsing clothing in fashion photographs. In: IEEE conference on computer vision and pattern recognition. IEEE; 2012. p. 3570–7.

[3] Jiang S, Wu Y, Fu Y. Deep bi-directional cross-triplet embedding for cross-domain clothing retrieval. In: Proceedings of the 2016 ACM on multimedia conference. ACM; 2016. p. 52–6.

[4] Liu S, Song Z, Liu G, Xu C, Lu H, Yan S. Street-to-shop: cross-scenario clothing retrieval via parts alignment and auxiliary set. In: IEEE conference on computer vision and pattern recognition. IEEE; 2012. p. 3330–7.

[5] Jiang S, Fu Y. Fashion style generator. In: Proceedings of the twenty-sixth international joint conference on artificial intelligence, IJCAI-17; 2017. p. 3721–7.

[6] Bossard L, Dantone M, Leistner C, Wengert C, Quack T, Van Gool L. Apparel classification with style. In: Asian conference on computer vision. Springer; 2013. p. 321–35.

[7] Kiapour MH, Yamaguchi K, Berg AC, Berg TL. Hipster wars: discovering elements of fashion styles. In: European conference on computer vision. Springer; 2014. p. 472–88.

[8] Chu WT, Chao YC. Line-based drawing style description for manga classification. In: ACM international conference on multimedia. ACM; 2014. p. 781–4.

[9] Goel A, Juneja M, Jawahar C. Are buildings only instances?: exploration in architectural style categories. In: Indian conference on computer vision, graphics and image processing. ACM; 2012.

[10] Van Gemert JC. Exploiting photographic style for category-level image classification by generalizing the spatial pyramid. In: ACM international conference on multimedia retrieval. ACM; 2011. p. 1–8.

[11] Xu Z, Tao D, Zhang Y, Wu J, Tsoi AC. Architectural style classification using multinomial latent logistic regression. In: European conference on computer vision. Springer; 2014. p. 600–15.

[12] Jiang S, Shao M, Jia C, Fu Y. Consensus style centralizing auto-encoder for weak style classification. In: Proceedings of the thirtieth AAAI conference on artificial intelligence. AAAI; 2016.

[13] Bengio Y. Learning deep architectures for AI. Foundations and Trends in Machine Learning 2009;2(1):1–127.

[14] Ding Z, Shao M, Fu Y. Deep low-rank coding for transfer learning. In: International joint conference on artificial intelligence; 2015. p. 3453–9.

[15] Vincent P, Larochelle H, Bengio Y, Manzagol PA. Extracting and composing robust features with denoising autoencoders. In: Proceedings of the 25th international conference on machine learning. ACM; 2008. p. 1096–103.

[16] Vincent P, Larochelle H, Lajoie I, Bengio Y, Manzagol PA. Stacked denoising autoencoders: learning useful representations in a deep network with a local denoising criterion. The Journal of Machine Learning Research 2010;11:3371–408.

[17] Kan M, Shan S, Chang H, Chen X. Stacked progressive auto-encoders (SPAE) for face recognition across poses. In: IEEE CVPR. IEEE; 2014. p. 1883–90.

[18] Rumelhart DE, Hinton GE, Williams RJ. Learning representations by back-propagating errors. Cognitive Modeling 1988;5:696–9.

[19] Chen M, Xu Z, Weinberger KQ, Sha F. Marginalized stacked denoising autoencoders. In: Learning workshop; 2012.

[20] Chapelle O, Scholkopf B, Zien A. Semi-supervised learning. IEEE Transactions on Neural Networks 2009;20(3):542.

[21] Zhao H, Fu Y. Dual-regularized multi-view outlier detection. In: Proceedings of international joint conference on artificial intelligence; 2015. p. 4077–83.

[22] Ding Z, Shao M, Fu Y. Deep robust encoder through locality preserving low-rank dictionary. In: European conference on computer vision. Springer; 2016. p. 567–82.

[23] Liu G, Lin Z, Yan S, Sun J, Yu Y, Ma Y. Robust recovery of subspace structures by low-rank representation. IEEE Transactions on Pattern Analysis and Machine Intelligence 2013;35(1):171–84.

[24] Liu G, Yan S. Latent low-rank representation for subspace segmentation and feature extraction. In: International conference on computer vision. IEEE; 2011. p. 1615–22.

[25] Bunea F, She Y, Wegkamp MH, et al. Joint variable and rank selection for parsimonious estimation of high-dimensional matrices. The Annals of Statistics 2012;40(5):2359–88.

[26] Chen M, Xu Z, Weinberger K, Sha F. Marginalized denoising autoencoders for domain adaptation. arXiv preprint arXiv:1206.4683, 2012.

[27] Yang Y, Ramanan D. Articulated pose estimation with flexible mixtures-of-parts. In: IEEE conference on computer vision and pattern recognition. IEEE; 2011. p. 1385–92.

[28] Varma M, Zisserman A. A statistical approach to texture classification from single images. International Journal of Computer Vision 2005;62(1–2):61–81.

[29] Yamaguchi K, Kiapour MH, Berg TL. Paper doll parsing: retrieving similar styles to parse clothing items. In: IEEE international conference on computer vision. IEEE; 2013. p. 3519–26.

[30] Barnes JA. Physical and social kinship. Philosophy of Science 1961;28(3):296–9.

[31] Fang R, Tang KD, Snavely N, Chen T. Towards computational models of kinship verification. In: ICIP. IEEE; 2010. p. 1577–80.

[32] Xia S, Shao M, Fu Y. Kinship verification through transfer learning. In: IJCAI. AAAI Press; 2011. p. 2539–44.

[33] Xia S, Shao M, Luo J, Fu Y. Understanding kin relationships in a photo. IEEE TMM 2012;14(4):1046–56.

[34] Zhou X, Lu J, Hu J, Shang Y. Gabor-based gradient orientation pyramid for kinship verification under uncontrolled environments. In: ACM-MM. ACM; 2012. p. 725–8.

[35] Xia S, Shao M, Fu Y. Toward kinship verification using visual attributes. In: International conference on pattern recognition. IEEE; 2012. p. 549–52.

[36] Lu J, Hu J, Zhou X, Shang Y, Tan YP, Wang G. Neighborhood repulsed metric learning for kinship verification. In: IEEE CVPR. IEEE; 2012. p. 2594–601.

[37] Fang R, Gallagher AC, Chen T, Loui A. Kinship classification by modeling facial feature heredity. In: ICIP. IEEE; 2013.

[38] Dibeklioglu H, Salah AA, Gevers T. Like father, like son: facial expression dynamics for kinship verification. In: IEEE ICCV. IEEE; 2013. p. 1497–504.

[39] Yan H, Lu J, Deng W, Zhou X. Discriminative multi-metric learning for kinship verification. IEEE TIFS 2014;9(7):1169–78.

[40] Lu J, Zhou X, Tan Y, Shang Y, Zhou J. Neighborhood repulsed metric learning for kinship verification. IEEE TPAMI 2014;36(2):331–45.

[41] Robinson JP, Shao M, Wu Y, Fu Y. Families in the wild (FIW): large-scale kinship image database and benchmarks. In: Proceedings of the 2016 ACM on multimedia conference. ACM; 2016. p. 242–6.

[42] Zhang J, Xia S, Shao M, Fu Y. Family photo recognition via multiple instance learning. In: Proceedings of the 2017 ACM on international conference on multimedia retrieval. ACM; 2017. p. 424–8.

[43] Zhao W, Chellappa R, Phillips PJ, Rosenfeld A. Face recognition: a literature survey. ACM Computing Surveys (CSUR) 2003;35(4):399–458.

[44] Li SZ, Jain AK. Handbook of face recognition. Springer; 2011.

[45] Lowe DG. Distinctive image features from scale-invariant keypoints. IJCV November 2004;60(2):91–110.

[46] Ahonen T, Hadid A, Pietikäinen M. Face recognition with local binary patterns. In: ECCV. Springer; 2004. p. 469–81.

[47] Dalal N, Triggs B. Histograms of oriented gradients for human detection. In: IEEE CVPR. IEEE; 2005. p. 886–93.

[48] Fei-Fei L, Perona P. A Bayesian hierarchical model for learning natural scene categories. IEEE CVPR, vol. 2. IEEE; 2005. p. 524–31.

[49] Yang J, Yu K, Gong Y, Huang T. Linear spatial pyramid matching using sparse coding for image classification. In: IEEE CVPR. IEEE; 2009. p. 1794–801.

[50] Bourlard H, Kamp Y. Auto-association by multilayer perceptrons and singular value decomposition. Biological Cybernetics 1988;59(4–5):291–4.

[51] Bengio Y, Lamblin P, Popovici D, Larochelle H. Greedy layer-wise training of deep networks. NIPS 2007.

[52] Coates A, Ng AY, Lee H. An analysis of single-layer networks in unsupervised feature learning. In: AISTATS; 2011. p. 215–23.

[53] Hinton GE, Salakhutdinov RR. Reducing the dimensionality of data with neural networks. Science 2006;313(5786):504–7.

[54] Luo P, Wang X, Tang X. Hierarchical face parsing via deep learning. In: IEEE CVPR. IEEE; 2012. p. 2480–7.

[55] Dehghan A, Ortiz EG, Villegas R, Shah M. Who do I look like? Determining parent–offspring resemblance via gated autoencoders. In: CVPR. IEEE; 2014. p. 1757–64.

[56] Candès EJ, Li X, Ma Y, Wright J. Robust principal component analysis? Journal of the ACM (JACM) 2011;58(3):11.

[57] Liu G, Lin Z, Yu Y. Robust subspace segmentation by low-rank representation. In: ICML; 2010. p. 663–70.

[58] Cheng B, Liu G, Wang J, Huang Z, Yan S. Multi-task low-rank affinity pursuit for image segmentation. In: ICCV. IEEE; 2011. p. 2439–46.

[59] Jhuo IH, Liu D, Lee D, Chang SF. Robust visual domain adaptation with low-rank reconstruction. In: CVPR. IEEE; 2012. p. 2168–75.

[60] Dudik M, Harchaoui Z, Malick J, et al. Lifted coordinate descent for learning with trace-norm regularization. In: AISTATS, vol. 22; 2012.

[61] Chen SS, Donoho DL, Saunders MA. Atomic decomposition by basis pursuit. SIAM Journal on Scientific Computing 1998;20(1):33–61.

[62] Kan M, Shan S, Xu D, Chen X. Side-information based linear discriminant analysis for face recognition. In: BMVC; 2011. p. 1–12.

[63] Chang CC, Lin CJ. LIBSVM: a library for support vector machines. ACM Transactions on Intelligent Systems and Technology (TIST) 2011;2(3):27.

CHAPTER 10

Image Dehazing: Improved Techniques

Yu Liu*, Guanlong Zhao[†], Boyuan Gong[†], Yang Li*, Ritu Raj[†], Niraj Goel[†],
Satya Kesav[†], Sandeep Gottimukkala[†], Zhangyang Wang[†], Wenqi Ren[‡],
Dacheng Tao[§]

*Department of Electrical and Computer Engineering, Texas A&M University, College Station, TX, United States
[†]Department of Computer Science and Engineering, Texas A&M University, College Station, TX, United States
[‡]Chinese Academy of Sciences, Beijing, China
[§]University of Sydney, Sydney, NSW, Australia

Contents

10.1.	Introduction	251
10.2.	Review and Task Description	252
	10.2.1 Haze Modeling and Dehazing Approaches	253
	10.2.2 RESIDE Dataset	253
10.3.	Task 1: Dehazing as Restoration	254
10.4.	Task 2: Dehazing for Detection	257
	10.4.1 Solution Set 1: Enhancing Dehazing and/or Detection Modules in the Cascade	257
	10.4.2 Solution Set 2: Domain-Adaptive Mask-RCNN	257
10.5.	Conclusion	260
	References	261

10.1. INTRODUCTION

In many emerging applications such as autonomous/assisted driving, intelligent video surveillance, and rescue robots, the performances of visual sensing and analytics are largely jeopardized by various adverse visual conditions, e.g., bad weather and illumination conditions from the unconstrained and dynamic environments. While most current vision systems are designed to perform in clear environments, i.e., where subjects are well observable without (significant) attenuation or alteration, a dependable vision system must reckon with the entire spectrum of complex unconstrained outdoor environments. Taking autonomous driving, for example, the industry players have been tackling the challenges posed by inclement weather; however, heavy rain, haze or snow will still obscure the vision of on-board cameras and create confusing reflections and glare, leaving the state-of-the-art self-driving cars in a struggle. Another illustrative example can be found in video surveillance cameras: even the commercialized cameras adopted by governments appear fragile in challenging weather conditions. Therefore, it

Deep Learning Through Sparse and Low-Rank Modeling
DOI: 10.1016/B978-0-12-813659-1.00019-6

251

is highly desirable to study to what extent, and in what sense, such adverse visual conditions can be coped with, for the goal of achieving robust computer vision systems in the wild that benefit security/safety, autonomous driving, robotics, and an even broader range of applications.

Despite the blooming research on relevant topics, such as dehazing and rain removal, a unified view towards these problems has been absent, so have collective efforts for resolving their common bottlenecks. On the one hand, such adverse visual conditions usually give rise to complicated, nonlinear and data-dependent degradations, which follow some parameterized physical models a priori. This will naturally motivate a combination of model-based and data-driven approaches. On the other hand, most existing research works confined themselves to solving image restoration or enhancement problems. In contrast, general image restoration and enhancement, known as part of low-level vision tasks, are usually thought as the preprocessing step for mid- and high-level vision tasks. The performance of high-level computer vision tasks, such as object detection, recognition, segmentation and tracking, will deteriorate in the presence of adverse visual conditions, and is largely affected by the quality of handling those degradations. It is thus highly meaningful to examine whether restoration-based approaches for alleviating adverse visual conditions would actually boost the target high-level task performance or not. Lately, there have been a number of works exploring the joint consideration of low- and high-level vision tasks as a pipeline and achieving superior performance.

This chapter takes image dehazing as a concrete example to illustrate the handling of the above challenges. Images taken in outdoor environments affected by air pollution, dust, mist, and fumes often contain complicated, nonlinear, and data-dependent noise, also known as haze. Haze complicates many high-level computer vision tasks such as object detection and recognition. Therefore, dehazing has been widely studied in the fields of computational photography and computer vision. Early dehazing approaches often required additional information such as the provision or capture of scene depth by comparing several different images of the same scene [1–3]. Many approaches have since been proposed to exploit natural image priors and to perform statistical analyses [4–7]. Most recently, dehazing algorithms based on neural networks [8–10] have delivered state-of-the-art performance. For example, AOD-Net [10] trains an end-to-end system and shows superior performance according to multiple evaluation metrics, improving object detection in the haze using end-to-end training of dehazing and detection modules.

10.2. REVIEW AND TASK DESCRIPTION

Here we study two haze-related tasks: (i) boosting single image dehazing performance as an image restoration problem; and (ii) improving object detection accuracy in the presence of haze. As noted by [11,10,12], the second task is related to, but is often unaligned with, the first.

While the first task has been well studied in recent works, we propose that **the second task is more relevant in practice and deserves greater attention**. Haze does not affect human visual perceptual quality as much as resolution, noise, and blur; indeed, some hazy photos may even have better aesthetics. However, haze in unconstrained outdoor environments could be detrimental to machine vision systems, since most of them only work well for haze-free scenes. Taking autonomous driving as an example, hazy and foggy weather will obscure the vision of on-board cameras and create confusing reflections and glare, creating problems even for state-of-the-art self-driving cars [12].

10.2.1 Haze Modeling and Dehazing Approaches

An atmospheric scattering model has been widely used to represent hazy images in haze removal works [13–15]:

$$I(x) = J(x)t(x) + A(1 - t(x)), \qquad (10.1)$$

where x indexes pixels in the observed hazy image, $I(x)$ is the observed hazy image, and $J(x)$ is the clean image to be recovered. Parameter A denotes the global atmospheric light, and $t(x)$ is the transmission matrix defined as

$$t(x) = e^{-\beta d(x)}, \qquad (10.2)$$

where β is the scattering coefficient, and $d(x)$ represents the distance between the object and camera.

Conventional single image dehazing methods commonly exploit natural image priors (for example, the dark channel prior (DCP) [4,5], the color attenuation prior [6], and the non-local color cluster prior [7]) and perform statistical analysis to recover the transmission matrix $t(x)$. More recently, convolutional neural networks (CNNs) have been applied for haze removal after demonstrating success in many other computer vision tasks. Some of the most effective models include the multi-scale CNN (MSCNN) which predicts a coarse-scale holistic transmission map of the entire image and refines it locally [9]; DehazeNet, a trainable transmission matrix estimator that recovers the clean image combined with estimated global atmospheric light [8]; and the end-to-end dehazing network, AOD-Net [10,16], which takes a hazy image as input and directly generates a clean image output. AOD-Net has also been extended to video [17].

10.2.2 RESIDE Dataset

We benchmark against the REalistic Single Image DEhazing (RESIDE) dataset [12]. RESIDE was the first large-scale dataset for benchmarking single image dehazing algorithms and includes both indoor and outdoor hazy images.[1] Further, RESIDE contains both synthetic and real-world hazy images, thereby highlighting diverse data sources and image contents. It is divided into five subsets, each serving different training or evaluation purposes. RESIDE contains 110,500 synthetic indoor hazy images (ITS) and 313,950 synthetic outdoor hazy images (OTS) in the training set, with an option to split them for validation. The RESIDE test set is uniquely composed of the synthetic objective testing set (SOTS), the annotated real-world task-driven testing set (RTTS), and the hybrid subjective testing set (HSTS) containing 1000, 4332, and 20 hazy images, respectively. The three test sets address different evaluation viewpoints including restoration quality (PSNR, SSIM and no-reference metrics), subjective quality (rated by humans), and task-driven utility (using object detection, for example).

Most notably, RTTS is the only existing public dataset that can be used to evaluate object detection in hazy images, representing mostly real-world traffic and driving scenarios. Each image is annotated with object bounding boxes and categories (person, bicycle, bus, car, or motorbike). Furthermore, 4807 unannotated real-world hazy images are also included in the dataset for potential domain adaptation.

For Task 1, we used the training and validation sets from ITS + OTS, and the evaluation is based on PSNR and SSIM. For Task 2, we used the RTTS set for testing and evaluated using mean average precision (MAP) scores.

10.3. TASK 1: DEHAZING AS RESTORATION

Most CNN dehazing models [8–10] refer to the mean-squared error (MSE) or ℓ_2 norm-based loss functions. However, MSE is well-known to be imperfectly correlated with human perception of image quality [18,19]. Specifically, for dehazing, the ℓ_2 norm implicitly assumes that the degradation is additive white Gaussian noise, which is oversimplified and invalid for haze. On the other hand, the ℓ_2 norm treats the impact of noise independently of the local image characteristics such as structural information, luminance and contrast. However, according to [20], the sensitivity of the Human Visual System (HVS) to noise depends on the local properties and structure of a vision.

Here we aimed to identify loss functions that better match human perception to train a dehazing neural network. We used AOD-Net [10] (originally optimized using MSE loss) as the backbone but replaced its loss function with the following options:

- ℓ_1 **loss.** The ℓ_1 loss for a patch P can be written as:

$$\mathcal{L}^{\ell_1}(P) = \frac{1}{N} \sum_{p \in P} |x(p) - \gamma(p)|, \tag{10.3}$$

---

[1] The RESIDE dataset was updated in March 2018, with some changes made to dataset organization. Our experiments were all conducted on the original RESIDE version, now called RESIDE-v0.

where N is the number of pixels in the patch, p is the index of the pixel, and $x(p)$ and $y(p)$ are the pixel values of the generated and the ground-truth images, respectively.

- **SSIM loss.** Following [19], we write the SSIM for pixel p as:

$$SSIM(p) = \frac{2\mu_x\mu_y + C_1}{\mu_x^2 + \mu_y^2 + C_1} \cdot \frac{2\sigma_{xy} + C_2}{\sigma_x^2 + \sigma_y^2 + C_2} \tag{10.4}$$
$$= l(p) \cdot cs(p).$$

The means and standard deviations are computed using a Gaussian filter with standard deviation σ_G. The loss function for SSIM can then be defined as

$$\mathcal{L}^{SSIM}(P) = \frac{1}{N}\sum_{p \in P} 1 - SSIM(p). \tag{10.5}$$

- **MS-SSIM loss.** The choice of σ_G would impact the training performance of SSIM. Here we adopt the idea of multi-scale SSIM [19], where M different values of σ_G are pre-chosen and fused:

$$\mathcal{L}^{MS-SSIM}(P) = l_M^\alpha(p) \cdot \prod_{j=1}^{M} cs_j^{\beta_j}(P). \tag{10.6}$$

- **MS-SSIM+ℓ_2 loss.** Using a weighted sum of MS-SSIM and ℓ_2 as the loss function,

$$\mathcal{L}^{MS-SSIM-\ell_2} = \alpha \cdot \mathcal{L}^{MSSSIM} + (1-\alpha) \cdot G_{\sigma_G^M} \cdot \mathcal{L}^{\ell_2}, \tag{10.7}$$

a point-wise product between $G_{\sigma_G^M}$ and $\mathcal{L}^{\ell_2}$ is added to the ℓ_2 loss function term, because MS-SSIM propagates the error at pixel q based on its contribution to MS-SSIM of the central pixel $\tilde{q}$, as determined by the Gaussian weights.

- **MS-SSIM+ℓ_1 loss.** Using a weighted sum of MS-SSIM and ℓ_1 as the loss function,

$$\mathcal{L}^{MSSSIM-\ell_1} = \alpha \cdot \mathcal{L}^{MSSSIM} + (1-\alpha) \cdot G_{\sigma_G^M} \cdot \mathcal{L}^{\ell_1}, \tag{10.8}$$

the ℓ_1 loss is similarly weighted by $G_{\sigma_G^M}$.

We selected 1000 images from ITS + OTS as the validation set and the remaining images for training. The initial learning rate and mini-batch size of the systems were set to 0.01 and 8, respectively, for all methods. All weights were initialized as Gaussian random variables, unless otherwise specified. We used a momentum of 0.9 and a weight

Table 10.1 Comparison of PSNR results (dB) for Task 1

Models	PSNR		
	Indoor	Outdoor	All
AOD–Net Baseline	**21.01**	24.08	22.55
ℓ_1	20.27	25.83	23.05
SSIM	19.64	26.65	23.15
MS-SSIM	19.54	**26.87**	23.20
MS-SSIM+ℓ_1	20.16	26.20	23.18
MS-SSIM+ℓ_2	20.45	26.38	**23.41**
MS-SSIM+ℓ_2 (fine-tuned)	20.68	26.18	**23.43**

Table 10.2 Comparison of SSIM results for Task 1

Models	SSIM		
	Indoor	Outdoor	All
AOD–Net Baseline	**0.8372**	0.8726	0.8549
ℓ_1	0.8045	0.9111	0.8578
SSIM	0.7940	0.8999	0.8469
MS-SSIM	0.8038	0.8989	0.8513
MS-SSIM+ℓ_1	0.8138	**0.9184**	0.8661
MS-SSIM+ℓ_2	0.8285	0.9177	**0.8731**
MS-SSIM+ℓ_2 (fine-tuned)	0.8229	**0.9266**	**0.8747**

decay of 0.0001. We also clipped the ℓ_2 norm of the gradient to be within $[-0.1, 0.1]$ to stabilize network training. All models were trained on an Nvidia GTX 1070 GPU for around 14 epochs, which empirically led to convergence. For SSIM loss, σ_G was set to 5; C_1 and C_2 in (10.4) were 0.01 and 0.03, respectively. For MS-SSIM losses, multiple Gaussian filters were constructed by setting $\sigma_G^i = \{0.5, 1, 2, 4, 8\}$; α was set as 0.025 for MS-SSIM+ℓ_1, and 0.1 for MS-SSIM+ℓ_2, following [19].

As shown in Tables 10.1 and 10.2, simply replacing the loss functions resulted in noticeable differences in performance. While the original AOD-Net with MSE loss performed well on indoor images, it was less effective on outdoor images, which are usually the images needing to be dehazed in practice. Of all the options, MS-SSIM-ℓ_2 achieved both the highest overall PSNR and SSIM results, resulting in 0.88 dB PSNR and 0.182 SSIM improvements over the state-of-the-art AOD-Net. We further fine-tuned the MS-SSIM-ℓ_2 model, including using a pre-trained AOD-Net as a warm initialization, adopting a smaller learning rate (0.002) and a larger minibatch size (16). Finally, the best achievable PSNR and SSIM were 23.43 dB and 0.8747, respectively. Note that the best SSIM represented a nearly 0.02 improvement over AOD-Net.

10.4. TASK 2: DEHAZING FOR DETECTION

10.4.1 Solution Set 1: Enhancing Dehazing and/or Detection Modules in the Cascade

In [10], the authors proposed a cascade of AOD-Net dehazing and Faster-RCNN [21] detection modules to detect objects in hazy images. We therefore considered it intuitive to try different combinations of more powerful dehazing/detection modules in the cascade. Note that such a cascade could be subject to further joint optimization, as many previous works [22,23,10]. However, **to be consistent with the results in** [12], all detection models used in this section were the original pre-trained versions, *without any retraining or adaptation.*

Our solution set 1 considered several popular dehazing modules including DCP [4], DehazeNet [8], AOD-Net [10], and the recently proposed densely connected pyramid dehazing network (DCPDN) [24]. Since hazy images tend to have lower contrast, we also included a contrast enhancement method called contrast limited adaptive histogram equalization (CLAHE). Regarding the choice of detection modules, we included Faster R-CNN [21],[2] SSD [26], RetinaNet [27], and Mask-RCNN [28].

The compared pipelines are shown in Table 10.3. In each pipeline, "X+Y" by default means applying Y directly on the output of X in a sequential manner. The most important observation is that simply applying more sophisticated detection modules is unlikely to boost the performance of the dehazing–detection cascade, due to the domain gap between hazy/dehazed and clean images (on which typical detectors are trained). The more sophisticated pre-trained detectors (RetinaNet, Mask-RCNN) may have overfitted the clean image domain, again highlighting the demand of handling domain shifts in real-world detection problems. Moreover, a better dehazing model in terms of restoration performance does not imply better detection results on its pre-processed images (e.g., DPDCN). Further, adding dehazing pre-processing does not always guarantee better detection (e.g., comparing RetinaNet versus AOD-Net + RetinaNet), consistent with the conclusion made in [12]. In addition, AOD-Net tended to generate smoother results but with lower contrast than the others, potentially compromising detection. Therefore, we created two three-stage cascades as in the last two rows of Table 10.3, and found that using DCP to process AOD-Net dehazed results (with greater contrast) further marginally improved results.

10.4.2 Solution Set 2: Domain-Adaptive Mask-RCNN

Motivated by the observations made on solution set 1, we next aimed to more explicitly tackle the domain gap between hazy/dehazed images and clean images for object detec-

---

[2] We have replaced the backbone of Faster R-CNN from VGG 16 as used by [12] with the ResNet101 model [25] to enhance performance.

Table 10.3 Solution set 1 mAP results on RTTS. Top 3 results are colored in red, green, and blue, respectively. (For interpretation of the colors in the tables, the reader is referred to the web version of this chapter)

Pipelines	mAP
Faster R–CNN	0.541
SSD	0.556
RetinaNet	0.531
Mask-RCNN	0.457
DehazeNet + Faster R–CNN	0.557
AOD-Net + Faster R–CNN	0.563
DCP + Faster R–CNN	0.567
DehazeNet + SSD	0.554
AOD-Net + SSD	0.553
DCP + SSD	0.557
AOD-Net + RetinaNet	0.419
DPDCN + RetinaNet	0.543
DPDCN + Mask-RCNN	0.477
AOD-Net + DCP + Faster R–CNN	0.568
CLACHE + DCP + Mask-RCNN	0.551

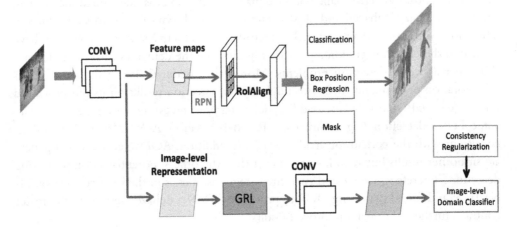

Figure 10.1 DMask-RCNN structure.

tion. Inspired by the recently proposed domain adaptive Faster-RCNN [29], we applied a similar approach to design a domain-adaptive mask-RCNN (DMask-RCNN).

In the model shown in Fig. 10.1, the primary goal of DMask-RCNN is to mask the features generated by feature extraction network to be as domain invariant as possible, between the source domain (clean input images) and the target domain (hazy images). Specifically, DMask-RCNN places a domain-adaptive component branch after the base feature extraction convolution layers of Mask-RCNN. The loss of the domain classifier is a binary cross entropy loss

$$-\sum_i (y_i \log(p_i) + (1 - y_i) \log(1 - p_i)), \tag{10.9}$$

where y_i is the domain label of the ith image, and p_i is the prediction probability from the domain classifier. The overall loss of DMask-RCNN can therefore be written as

$$\begin{aligned} L(\theta_{res}, \theta_{head}, \theta_{domain}) = &\ L_{C,B}(C, B | \theta_{res}, \theta_{head}, x \in D_s) \\ &- \lambda L_d(G_d | \theta_{res}, x \in D_s, D_t) \\ &+ \lambda L_d(G_d | \theta_{domain}, x \in D_s, D_t), \end{aligned} \tag{10.10}$$

where x is the input image, and D_s and D_t represent the source and target domain, respectively; θ denotes the corresponding weights of each network component; G represents the mapping function of the feature extractor; I is the feature map distribution; B is the bounding box of an object; and C is the object class. Note that when calculating the $L_{C,B}$, only source domain inputs will be counted in since the target domain has no labels.

As seen from Eq. (10.10), the negative gradient of the domain classifier loss needs to be propagated back to ResNet, whose implementation relies on the gradient reverse layer [30] (GRL, Fig. 10.1). The GRL is added after the feature maps generated by the ResNet and feeds its output to the domain classifier. This GRL has no parameters except for the hyper-parameter λ, which, during forward propagation, acts as an identity transform. However, during back propagation, it takes the gradient from the upper level and multiplies it by $-\lambda$ before passing it to the preceding layers.

Experiments

To train DMask-RCNN, MS COCO (clean images) were always used as the source domain, while **two target domain options** were designed to consider two types of domain gap: (i) all unannotated realistic haze images from RESIDE; and (ii) dehazed results of those unannotated images, using MSCNN [9]. The corresponding DMask-RCNNs are called DMask-RCNN1 and DMask-RCNN2, respectively.

We initialized the Mask-RCNN component of DMask-RCNN with a pre-trained model on MS COCO. All models were trained for 50,000 iterations with learning rate 0.001, then another 20,000 iterations with learning rate 0.0001. We used a naive batch size of 2, including one image randomly selected from the source domain and the other from the target domain, noting that larger batches may further benefit performance. We also tried to concatenate dehazing pre-processing (AOD-Net and MSCNN) with DMask-RCNN models to form new dehazing–detection cascades.

Table 10.4 Solution set 2 mAP results on RTTS. Top 3 results are colored in red, green, and blue, respectively

Pipelines	mAP
DMask-RCNN1	0.612
DMask-RCNN2	0.617
AOD-Net + DMask-RCNN1	0.602
AOD-Net + DMask-RCNN2	0.605
MSCNN + Mask-RCNN	0.626
MSCNN + DMask-RCNN1	0.627
MSCNN + DMask-RCNN2	0.634

Table 10.4 shows the results of solution set 2 (the naming convention is the same as in Table 10.3), from which we can conclude that:

- The domain-adaptive detector presents a very promising approach, and its performance significantly outperforms the best results in Table 10.3[3];
- The power of strong detection models (Mask-RCNN) is fully exploited, given the proper domain adaptation, in contrast to the poor performance of vanilla Mask RCNN in Table 10.3;
- DMask-RCNN2 is always superior to DMask-RCNN1, showing that the choice of dehazed images as the target domain matters. We make the reasonable hypothesis that the domain discrepancy between dehazed and clean images is smaller than that between hazy and clean images, so DMask-RCNN performs better when the existing domain gap is narrower; and
- The best result in solution set 2 is from a dehazing–detection cascade, with MSCNN as the dehazing module and DMask-RCNN as the detection module and highlighting: **the joint value of dehazing pre-processing and domain adaption**.

10.5. CONCLUSION

This chapter tackles the challenge of single image dehazing and its extension to object detection in haze. The solutions are proposed from diverse perspectives ranging from novel loss functions (Task 1) to enhanced dehazing–detection cascades, as well as domain-adaptive detectors (Task 2). By way of careful experiments, we significantly improve the performance of both tasks, as verified on the RESIDE dataset. We expect further improvements as we continue to study this important dataset and tasks.

---

[3] By saying this, we also emphasize that Table 10.3 results have not undergone joint tuning as in [31,10], so there is potential for further improvements.

REFERENCES

[1] Tan K, Oakley JP. Enhancement of color images in poor visibility conditions. In: Image processing, 2000. Proceedings. 2000 International conference on, vol. 2. IEEE; 2000. p. 788–91.

[2] Schechner YY, Narasimhan SG, Nayar SK. Instant dehazing of images using polarization. In: Computer vision and pattern recognition, 2001. CVPR 2001. Proceedings of the 2001 IEEE computer society conference on, vol. 1. IEEE; 2001.

[3] Kopf J, Neubert B, Chen B, Cohen M, Cohen-Or D, Deussen O, et al. Deep photo: model-based photograph enhancement and viewing, vol. 27. ACM; 2008.

[4] He K, Sun J, Tang X. Single image haze removal using dark channel prior. IEEE Transactions on Pattern Analysis and Machine Intelligence 2011;33(12):2341–53.

[5] Tang K, Yang J, Wang J. Investigating haze-relevant features in a learning framework for image dehazing. In: Proceedings of the IEEE conference on computer vision and pattern recognition; 2014. p. 2995–3000.

[6] Zhu Q, Mai J, Shao L. A fast single image haze removal algorithm using color attenuation prior. IEEE Transactions on Image Processing 2015;24(11):3522–33.

[7] Berman D, Avidan S, et al. Non-local image dehazing. In: Proceedings of the IEEE conference on computer vision and pattern recognition; 2016. p. 1674–82.

[8] Cai B, Xu X, Jia K, Qing C, Tao D. DehazeNet: an end-to-end system for single image haze removal. IEEE Transactions on Image Processing 2016;25(11):5187–98.

[9] Ren W, Liu S, Zhang H, Pan J, Cao X, Yang MH. Single image dehazing via multi-scale convolutional neural networks. In: European conference on computer vision. Springer; 2016. p. 154–69.

[10] Li B, Peng X, Wang Z, Xu J, Feng D. AOD-Net: all-in-one dehazing network. In: Proceedings of the IEEE international conference on computer vision; 2017. p. 4770–8.

[11] Wang Z, Chang S, Yang Y, Liu D, Huang TS. Studying very low resolution recognition using deep networks. In: Proceedings of the IEEE conference on computer vision and pattern recognition; 2016. p. 4792–800.

[12] Li B, Ren W, Fu D, Tao D, Feng D, Zeng W, et al. RESIDE: a benchmark for single image dehazing. arXiv preprint arXiv:1712.04143, 2017.

[13] McCartney EJ. Optics of the atmosphere: scattering by molecules and particles. New York: John Wiley and Sons, Inc.; 1976. 421 p.

[14] Narasimhan SG, Nayar SK. Chromatic framework for vision in bad weather. In: Computer vision and pattern recognition, 2000. Proceedings. IEEE conference on, vol. 1. IEEE; 2000. p. 598–605.

[15] Narasimhan SG, Nayar SK. Vision and the atmosphere. International Journal of Computer Vision 2002;48(3):233–54.

[16] Li B, Peng X, Wang Z, Xu J, Feng D. An all-in-one network for dehazing and beyond. arXiv preprint arXiv:1707.06543, 2017.

[17] Li B, Peng X, Wang Z, Xu J, Feng D. End-to-end united video dehazing and detection. arXiv preprint arXiv:1709.03919, 2017.

[18] Zhang L, Zhang L, Mou X, Zhang D. A comprehensive evaluation of full reference image quality assessment algorithms. In: Image processing (ICIP), 2012 19th IEEE international conference on. IEEE; 2012. p. 1477–80.

[19] Zhao H, Gallo O, Frosio I, Kautz J. Loss functions for image restoration with neural networks. IEEE Transactions on Computational Imaging 2017;3(1):47–57.

[20] Wang Z, Bovik AC, Sheikh HR, Simoncelli EP. Image quality assessment: from error visibility to structural similarity. IEEE Transactions on Image Processing 2004;13(4):600–12.

[21] Ren S, He K, Girshick R, Sun J. Faster R-CNN: towards real-time object detection with region proposal networks. In: Advances in neural information processing systems; 2015. p. 91–9.

[22] Liu D, Wen B, Liu X, Wang Z, Huang TS. When image denoising meets high-level vision tasks: a deep learning approach. arXiv preprint arXiv:1706.04284, 2017.

[23] Cheng B, Wang Z, Zhang Z, Li Z, Liu D, Yang J, et al. Robust emotion recognition from low quality and low bit rate video: a deep learning approach. arXiv preprint arXiv:1709.03126, 2017.

[24] Zhang H, Patel VM. Densely connected pyramid dehazing network. In: The IEEE conference on computer vision and pattern recognition (CVPR); 2018.

[25] He K, Zhang X, Ren S, Sun J. Deep residual learning for image recognition. In: Proceedings of the IEEE conference on computer vision and pattern recognition; 2016. p. 770–8.

[26] Liu W, Anguelov D, Erhan D, Szegedy C, Reed S, Fu CY, et al. SSD: single shot multibox detector. In: European conference on computer vision. Springer; 2016. p. 21–37.

[27] Lin TY, Goyal P, Girshick R, He K, Dollár P. Focal loss for dense object detection. arXiv preprint arXiv:1708.02002, 2017.

[28] He K, Gkioxari G, Dollár P, Girshick R. Mask R-CNN. In: Computer vision (ICCV), 2017 IEEE international conference on. IEEE; 2017. p. 2980–8.

[29] Chen Y, Li W, Sakaridis C, Dai D, Van Gool L. Domain adaptive faster R-CNN for object detection in the wild. In: Proceedings of the IEEE conference on computer vision and pattern recognition; 2018. p. 3339–48.

[30] Ganin Y, Lempitsky V. Unsupervised domain adaptation by backpropagation. arXiv preprint arXiv:1409.7495, 2014.

[31] Liu D, Cheng B, Wang Z, Zhang H, Huang TS. Enhanced visual recognition under adverse conditions via deep networks. arXiv preprint arXiv:1712.07732, 2017.

CHAPTER 11

Biomedical Image Analytics: Automated Lung Cancer Diagnosis

Steve Kommrusch, Louis-Noël Pouchet
Colorado State University, Fort Collins, CO, United States

Contents

11.1.	Introduction	263
11.2.	Related Work	264
11.3.	Methodology	265
11.4.	Experiments	268
11.5.	Conclusion	270
	Acknowledgments	271
	References	271

11.1. INTRODUCTION

Worldwide in 2017, lung cancer remained the leading cause of cancer deaths [6]. Computer aided diagnosis, where a software tool analyzes the patient's medical imaging results to suggest a possible diagnosis, is a promising direction: from an input low-resolution 3D CT scan, image processing techniques can be used to classify nodules in the lung scan as potentially cancerous or benign. But such systems require quality 3D training images to ensure the classifiers are adequately trained with sufficient generality. Cancerous lung nodule detection still suffers from a dearth of training images which hampers the ability to effectively automate and improve the analysis of CT scans for cancer risks [7]. In this work, we propose to address this problem by automatically generating synthetic 3D images of nodules, to augment the training dataset of such systems with meaningful (yet computer-generated) lung nodule images.

Li et al. showed how to analyze nodules using computed features from the 3D images (such as volume, degree of compactness and irregularity, etc.) [8]. These computed features are then used as inputs to a nodule classification algorithm. 2D lung nodule image generation has been investigated using generative adversarial networks (GANs) [9], reaching sufficient quality to be classified by radiologists as actual CT scan images. In our work, we aim to generate 3D lung nodule images which match the feature statistics of actual nodules as determined by an analysis program. We propose a new system inspired from autoencoders, and extensively evaluate its generative capabilities.

Deep Learning Through Sparse and Low-Rank Modeling
DOI: 10.1016/B978-0-12-813659-1.00020-2

Precisely, we introduce LuNG: a synthetic lung nodule generator, which is a neural network trained to generate new examples of 3D shapes that fit within a broad learned category.

Our work is aimed at creating synthetic images in cases where input images are difficult to get. For example, the Adaptive Lung Nodule Screening Benchmark (ALNSB) from the NSF Center for Domain-Specific Computing uses a flow that leverages compressive sensing to reconstruct images from low-dose CT scans. These images are slightly different than those built from filtered backprojection, a technique which has more samples readily available (such as LIDC/IDRI [10]). To evaluate our results, we integrate our work with the ALNSB system [11] that automatically processes a low-dose 3D CT scan, reconstructs a higher-resolution image, isolates all nodules in the 3D image, computes features on them, and classifies each nodule as benign or suspicious. We use original patient data to train LuNG, and then use it to generate synthetic nodules that are processed by ALNSB. We create a network which optimizes 3 metrics: (i) increase the percentage of generated images accepted by the nodule analyzer; (ii) increase the variation of the generated output images relative to the limited seed images; and (iii) decrease the error of the seed images with themselves when input to the autoencoder. We make the following contributions:

- A new 3D image generation system, which can create synthetic images that resemble (in terms of features) the training images. The system is fully implemented and automated.
- Novel metrics, which allow for numerical evaluation of 3D image generation aligned with qualitative goals related to lung nodule generation.
- An extensive evaluation of this system to generate 3D images of lung nodules, and its use within an existing computer-aided diagnosis benchmark application.
- The evaluation of iterative training techniques coupled with the ALNSB nodule classifier software, to further refine the quality of the image generator.

11.2. RELATED WORK

Improving automated CT lung nodule classification techniques and 3D image generation are areas that are receiving significant research attention.

Recently, Valente et al. provided a good overview of the requirements for CADe (Computer Aided Detection) systems in medical radiology and they evaluate the status of recent approaches [7]. Our aim is to provide a tool which can be used to improve the results of such CADe systems by both increasing the true positive rate (sensitivity) and decreasing the false positive rate of CADe classifiers through the use of an increase in nodules for analysis and training. Their survey paper discusses in detail the preprocessing, segmentation, and nodule detection steps similar to those used in the ALNSB nodule analyzer/classifier which we used in this project.

Li et al. provide an excellent overview of recent approaches to 3D shape generation in their paper "GRASS: generative recursive autoencoders for shape structures" [12]. While we do not explore the design of an autoencoder with convolutional and deconvolutional layers, the same image generation quality metrics that we teach could be used to evaluate such designs. Similar tradeoffs between overfitting and low error rates with seed images would have to be considered when setting the depth of the network and number of feature maps in the convolutional layers.

Durugkar et al. describe the challenges of training GANs well and discuss the advantages of multiple generative networks trained with multiple adversaries to improve the quality of images generated [13]. LuNG explored using multiple networks during image feedback experiments. Larsen et al. [14] teach a system which combines an autoencoder with a GAN which could be a basis for future work introducing GAN methodologies into the LuNG system by preserving our goal of generating shapes similar to existing seed shapes.

11.3. METHODOLOGY

To begin, guided training is used in which each nodule is modified to create 15 additional training samples. We call the initial nodule set, of which we were provided 51 samples, the "seed" nodules, examples of which are shown in Fig. 11.2. The "base" nodules include image reflections and offsets to create 15 modified samples per seed nodule for a total of 816 samples. Fig. 11.1 shows the general structure of the LuNG system. The base nodules are used to train an autoencoder neural network with 3 latent feature neurons in the bottleneck layer. Images output by the neural network pass through a reconnection algorithm to guarantee that viable fully connected nodules are being generated for analysis by the nodule analyzer. A nodule analyzer program then extracts relevant 3D features from the nodules and prunes away nodules which can be rejected as not interesting for classification (definitely not a cancerous nodule). The analyzer has range checking on features which include nodule size, elongation, and surface area to volume ratio. The original CT scan nodule candidates pass through these checks before creating our 51 seed nodules, and the LuNG generated images are processed with the same criteria. The "analyzer accepted images" are the final output of LuNG and can be used to augment the image training set for a classifier. We use the ALNSB [11] nodule analyzer and classifier code for the LuNG project, but similar analyzers compute similar 3D features to aid in classification. The support vector machine is an example classifier to which LuNG can provide augmented data. Such augmented data is helpful in overcoming drawbacks in current lung nodule classification work [7]. To evaluate generated nodules, we develop a statistical distance metric similar to the Mahalanobis distance. Given the set of nodules output by LuNG, we explore adding them to the autoencoder training set to improve the generality of the generator. We also used the Score to evaluate various nodule reconnection options and network hyperparameters.

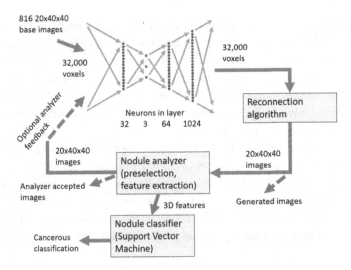

Figure 11.1 Interaction between autoencoder, nodule analyzer, and support vector machine. Analyzer accepted images are suitable for augmented the training dataset for classifier.

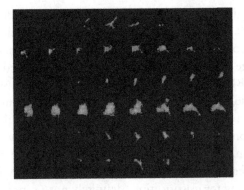

Figure 11.2 6 of 51 original seed nodules showing middle 8 2D slices of 3D image from the CT scan.

We chose an autoencoder architecture for LuNG because it can be split into both an encoder and decoder network for different use models. The encoder (or "feature network") is the portion of the autoencoder that takes a 32,000 voxel image as input and outputs 3 bottleneck feature neuron values between −1 and 1. The decoder (or "generator network") is the portion of the autoencoder that takes 3 feature neuron values between −1 to 1 as input and generates a 32,000 voxel output image. Thus, given 2 seed nodules, one can use the feature network to find their latent space coordinates and then step from one nodule to another with inputs to the generator network. We analyze using a uniform random value from −1 to 1 on the 3 feature neurons to generate random nodules which we analyze for augmenting the classifier training set.

While the LuNG use model relies on having both an encoder and decoder network as provided by autoencoder training, future work could merge our technique with a generative adversarial network to enhance the generator or test whether convolution/deconvolution layers can help improve our overall quality metrics [12].

Metrics for Scoring Images

Our goals for LuNG are to generate images that have a high acceptance rate for the analyzer and a high variation relative to the seed images while minimizing the error of the network when a seed image is reproduced. We track the acceptance rate simply as the percentage of randomly generated nodules that are accepted by the analyzer. For a metric of variation, we compute a feature distance *FtDist* based on the 12 3D image features used in the analyzer. To track how well the distribution of output images matches the seed image means, we compute an *FtMMSE* based on the distribution means. The ability of the network to reproduce a given seed image is tracked with the mean squared error of the image output voxels, as is typical for autoencoder image training.

FtDist has some similarity to Mahalanobis distance, but finds the average over all the accepted images of the distance from the image to the closest seed image in the 12-dimensional analyzer feature space. As *FtDist* increases, the network is generating images that are less similar to specific samples in the seed images, hence it is a metric we want to increase with LuNG. Given an accepted set of n images Y and a set of 51 seed images S, and given y_i denotes the value of feature i for an image and σ_{Si} denotes the standard deviation of feature i within S,

$$FtDist = \frac{1}{n} \sum_{y \in Y} \min_{s \in S} \sqrt{\sum_{i=1}^{12} \left(\frac{y_i - s_i}{\sigma_{Si}}\right)^2}.$$

FtMMSE tracks how much the 12 3D features have the same mean between the accepted set of images X and the seed images S. As *FtMMSE* increases, the network is generating average images that are increasing outside the typical seed image distribution, hence it is a metric we want to decrease with LuNG. Given μ_{Si} is the mean of feature i in the set of seed images and μ_{Yi} is the mean of feature i in the final set of accepted images,

$$FtMMSE = \frac{1}{12} \sum_{i=1}^{12} (\mu_{Yi} - \mu_{Si})^2.$$

Score is our composite network scoring metric used to compare different networks, hyperparameters, feedback options, and reconnection options. In addition to *FtDist* and *FtMMSE*, we use *AC*, which is the fraction of generated images which the analyzer

Figure 11.3 Generated images of 6 steps through latent feature space from seed nodules 2 to 4.

accepted, and *MSE* which is the traditional mean squared error which results when the autoencoder is used to regenerate the 51 seed nodule images:

$$Score = \frac{FtDist - 1}{(FtMMSE + 0.1) * (MSE + 0.1) * (1 - AC)}.$$

Score increases as *FtDist* or *AC* increase and decreases when *FtMMSE* or *MSE* increase. The constants in the equation are based on qualitative assessments of network results; for example, using $MSE + 0.1$ means that *MSE* values below 0.1 don't override the contribution of other components and mathematically aligns with the qualitative statement that an *MSE* of 0.1 yielded acceptable images visually in comparison with the seed images.

11.4. EXPERIMENTS

By using the trained encoder network to find the latent feature coordinates for seed nodules 2 and 4, Fig. 11.3 shows 6 steps between these nodules. The top and bottom nodules in the image can be seen to accurately reproduce seed nodules 2 and 4 from Fig. 11.2. The 4 intermediate nodules are novel images from the generator which can be used to improve an automated classifier system.

In addition to stepping through latent feature space values, we use our *Score* metric to evaluate a full nodule generation system that includes using uniform random values between -1 and 1 as inputs to the generator network, then sends images through the reconnection algorithm and analyzer for acceptance of the images for analysis. Fig. 11.4 shows the components of the score for the final parameter analysis we did on the network. Note that the MSE metric (mean squared error of the network on training set) continues to decrease with larger networks, but the maximum *Score* we are measuring occurs with 3 bottleneck latent feature neurons.

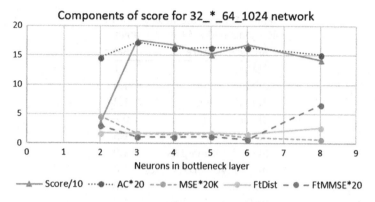

Figure 11.4 There are 4 components used to compute the network score. The component values are scaled as shown so that they can all be plotted on the same scale.

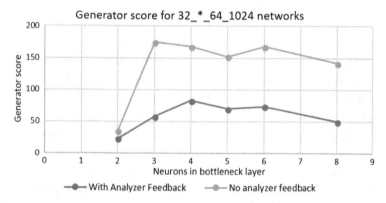

Figure 11.5 This figure compares results between a network that used 816 base images with no analyzer feedback for 150,000 iterations of training and a network that trained for 25,000 iterations on the base images, then added 302 generated nodules to train for 25,000 iterations, then added a different 199 generated nodules to train for 25,000 iterations, and then trained for a final 75,000 iterations on the base images.

Our *Score* metric is used to evaluate our system as a whole and it can be used to explore variations to autoencoder training. We tested multiple methodes of using accepted images to augment the autoencoder training set, but ultimately found that such images resulted in confirmation bias that diminished the total system score as shown in Fig. 11.5. Exploring such system options resulted in a system with a *Score* that incorporates knowledge of actual nodule shapes from the seed nodule set, expert domain knowledge as encoded in the analyzer acceptance criteria, and machine learning researcher knowledge as represented by network size and system features for LuNG.

As an example of the nodule metrics we use to evaluate network architectures and interface options. We analyze the 12 3D features computed by the nodule analyzer

Table 11.1 Feature means ratios for 400 generated vs 51 seed nodules

Analyzer feature	Ratio
2D area	1.1
2D max(xL,yL)	1.0
2D perimeter	1.2
2D area/peri^2	0.8
3D volume	1.3
3D rad/MeanSqDis	1.0
min(xL,yL)/max(xL,yL)	1.0
minl/maxl	1.0
surface area^3/volume^2	1.2
mean breadth	1.3
euler3D	1.1
maskTem area/peri^2	1.0

(ALNSB). Table 11.1 shows that when 400 novel random images are generated by LuNG, the mean feature value for all 12 3D features stays within 30% of the seed nodules. (When image feedback is used in the training set, we see the mean of the generated images tend to be further from the seed nodule mean for any given feature; hence our conclusion that confirmation bias harms the *Score* for systems that used image feedback.) Based on this alignment, we plot SVM distance values for 1000 nodules and the 51 seed nodules in Fig. 11.6. After the support vectors are applied, nodules closer to the positive than the negative centroid are classified as suspicious. The results show that LuNG generated nodules augment the available nodules for analysis well, including providing many nodules which are near the existing boundary and can be useful for improving the sensitivity of the classifier. For example, by having a trained radiologist classify a subset of 100 of the 1000 generated nodules which are near the current classifier boundary or on the "cancerous" side of the boundary, a more balanced and varied classifier training set could be produced which would improve classification.

11.5. CONCLUSION

To produce quality image classifiers, machine learning requires a large set of training images. This poses a challenge for application areas where such training sets are rare if they exist, such as for computer-aided diagnosis of cancerous lung nodules.

In this work we developed LuNG, a lung nodule image generator, allowing us to augment the training dataset of image classifiers with meaningful (yet computer-generated) lung nodule images. Specifically, we have developed an autoencoder-based system that learns to produce 3D images that resembles the original training set, while

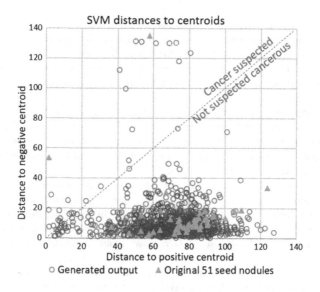

Figure 11.6 Support Vector Machine (SVM) coordinates for 1000 generated and 51 seed nodules.

covering adequately the feature space. Our tool, LuNG, was developed using PyTorch and is fully implemented. We have shown that the 3D nodules generated by this process visually and numerically align well with the general image space presented by the limited set of seed images.

ACKNOWLEDGMENTS

This work was supported in part by the U.S. National Science Foundation award CCF-1750399.

REFERENCES

[1] Huang S, Zhou J, Wang Z, Ling Q, Shen Y, Biomedical informatics with optimization and machine learning. 2016.
[2] Samareh A, Jin Y, Wang Z, Chang X, Huang S. Predicting depression severity by multi-modal feature engineering and fusion. arXiv preprint arXiv:1711.11155, 2017.
[3] Samareh A, Jin Y, Wang Z, Chang X, Huang S. Detect depression from communication: how computer vision, signal processing, and sentiment analysis join forces. IISE Transactions on Healthcare Systems Engineering 2018:1–42.
[4] Sun M, Baytas IM, Zhan L, Wang Z, Zhou J. Subspace network: deep multi-task censored regression for modeling neurodegenerative diseases. arXiv preprint arXiv:1802.06516, 2018.
[5] Karimi M, Wu D, Wang Z, Shen Y. DeepAffinity: interpretable deep learning of compound-protein affinity through unified recurrent and convolutional neural networks. arXiv preprint arXiv:1806.07537, 2018.
[6] Siegel RL, Miller KD, Jemal A. Cancer statistics, 2017. CA: A Cancer Journal for Clinicians 2017;67(1):7–30. https://doi.org/10.3322/caac.21387.

[7] Valente IRS, Cortez PC, Neto EC, Soares JM, de Albuquerque VHC, Tavares JaMR. Automatic 3D pulmonary nodule detection in CT images. Computer Methods and Programs in Biomedicine 2016;124(C):91–107. https://doi.org/10.1016/j.cmpb.2015.10.006.

[8] Li Q, Li F, Doi K. Computerized detection of lung nodules in thin-section CT images by use of selective enhancement filters and an automated rule-based classifier. Academic Radiology 2008;15(2):165–75.

[9] Chuquicusma MJM, Hussein S, Burt J, Bagci U. How to fool radiologists with generative adversarial networks? A visual turing test for lung cancer diagnosis. ArXiv e-prints arXiv:1710.09762, 2017.

[10] Rong J, Gao P, Liu W, Zhang Y, Liu T, Lu H. Computer simulation of low-dose CT with clinical lung image database: a preliminary study. Society of Photo-Optical Instrumentation Engineers (SPIE) Conference Series, vol. 10132. 2017. p. 101322U.

[11] Shen S, Rawat P, Pouchet LN, Hsu W. Lung nodule detection C benchmark. URL: https://github.com/cdsc-github/Lung-Nodule-Detection-C-Benchmark, 2015.

[12] Li J, Xu K, Chaudhuri S, Yumer E, Zhang H, Guibas L. GRASS: generative recursive autoencoders for shape structures. ACM Transactions on Graphics (Proceedings of SIGGRAPH 2017) 2017;36(4).

[13] Durugkar I, Gemp I, Mahadevan S. Generative multi-adversarial networks. ArXiv e-prints arXiv:1611.01673, 2016.

[14] Boesen Lindbo Larsen A, Kaae Sønderby S, Larochelle H, Winther O. Autoencoding beyond pixels using a learned similarity metric. ArXiv e-prints arXiv:1512.09300, 2015.

INDEX

A

Action recognition, 4, 183, 198–201, 203, 205, 209, 210
  conventional, 209
Adaptive Lung Nodule Screening Benchmark (ALNSB), 264
Anchor graph hashing (AGH), 172
Anchored neighborhood regression (ANR), 73
Approximated message passing (AMP), 43
Architecture style classification, 214, 216, 230
  dataset, 227
Area under curve (AUC), 242
Augmented Lagrange multipliers (ALM), 146
Autoencoder (AE), 102, 184, 187, 216, 217, 232, 235, 238, 240, 263–266, 268
  rPAE, 242
Automatic speech recognition (ASR), 130
Auxiliary loss, 106
AVIRIS Indiana Pines, 20, 40
  classification, 20
  dataset, 41

B

Baseline encoder (BE), 39, 109
Bicubic interpolation, 48, 55, 57–61, 72, 76, 77
Bounded Linear Unit (BLU), 165, 167, 169

C

Canonical correlation analysis (CCA), 185
Cascade, 48, 55, 57, 63, 199, 201, 257
Cascade of SCNs (CSCN), 49, 61, 63, 64, 69, 72
  deeper network, 55
Classification, 4, 9, 11–13, 17, 18, 22, 23, 38, 40, 43, 106, 116, 145, 149, 155, 170, 183, 196–198, 205, 214, 227, 246, 265
Classifier, 11, 12, 14, 15, 18, 19, 22, 91, 192, 201, 227, 264, 265, 270
  domain, 259
  hyperspectral image, 9
Cluster label
  predicted, 38, 90, 92, 105
  true, 38, 90

Clustering, 4, 17, 31, 32, 38, 42, 43, 87–89, 92, 97, 98, 101, 103, 104, 106, 107, 112, 114, 116, 117
  discriminative, 87, 88, 98, 102, 109
  joint optimization, 88
  loss of, 90, 91
CMU MultiPIE, 108, 109, 114
CMU PIE, 42, 157, 158, 161
  dataset, 42, 161
Collective matrix factorization (CMF), 185
Color attenuation prior, 253
Compressive sensing (CS), 121
Consensus style centralizing autoencoder (CSCAE), 216, 222, 225, 229
Contrast limited adaptive histogram equalization (CLAHE), 257
Convolutional layers, 4, 53, 76, 77, 265
Convolutional neural networks (CNN), 2, 48, 125, 253
Corrupted version, 151, 187–189, 221
Cutting plane maximum margin clustering (CPMMC), 91

D

Dark channel prior (DCP), 253
Datasets, 19, 23, 87, 93, 107, 108, 159, 163, 164, 167, 226, 227, 254
  3D, 200, 206
  AR, 157
  architecture style, 230
  BSD500, 63
  CIFAR10, 172
  CMU MultiPIE, 108
  COIL-100, 157, 162
  COIL20, 108
  daily and sports activities (DSA), 191, 210
  Extend YaleB, 157
  Extended YaleB, 159
  fashion, 214
  ILSVRC2013, 63
  IXMAS, 210
  MNIST, 38, 40, 107–109, 111, 112, 162, 163

MSCNN, 253, 259, 260
MSR-Action3D, 206–208
MSR-Gesture3D, 206, 207
ORL, 157
RESIDE, 253, 260
SUN, 228
Decoder, 123, 125, 146, 151, 187, 266
Deep
    architectures, 38, 43, 48, 103, 109, 125, 191,
        201
    clustering models, 117
    encoders, 38–40, 42
    learning, 1–4, 32, 43, 48, 57, 58, 63, 101, 102,
        129, 146, 165–167, 187, 213, 216, 217,
        226–229, 242, 246
    learning for clustering, 101, 102
    learning for speech denoising, 130
    models, 3, 109, 111, 126, 200, 201
    network cascade, 64
    networks, 3, 4, 32, 36, 41, 47–49, 72, 73, 81,
        106, 130, 170, 184
    neural network, 2, 130, 132, 139, 210
"deep" competitors, 173
Deep network cascade (DNC), 64
Deep neural networks (DNN), 131
Deep nonlinear regression, 132
"deep unfolding", 34
Deeply Optimized Compressive Sensing (DOCS),
        123–127, 129
    loss function, 125
Deeply-Task-specific And Graph-regularized
        Network (DTAGnet), 106, 115
Dehazing, 252, 254, 257
Denoising autoencoder (DAE), 131, 221, 228
Densely connected pyramid dehazing network
        (DCPDN), 257
Depth videos, 198–200, 207
Dictionary, 2, 10, 13–15, 18, 19, 32, 39, 43, 51,
        60, 88, 89, 91, 95, 97, 108, 121, 144, 145,
        147, 149, 150, 152, 155, 166
    atoms, 11, 12, 19, 31, 146, 157
    for feature extraction, 26
    learned, 13, 19, 32, 89, 102, 145, 146
    learning, 13, 50, 88, 122, 143–145, 147–149,
        152, 157, 165
Discrete cosine transform (DCT), 59, 122
Discriminative, 87, 88, 106, 148, 152, 165, 188
    clustering, 87, 88, 98, 102, 109
    coefficients, 152

effects, 10
    information, 146, 148, 165, 188, 189
Discriminative and robust classification, 11, 12
Domain expertise, 48

E
Effectiveness, 22, 38, 78, 81, 101, 105, 117, 144,
        147, 157, 160, 172, 193, 196–198, 229, 239
Encoder, 42, 123, 146, 150, 170, 171, 173, 187,
        217, 232, 266, 267
Encoder and decoder network, 266, 267

F
Face recognition (FR), 200, 244
Facial images, 93, 107, 231, 232, 234, 239, 244
Familial features, 231–233, 235, 243, 244
    in kinship verification, 232
Family faces, 232
"family faces", 232
Family faces, 233–235
Family membership recognition (FMR), 231, 239,
        243, 244
Fashion style classification, 214, 216, 229
Fashion style classification dataset, 226
Features
    descriptors, 214–216, 221, 222
    extraction, 11, 12, 17–19, 102, 143, 146, 259
    extraction network, 258
    for clustering, 88, 89, 101
    learning, 103, 109, 146, 147, 167, 184, 200, 242
    learning layers, 102
    maps, 36, 202, 205, 259, 265
    network, 266
    representation, 3, 192, 214, 217, 226, 227, 232
    robustness, 202
    space, 10, 49, 50, 73, 186, 210, 267, 271
    vector, 12, 192, 193, 199, 202, 232, 234, 240

G
Gated recurrent unit (GRU), 2
Gaussian mixture model (GMM), 87, 101
Generalized multiview analysis (GMA), 185
Generative, 87
Generative adversarial networks (GAN), 263
Generator network, 266, 268
Generic SCN, 49, 57, 58, 67, 78
Gradient descent, 32, 167

H

Hard thrEsholding Linear Unit (HELU), 35, 36, 170
Hashing codes, 176
Hashing methods
    neural-network hashing (NNH), 172
    sparse neural-network hashing (SNNH), 172
Hazy images, 253, 254, 257, 259
Hidden
    layer features, 219
    layer neural network, 146
    layers, 2, 39, 106, 109, 131, 134–136, 146, 151, 171, 173, 217, 232, 235, 240
    units, 151, 165, 187, 240, 242
HR image, 49–51, 53, 55, 73–76
Human Visual System (HVS), 254
Hybrid convolutional-recursive neural network (HCRNN), 199, 201, 206–209
Hybrid neural network for action recognition, 198
Hybrid subjective testing set (HSTS), 254

I

Inputs, 1, 133, 135, 187, 191, 204, 218, 228, 232, 259, 263, 266, 268
Iterative shrinkage and thresholding algorithm (ISTA), 33, 43, 52, 124

J

Joint
    EMC, 95, 97
    MMC, 95, 97
    optimization, 11, 13, 18, 26, 37, 54, 95, 97, 99, 101, 103, 123, 167, 257

K

Kernelized supervised hashing (KSH), 172
Kinship pairs, 239, 243
    negative, 240
    positive, 240
Kinship verification, 230–233, 239, 243, 244

L

Label information, 184, 189, 190, 193, 197, 198
Layer, 4, 48, 52–54, 57, 60, 102, 104, 106, 107, 116, 131–133, 135, 169, 191, 192, 202, 204, 218, 219, 229, 235, 240, 242, 259
    biases, 132
    convolution, 53, 267
    deconvolution, 267

linear, 53
    weight, 132, 173
LDA Hash (LH), 172
Learned
    atoms, 13, 32, 89, 102
    dictionary, 13, 19, 32, 89, 102, 145, 146
    shared features, 193, 196, 198
Learned iterative shrinkage and thresholding algorithm (LISTA), 4, 33, 35, 43, 48, 52, 75, 124, 173, 176
Learning, 2, 18, 19, 39, 58, 73, 74, 81, 101, 102, 123, 143, 147, 149, 165, 186, 187, 189, 190, 196, 197, 199, 206, 209, 218, 228
    capacity, 32, 48, 101, 109, 172
    deep, 1–4, 32, 43, 48, 57, 58, 63, 101, 102, 129, 146, 165–167, 187, 213, 216, 217, 226–229, 242, 246
    dictionary, 13, 50, 88, 122, 143–145, 147–149, 152, 157, 165
    features, 103, 109, 146, 147, 167, 184, 200, 242
    rate, 16, 19, 38, 77, 108, 172, 255, 256, 259
Linear discriminant analysis (LDA), 88
Linear layers, 53, 60
Linear regression classification (LRC), 157
Linear scaling layers, 35, 53, 104
Linear search hashing (LSH), 170
Local contrast normalization (LCN), 202
Local Coordinate Coding (LCC), 149
Locality constraint, 143, 144, 147–150, 164, 165
Logistic regression classifier, 26
Loss, 73, 76, 93, 105, 254, 255, 259
Loss function, 10, 14, 16, 38, 88, 90, 105, 235, 254–256, 260
    classical, 13
    for SSIM, 255
    joint, 14
    quadratically smoothed, 93
Low-rankness, 2–4, 143
LR
    feature space, 49, 50, 73
    image, 47, 48, 50, 57, 58, 67, 73–76
    patch, 50–53, 60, 73
Lung nodule, 263, 264, 270
Lung nodule classification, 265

M

Manga dataset, 226, 227, 229
Manga style classification, 214, 216, 219, 222, 229
    dataset, 226

Marginalized denoising dictionary learning (MDDL), 143, 144, 148, 152, 154, 155, 157, 159, 160, 162, 164
Marginalized denoising transformation, 148, 152, 165
Marginalized stacked denoising autoencoder (mSDA), 147, 187, 197, 198, 228
Max_M
    pooling, 36, 37, 43
    unpooling, 36, 37, 43
Maximum margin clustering (MMC), 91, 105
Mean average precision (MAP), 173, 254
Mean precision (MP), 173
Mean squared error (MSE), 53, 133

N

Nearest neighbor, 218, 221, 229, 232–234, 244, 245
Negative matrix factorization (NMF), 130
Network, 33, 36, 39, 48, 51, 53, 54, 60, 74, 75, 77, 78, 81, 83, 108, 124, 132–135, 253, 264–268
    compression, 3
    formulation, 32
    implementation, 33, 51
    unified, 74
Network cascade, 54, 55, 72
    deep, 64
Network in network, 53
Neural network, 1, 10, 48, 51, 57, 73–76, 106, 122, 124, 130, 131, 151, 199, 204, 252, 265
    deep, 2, 130, 132, 139, 210
Neurons, 3, 32, 125, 146, 151, 170, 265, 268
Nodules, 263–265, 268
    analyzer, 264, 265, 269
    base, 265
    random, 266, 267
    "seed", 265
Noise, 31, 49, 57, 59, 69, 88, 112, 134, 137–139, 144, 145, 147, 148, 150, 159, 160, 164–166, 223, 224, 228, 253, 254
Nonnegative matrix factorization (NMF), 139
Normalized mutual information (NMI), 94, 95, 108

O

Object detection, 1, 146, 252, 254, 258, 260
    in hazy images, 254

Overall accuracy (OA), 20

P

Parallel autoencoders, 231, 232, 234, 235, 238, 242
Parallel autoencoders learning, 239
Parameter-sensitive hashing (PSH), 172
Patch, 38, 39, 51, 53, 59, 61, 74, 199, 201–205, 209, 216, 218, 222, 223, 230, 255
    LR, 50–53, 60, 73
Peak signal-to-noise ratio (PSNR), 62, 64, 77, 78, 254, 256
Principal component analysis (PCA), 88
Private features, 188, 193, 196–198

R

Rank-constrained group sparsity autoencoder (RCGSAE), 216, 222, 224
Real-world task-driven testing set (RTTS), 254
Reconstruction loss, 14, 107
Rectified linear activation, 132, 135, 136
Rectified linear unit (ReLU), 4, 35, 54, 76, 170
Recurrent neural networks (RNN), 2
Regularized parallel autoencoders (rPAE), 231, 238–240, 243
    hidden layer, 239
Restricted Boltzmann machines (RBM), 131
Restricted isometry property (RIP), 43
Robustness, 9, 57, 67, 72, 117, 137, 160, 171, 184
    features, 202
ROC curves, 242

S

Seed images, 264, 265, 267, 268
    limited, 264, 271
Seed nodules, 265, 266, 268, 270
Shallow neural networks (SNN), 130
Shojo, 214, 226
Siamese network, 170, 171, 173
Sparse
    approximation, 31, 32, 38, 43, 125, 167
    codes, 10–12, 14, 31–33, 39, 51, 53, 88, 89, 91, 97, 98, 102, 104, 106, 112, 167
    coding, 2–4, 13, 16, 31–33, 36, 38, 41–43, 48, 50–52, 54, 60, 61, 64, 67, 72, 73, 75, 77, 87–89, 97, 98, 101, 105, 106, 109, 111, 149
    coding domain expertise, 103, 117
    coding for clustering, 102
    coding for feature extraction, 12

representation, 10, 13, 48, 50, 54, 72, 73, 77, 89, 121, 126, 144–146, 149
Sparse coding based network (SCN), 54, 55, 57
  cascade, 49, 54, 55
  model, 49, 53–55, 57, 60, 72
Sparsity, 2–4, 10, 19, 93, 143–145, 149, 150, 176
Spectral hashing (SH), 170
Speech denoising, 130, 131, 140, 144
SR
  inference, 75–78
  inference modules, 74–78, 81
  methods, 59, 64, 69, 71, 72, 77, 78
  performance, 67, 74, 77, 81
  results, 49, 54, 55, 64, 70, 81
Stochastic gradient descent (SGD), 11, 15, 16, 34, 37, 91, 101, 106, 147
Structural similarity (SSIM), 64, 77, 78, 254–256
  performance, 255
Style
  centralizing, 220, 229
  centralizing autoencoder, 216, 217, 221, 228, 246

classification, 213, 214, 216, 229, 246
descriptor, 214, 215, 227
level, 216, 218–220, 226, 229
Style centralizing autoencoder (SCAE), 216–219, 221, 222, 228, 229, 246
  inputs, 216, 228
Superresolution, 4
Support vector machine (SVM), 10, 90
Synthetic objective testing set (SOTS), 254

T
Task-specific And Graph-regularized Network (TAGnet), 103–106, 109, 112

U
Unlabeled samples, 11, 12, 14, 20
Upscaling factors, 49, 63, 73, 77, 78, 81

W
Weak style, 214, 216, 220, 221, 246
Weighted sum of loss, 14